Speech Signal Processing

Speech Signal Processing

V. Udayashankara
Professor
Department of Electronics and Instrumentation Engineering
Sri Jayachamarajendra College of Engineering
JSS Science and Technology University
Mysore, Karnataka

TECHSAR™
NEW DELHI

Published and Printed by
Techsar LLP
4435-36/7, Ansari Road, Daryaganj,
New Delhi–110 002 (India)
E-mail: tech_info@techsarworld.com
Website: www.techsarworld.com

ISBN 978-93-90620-05-0

Reprint 2026

Dedicated to

My Beloved Parents

Sri N.N. Veerappa

Smt. Susheela Veerappa

Preface

The aim of this book is to introduce, the most fundamental and important speech processing algorithms with DSP knowledge for speech production, analysis, coding, enhancement, speech and speaker recognition. It includes DSP algorithms like short time Fourier transform, Mel cepstral analysis, linear predictive coding to solve several complicated problems in speech processing.

The book has 13 chapters. First chapter gives an introduction to speech signals. Chapter 2 covers speech production process in human beings and the associated anatomy. This chapter also covers phonetics and digital models for speech signals. Chapter 3 includes basics of digital signal processing and wavelets. Chapter 4-6 covers the key elements of speech analysis like time domain analysis, frequency domain analysis, spectrograms, cepstral analysis, mel frequency cepstrum analysis and LPC analysis. Speech coding finds applications in cellular telephony, multimedia, voice mail, audio video teleconferencing and the Internet and hence three types of speech coders: waveform coders, vocoders and hybrid coders are explained in Chapter 7. Chapter 8 is dedicated to speech synthesis based on waveform coding, analysis-synthesis method, production mechanism and text-to-speech conversion. Speech recognition is another important area and is covered in Chapter 9. This chapter also covers difficulties in speech recognition, classification and steps for speech recognition. Chapter 10 consists of different types of speech enhancement algorithms that can be used to improve the quality of degraded speech. Chapter 11 introduces anatomy and functioning of the ear, types of hearing impairments and hearing aids. This chapter also includes analog hearing aid, digital hearing aid, DSP based hearing aid, ASIC based hearing aid and cochlear implant. Chapter 12 covers features of digital signal processing (DSP) chip and its applications. PRAAT and wave surfer are computer programs used for speech processing, Chapter 13 covers Sound Editor of PRAAT and wave surfer.

This book has been designed primarily for UG and PG programmes. It can also been used as a reference book by research scholars. A large number of MATLAB programs given in this book will be useful to conduct laboratory courses for speech processing. This book also serves as a useful reference for competitive examinations.

V. Udayashankara

Acknowledgements

I express my sincere gratitude to JSS Mahavidyapeetha for encouraging me to achieve this accomplishment. I thank Dr. Siddaramaiah (acting VC), JSS Science & Technology University and Dr. S.B. Kivade, Principal, Sri Jayachamarajendra College of Engineering, Mysore, India for their encouragement and support. I thank my research students Pradeep M.N., Dr. Shreekanth T., Dr. Amaresh P. Kandagal and Dr. S. Rathnakara for their valuable contributions to the book.

I am indebted to my colleagues and students Bindu Shree C., Navyabharathi K.S. and Shreeranga R.B. for their timely help, in planning, execution and for providing valuable suggestions for this book.

I am thankful to my wife Renuka and son Suveer for their patience and support while writing this book.

V. Udayashankara

Contents

1

CHAPTER

Introduction

This chapter provides insights on

- Characteristics of speech signal
- Speech signal representations
- Basic elements function of discrete time speech processing system
- Application areas of speech signal processing
- Outline of the book

1.1 INTRODUCTION

Speech is the most natural form of human communication. Because of its manifold applications, speech processing has been one of the most exciting areas of signal processing. Human speakers and machine synthesizers produce speech signals, while human listeners and machine recognizers receive and analyze speech signals. The topic discrete time speech signal processing can be defined as the manipulation of sampled speech signals by a digital processor to obtain new signals with some desired properties. The analysis, processing and synthesis of signals are the necessary keys of speech processing and this book develops tools to understand signals and their manipulation. In the section below, we describe characteristic of speech signal. Section 1.3 discusses about speech signal processing. Discrete time speech signal processing is discussed in section 1.4. Finally, a brief introduction to applications of speech signal processing is given in section 1.5.

1.2 CHARACTERISTICS OF SPEECH SIGNAL

The speech signal is commonly considered a communication signal; however it can serve as a diagnostic signal when speech and vocal tract disorders need to be investigated. They are generated naturally; and hence they are non-stationary signals. Speech signals are generated by important parts of human vocal system like vocal cords and vocal tract. Hence, many researchers represent the speech signal as output of vocal cord and vocal tract response parameters. Throughout this book, speech representation is based on the waveform and parametric (parameters of the model for speech production) representation. In waveform representation, as the name implies, we preserve the wave shape of the analog speech signal. Speech signal can be represented as a waveform as shown in Figure 1.1.

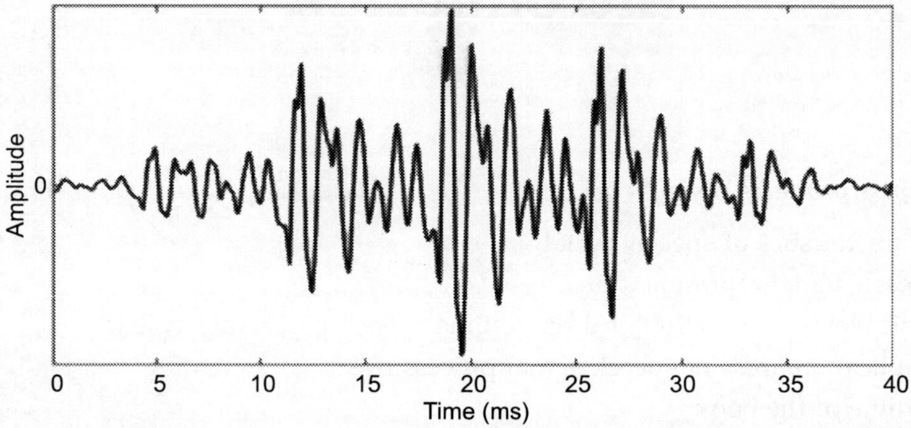

Figure 1.1 Speech signal

Let us assume speech signal is characterized by the following features

- Sampling frequency F_S : 10 kHz
- Length L of the file : 8000 samples

Then sampling period of the signal, is given by $T_S = \frac{1}{F_S} = \frac{1}{10000} = 100\ \mu s$

Then total playing time = $L \times T_S = 8000 \times 100\ \mu s = 0.8$ s

The speech signal is a slowly time varying signal in the sense that in short period of time (between 5 and 10 ms) its characteristics are fairly stationary; however over long periods of time the signal characteristics change to reflect the different speech sounds being spoken and they are non-stationary.

1.3 SPEECH SIGNAL PROCESSING

In case of speech signals, the speaker is the information source. The speaker produces a speech signal in the form of pressure waves that travel from speaker to listener as shown in Figure. 1.2. The measurement or observation is the acoustic waveform.

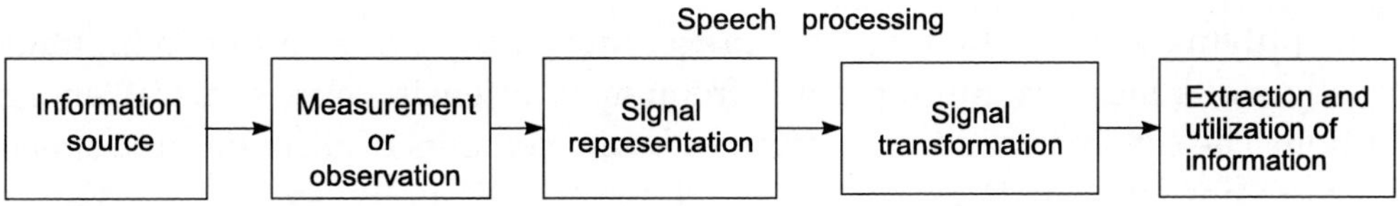

Figure 1.2 General view of speech signal processing

Speech processing involves two tasks, they are:

- Signal representation
- Signal transformation

Signal Representation: Representation of speech signal is either waveform or parametric form as shown in Figure 1.3.

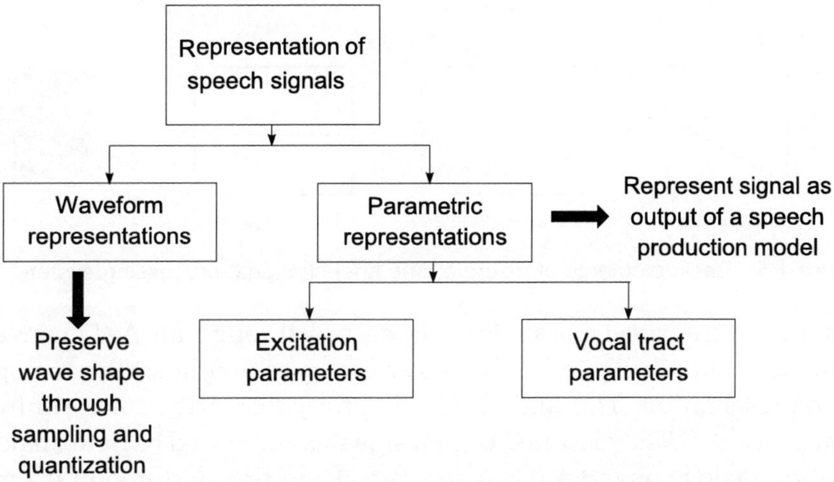

Figure 1.3 Representation of speech signals

In waveform representation speech signal is represented in digital form. The analog speech signal is converted to digital form through sampling and quantization process. In parametric representation, first speech signal is sampled and quantized and then further processed to obtain the parameters of the model for speech production. Parameters of the model are classified as excitation parameter (related to source of speech sound) or vocal tract parameter (related to individual speech sound). Second process in signal processing is signal transformation, i.e., transforming the signal representation into alternate forms, which are more appropriate to specific applications. The last stage in the process is the extraction and utilization of message information. Generally, it is human listeners or machines.

1.4 DISCRETE TIME SPEECH SIGNAL PROCESSING

Speech is produced by a variation of the air pressure, caused and emitted by the articulator system. Speech can be analyzed, first by digitizing the speech waveform and then applying various digital signal processing operations in order to highlight its acoustic traits such as fundamental frequency, intensity and spectral energy distribution. The schematic representation of basic elements of discrete time speech signal processing system is as shown in Figure 1.4. Usually, system consists of sensors or transducers; this converts speech signal into an electrical signal (ex microphone, output voltage of microphone is 20 millivolts). The output of sensors is usually connected to a preprocessing circuit. The basic preprocessing circuit is called signal conditioning circuit. The signal conditioning circuit consists of preamplifier and anti-aliasing filter. The preamplifier amplifies the output of sensor to match the amplitude of the signal to the range of the analog to digital converter specifications and anti-aliasing filters are used to reduce the portion of undesired noise.

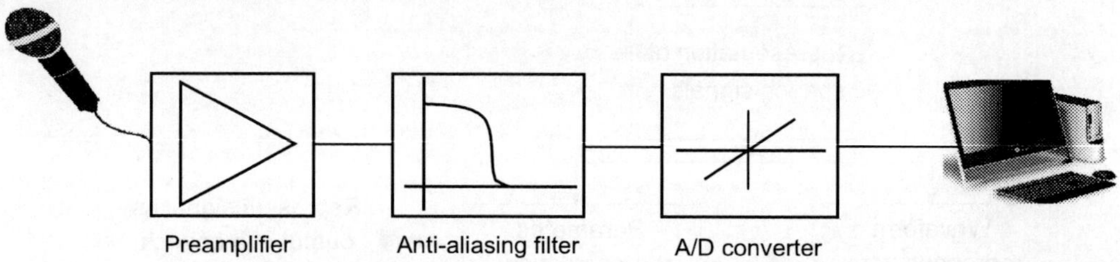

Figure 1.4 Basic elements of discrete time speech signal processing system

The continuous-time speech waveform is passed through an A/D converter to obtain a sequence of numbers, i.e., A/D converter serves to convert the analog signal to a digital representation. The digital data is processed in the microcontroller or digital signal processor. The modified digital signal is converted back to analog form with a digital-to-analog converter (D/A converter) and finally stored in the memory or played directly through an amplifier and speaker. Modern medical instruments use standard interfaces to exchange data with computers. The processor is connected to a computer or to other instruments through RS-232, USB or Ethernet. When the results of the processing are not required immediately following the signal, off-line processing methods are used. Off-line processing can be performed on general-purpose computers using microprocessor/microcontroller. When results are required during and immediately after the signal sample has been acquired, real-time or on-line processing is performed using digital signal processor based systems. Real-time processing application requires mathematical operations to be performed quickly. The digital signal processor is optimized to accomplish these tasks fast enough to maintain real-time operations. Digital signal processor (DSP) is designed to support several speech and audio applications; they are automatic speech recognition, speech synthesis, speaker verification, noise and echo cancellation.

1.5 OUTLINE OF THE BOOK

In this section, we will have a brief discussion of the topics which will be discussed in detail in subsequent chapters. Figure 1.5 shows few application areas of speech processing which will be covered in detail in this book.

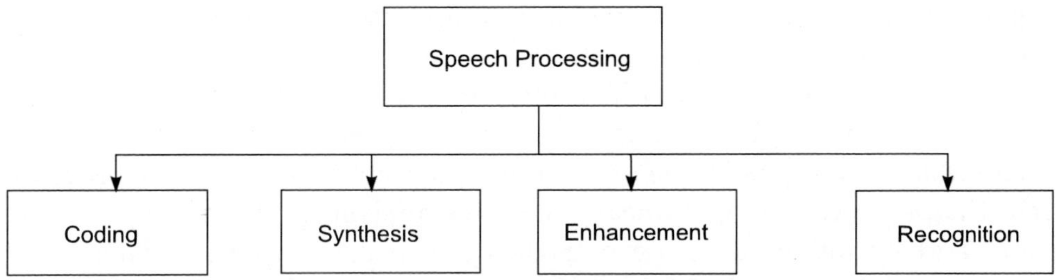

Figure 1.5 Application areas of speech processing

The speech signal is non-stationary (time-varying signal), changing its characteristics as the muscles of the vocal tract contract and relax. In chapter 2, we review speech production process in human beings and the associated anatomy. Speech sounds are divided into two broad classes, they are vowel and consonants. This chapter also introduces study of phonetics. We discuss the relationship between the positions and movements of speech organs and the sound of a language. We conclude this chapter with modeling aspects of human behavior during speech communication. First, we discuss a simple model for speech production. This model is then used to develop a lossless tube models and digital models for speech signals.

It is essential that a reader have a good understanding of fundamental of digital signal processing, chapter 3 covers basics of linear time invariant system, correlation, DFT, FFT, design of FIR filters using windows, design of butterworth IIR filters, decimation & interpolation concepts in multirate digital signal processing and wavelets.

Chapter 4-6 cover the key elements of speech analysis. The process of extracting properties or features of speech signal is called speech analysis. The following analyses are discussed in detail in chapters 4-6.

- Time domain analysis
- Frequency domain analysis
- Spectrograms
- Cepstral analysis
- Mel frequency cepstrum analysis
- LPC analysis

Chapter 4 covers time domain analysis. In time domain analysis, we compute short time energy and zero crossing rate, these are useful in detecting the starting and end point of speech and also we derive the formula for short time auto correlation function which is useful in computing fundamental frequency of speech signal.

In chapter 5, we introduce the short time Fourier transform (STFT) and its magnitude for analyzing the spectral properties of time varying speech waveform. This chapter covers spectrograms, which is used to obtain frequency-time analysis. This chapter also covers cepstral analysis, it describes an alternate means of de-convolution of the source and system components and it is referred to as homomorphic filtering. This chapter concludes with Mel frequency cepstrum analysis.

Linear predictive analysis is one of the most popular approaches for processing of speech. Chapter 6 covers how linear predictive analysis first extracts the system component and by inverse filtering extracts the source component. This chapter concludes with application of LPC analysis.

Chapter 7 investigates how speech signals coded for efficient transmission or storage. The goal of this chapter is to reduce information rate, measured in bits per second, while maintaining the quality of the original speech waveform. This chapter covers three types of speech coders: waveform coders, vocoders and hybrid coders. Waveform coders, represents the speech waveform directly and do not rely on a speech production model. Vocoders are largely speech model based and rely on a small set of model parameters. While the hybrid coders are partly waveform based and partly speech model based. Speech coding finds diverse range of applications in cellular telephony, multimedia, voice mail, audio video teleconferencing and the internet.

Speech synthesis is a process which artificially produces speech for various applications. The quality of speech synthesis is affected by two factors: intelligibility and naturalness. Speech intelligibility depends on the synthesizer's ability to produce formants, formant bandwidth and formant transitions, whereas speech naturalness depends on the excitation waveform characteristics. Chapter 8 covers synthesis based on waveform coding, analysis-synthesis method, production mechanism and text-to-speech conversion.

Speech recognition is the process of automatically extracting and determining linguistic information conveyed by a speech wave using computers or electronic circuits. Chapter 9 covers difficulties in speech recognition, classification and steps for speech recognition. Chapter concludes with speaker recognition. The technology closely related to speech recognition, i.e., speaker recognition. Speaker recognition is the automatic recognition of speaker (talker) with the help of characteristics of speaker voice signal.

The goal of chapter 10 is to improve the quality of degraded speech. This chapter covers different algorithms for speech enhancement, i.e., different algorithms to improve the quality of degraded speech by reduction of additive noise.

In chapter 11, we introduce anatomy and functioning of the ear. This chapter covers types of hearing impairments and hearing aids. This chapter also covers analog hearing aid, digital hearing aid, DSP based hearing aid and ASIC based hearing aid. This chapter concludes with cochlear implant.

Floating point digital signal processing chips are used to implement all speech application programs in real time. Chapter 12 covers features of digital signal processing (DSP) chip, TMS 320C6713 floating point processor, its architecture, addressing modes and its applications. Speech processing chips are powerful, specialized integrated circuits used to provide synthesized voice output and/or voice recognition input for many different types of computerized systems. Chapter 12 also covers the speech processing chips and summarizes the important characteristic of each.

PRAAT and wave surfer are computer programs used for speech processing purposes. Chapter 13 covers Sound Editor of PRAAT and wave surfer.

1.6 SUMMARY

This chapter summarizes the characteristics of speech signal, classification and representation of speech, basic elements of discrete time speech processing systems. Various application areas and the outline of the book is also discussed.

[illegible] This chapter [illegible] covers types of [illegible] algorithms [illegible] learning [illegible]. This chapter also [illegible] [illegible] and [illegible] [illegible].

[illegible] all speech applications [illegible] [illegible] [illegible] [illegible] [illegible] Chapter [illegible] each.

[illegible] [illegible] MATLAB [illegible] used for speech processing [illegible] [illegible].

1.6 SUMMARY

This chapter summarizes [illegible] speech signal classification [illegible] [illegible] and the organization of the book is also discussed.

2

CHAPTER

Fundamentals of Speech Signal

This chapter provides insights on

- Speech production mechanism
- Fundamental and formant frequencies
- Distinguishing voiced and unvoiced speech signals
- Phonemes classification
- Prosody features
- Basic and digital model of speech production

2.1 INTRODUCTION

The purpose of this chapter is to give the reader an understanding of how human speech is produced. Speech is generated by human respiratory and articulator systems that consist of different organs and muscles. Coordinated action of these organs results in sound waves that propagate in the air. This chapter gives introduction to phonetics. The study of sound variations is called phonetics. This chapter covers the summary of the phoneme of English and manner of articulation for each of the major phoneme classes. It also studies the linear time invariant model for speech production. This approach is used to obtain discrete time models for the speech signal. The chapter concludes with digital model for speech production.

2.2 SPEECH PRODUCTION MECHANISM

The speaker produces speech signal in the form of pressure waves that travel from the speaker to the listener. This signal consists of variations in pressure as a function of time. Figure 2.1 shows the important parts of human vocal system. The organs that are involved in producing speech can be divided into three main groups. They are:

- Lungs

- Larynx
- Parts of the mouth and the nose

Speech sounds are produced as air is pushed from the lungs and converted into fluctuating energy. The air source and nature of its flow are referred to as *excitation signal*. The lungs are the source of airflow that passes through the larynx and oral/nasal tract. Breathing is accomplished primarily with the diaphragm at the bottom of the thorax; the contraction of the diaphragm expands the lungs and the relaxation of the diaphragm expels air from the lungs

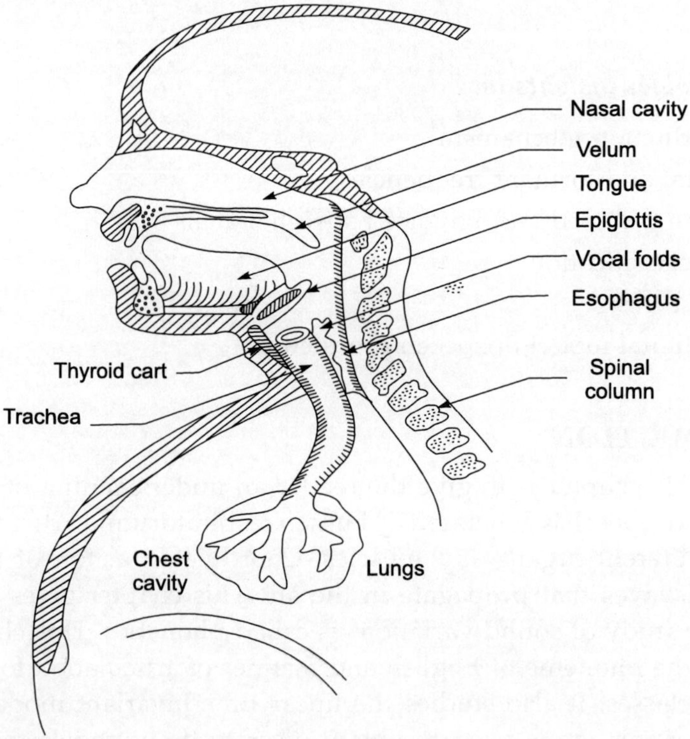

Figure 2.1 Schematic view of the human vocal mechanism

The diaphragm expands and contracts assisting the lungs in forcing the air through the trachea, across the vocal cord or vocal fold and finally through the nasal and oral cavities. The sources of most speech occur in the larynx, where vocal cord can partially or completely obstruct the airflow from the lungs. If the vocal cord or glottis is completely closed, the airflow ceases and no sound emerges or when it opens, it allows the air from the lungs through the larynx and oral/nasal cavity. By opening and closing it breaks the airflow into periodic pulses. The velum can be raised or lowered to allow acoustic coupling of nasal cavity as shown in Figure 2.2. If the velum is lowered air can pass through the nasal cavity, if the velum is raised, thereby closing off the nasal cavity air passes through the oral cavity. The speech organs in the vocal tract that are involved in the production of different sounds are called *articulators.* The tongue, lips, teeth, velum and jaw are grouped under articulators. Tongue and lip positioning creates a large number of vocal tract shapes necessary to produce different speech sounds.

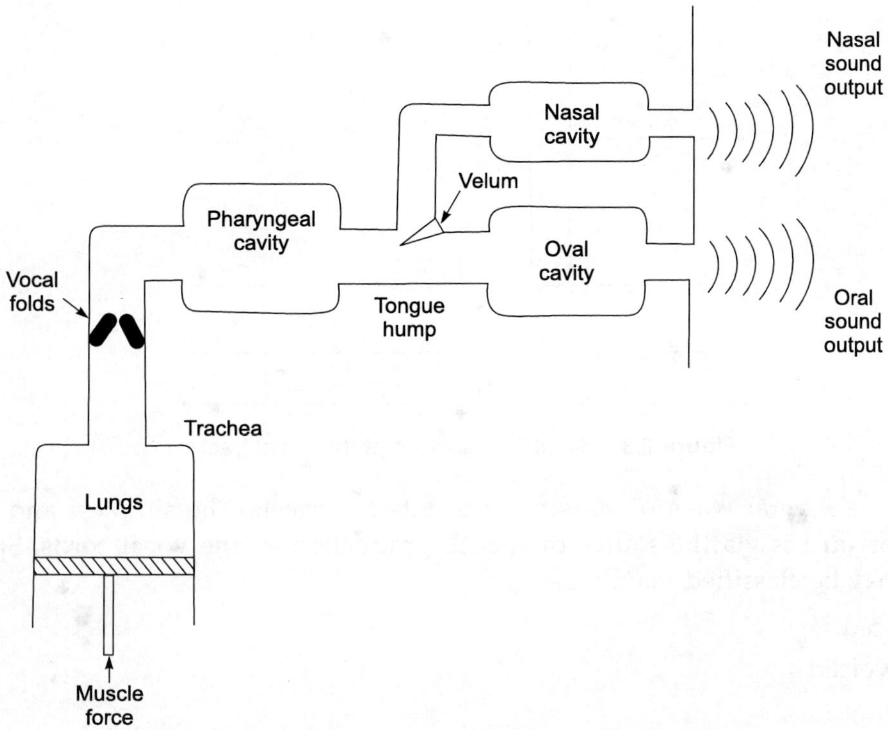

Figure 2.2 Schematic representation of the complete physiological mechanism of speech production

Thus, speech can be described as the result of the coordinated action of a number of muscles. The respiratory organs provide the energy needed to produce speech sound forcing airflow in the trachea and through the vocal cords. Thus, the speech sounds are produced as air is pushed from the lungs and converted into the fluctuating energy. The air source and nature of its flow are referred to as *excitation signal.* Vocal tract is a tube of non-uniform cross section and is about 17 cm long. It branches after 8.5 cm to give the nasal cavity, a tube about 13 cm long as shown if Figure 2.3. The resonance frequencies of the vocal tract tube are called *formant frequencies* or simply, the *formants.* If the vocal tract were of uniform cross section, many natural frequencies would occur at

$$F_n = (2n-1)\frac{c}{4L} \tag{2.1}$$

here, $n = 1,2,3....$, in air $c = 350\,\text{m/s}$, so, for a tube of length $L = 17\,\text{cm}$, these frequencies occur at odd multiples of almost 500 Hz. Since the cross section is not uniform; as a result the resonances are not equally spaced.

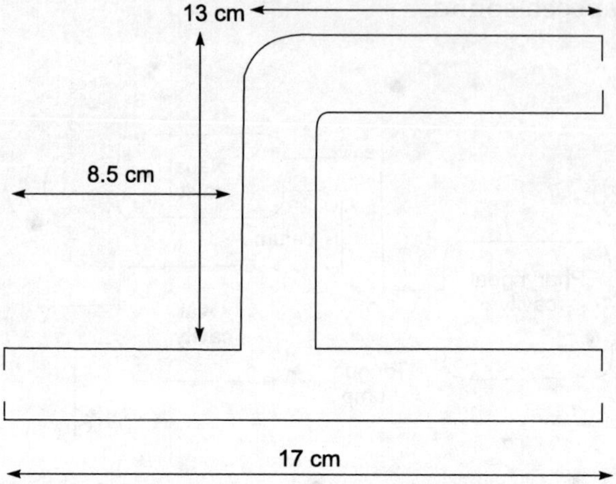

Figure 2.3 Simplified diagram of the vocal tract

There are several ways of classifying events in speech. The simplest and most straightforward is via the source of speech production — the vocal cords. Speech sounds may be classified mainly as:

- Voiced
- Unvoiced
- Silence

Voiced sound is one in which the vocal folds or glottis vibrate, i.e., voiced sounds involve the vocal cord or the glottis. Vocal cord are tensed and therefore, vibrate periodically and break the airflow into the periodic pulses. The input to the vocal tract may be treated as an impulse train that is almost periodic. The two cycles of the glottal pulse waveform is shown in Figure 2.4.

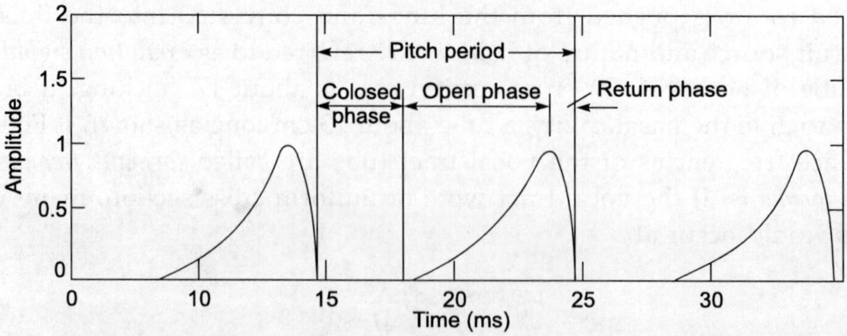

Figure 2.4 Glottal pulse waveform

In Figure 2.4, if T is the time during which vocal folds are open and T_0 is the total duration of the cycle, then $(T_0 - T)$ marks the time when the vocal folds are closed. Thus, the air leaving the lungs is interrupted by periodic closing and opening of the vocal folds; the rate of vibration is called the *pitch frequency* (F_0). The time span between successive vocal fold closures is referred to as **pitch period** or **fundamental period** (T_0). **Pitch frequency** is the reciprocal of pitch period, $F_0 = \frac{1}{T_0}$. Pitch frequency is also referred to as **fundamental frequency**. Filtering the vocal cord output is done in the vocal tract, which modifies the spectral properties of the signal and results in speech signal. The human vocal tract can be seen as an acoustic tube whose shape can be altered, and then depending on the current shape it has a tendency to amplify signal components of certain frequencies. Vocal tract is a tube of non-uniform cross section approximately 17 cm long. The vocal tract acts as a resonator whose behaviour can be described by a transfer function which depends on the shape of the vocal tract. The resonator has certain frequencies which it tends to amplify over other frequencies. These resonant frequencies are called **formant frequency** or **formants**. Formants are denoted by $F_1, F_2, \ldots F_n$. Hence, the vocal tract filter is characterized by formant frequencies (or formants). Where the number indicates which formant has the lowest frequency and which one has the highest frequency. F_1 stands for the lowest formant frequency.

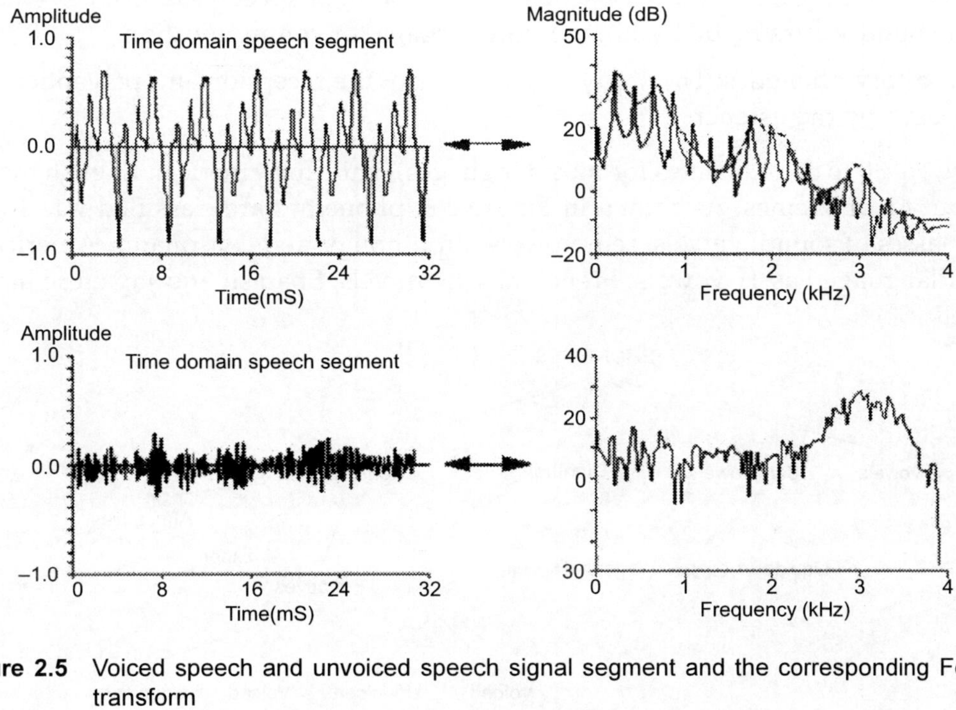

Figure 2.5 Voiced speech and unvoiced speech signal segment and the corresponding Fourier transform

Figure 2.5 shows a voiced speech signal segment. The signal is periodic in nature. All vowels are voiced sounds. In unvoiced sounds, the vocal cords do not vibrate, so the resulting waveform is random in nature. Figure 2.5 also shows an unvoiced speech signal segment. The voiced speech has higher magnitude compared to unvoiced speech signal. Silence involves the complete closure of the vocal tract, hence no speech is produced.

2.3 ACOUSTIC PHONETIC

Linguistics is the study of the rules by which speech sounds are assembled in a language. Linguists study the units of language and their functioning. The basic unit of linguistics is the *phoneme*, an articulated sound. Speech sounds are produced as air is pushed from the lungs and converted into fluctuating energy. For a spoken language, the acoustic signal radiated from the lips consists of a string of sound elements called **phonemes**. Any language can be described in terms of a set of distinctive sounds called **phonemes**. The classification of sounds of speech is called **phonetics**. The study of phonetics focuses on speech. The discipline of phonetics includes three basic areas of study, they are:

- Articulator phonetics: Study of physiology of speech production and the movement of vocal organs by the speaker.
- Acoustic phonetics: Study of acoustic description of speech sounds, i.e., study of sound waves made by human vocal organs for communication.
- Auditory phonetics: The study of phonetics of the reception and perception of speech by the listener.

There are 30 to 50 phonemes for most languages, while for American English there are about 41 phonemes. As shown in Figure 2.6, phonemes are classified into four broad classes of sound: vowels, semivowels, diphthongs, and consonants. American English language has 11 vowels, French has 15 vowels, Spanish and Japanese have 5 vowels.

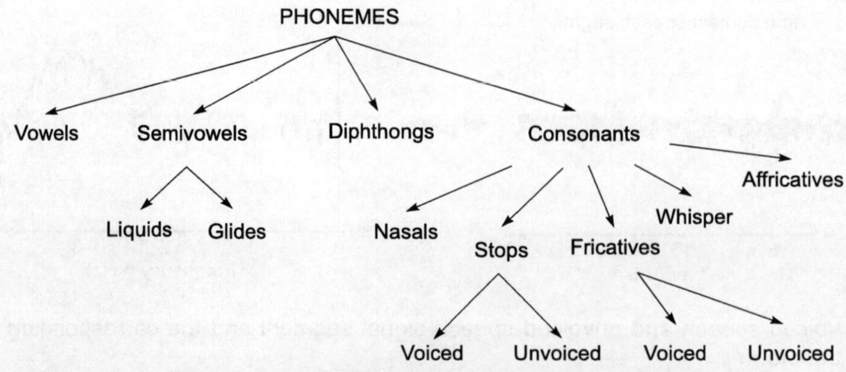

Figure 2.6 Phonemes of American English language

Phonemes may be combined into larger units called *syllables*. A common definition of a syllable is that it has a central vowel that is preceded and succeeded by one or more consonants. A combination of syllables forms a word, which is usually a combination of two or five phonemes. Each of these classes may be further broken down into subclasses which are related to the type of excitation and place of articulation of the sound within the vocal tract.

Hence, subclass classification is according to the type of excitation, for example, voiced and unvoiced sounds. Sounds generated when the vocal cords or glottis are in the relaxed state are called **unvoiced sounds** (without the aid of vocal cords). Sounds such as S and P are unvoiced sounds. Sounds generated when the vocal cords are tense and air passes across them are called **voiced sounds**. Hence, sounds created by the spectral shaping of the glottal pulses are called voiced sounds. All the vowels and some consonants in the English language are voiced sounds. Another subclass classification is by the manner and place of articulation. Typical place of articulation includes place of tongue, lips, teeth, alveolar, palate and glottis.

Vowels: Vowels are voiced speech sounds and are produced without significant movement of articulators except the tongue. The vowel sound produced is determined primarily by the position of the tongue, but to a small extent articulators, jaw and lips, also influence the resulting sound. Vowels are subdivided into three groups defined by the tongue position: front, mid (central), and back as shown in Table 2.1.

TABLE 2.1 Constriction/tongue position in vowels

Constriction/Tongue position	Front	Mid	Back
High	$\|i\|$, beet $\|I\|$, bit	$\|ER\|$, bird	$\|u\|$, Foot
Medium	$\|e\|$, bet	$\|UH\|$, but	$\|o\|$, bought
Low	$\|ae\|$, bat	$\|a\|$, Father	

Vowels are produced by fixed vocal tract shape with quasi-periodic pulses of air caused by the vibration of vocal cords. The cross-sectional area varying along the vocal tract determines the resonance frequencies of the tract (formants) and thereby the sound is produced. The vowels are generally long in duration and spectrally well defined. Vowels are voiced and have the largest amplitude among phonemes and range in duration from 50 to 400 ms in normal speech. Vowel energy is primarily concentrated below 1 kHz. The signals are quasi-periodic due to repeated excitation of the vocal tract by vocal fold closures. Vowels have line spectra and energy is concentrated at multiples of F_0. The largest harmonic amplitude is near the low formant frequency.

The vowels may be divided into tense and lax classes depending on the duration how far the tongue is displaced from the neutral, central position. In American English some vowels are intrinsically longer than others, this is because the tongue reaches a more extreme position. The vowels longer duration are termed *tense vowels*, for example, see, saw, sew, say, sow, law, etc. Shorter vowels are called *lax vowels* because they are produced with less extreme movements, examples are: sing, strength, sang, song, sung, etc.

Speech signals of the three vowels ($/a/$ $/i/$ $/u/$) are presented in glottal shape and frequency domain in Figure 2.7. As shown in Figure 2.7, the first formant of $|a|$ falls roughly at 800 Hz, while the weaker ones, roughly at 1300 Hz, 2300 Hz and 3300 Hz respectively. The first formant of $|i|$ falls roughly at about 400 Hz, while the second, third and fourth format at 2100 Hz, 3100 Hz and 3800 Hz respectively and the first formant of $|u|$ falls roughly at about 300 Hz, while the second, third and fourth format at 900 Hz, 2100 Hz and 2600 Hz respectively. The formant frequencies for different vowels are shown in Table 2.2.

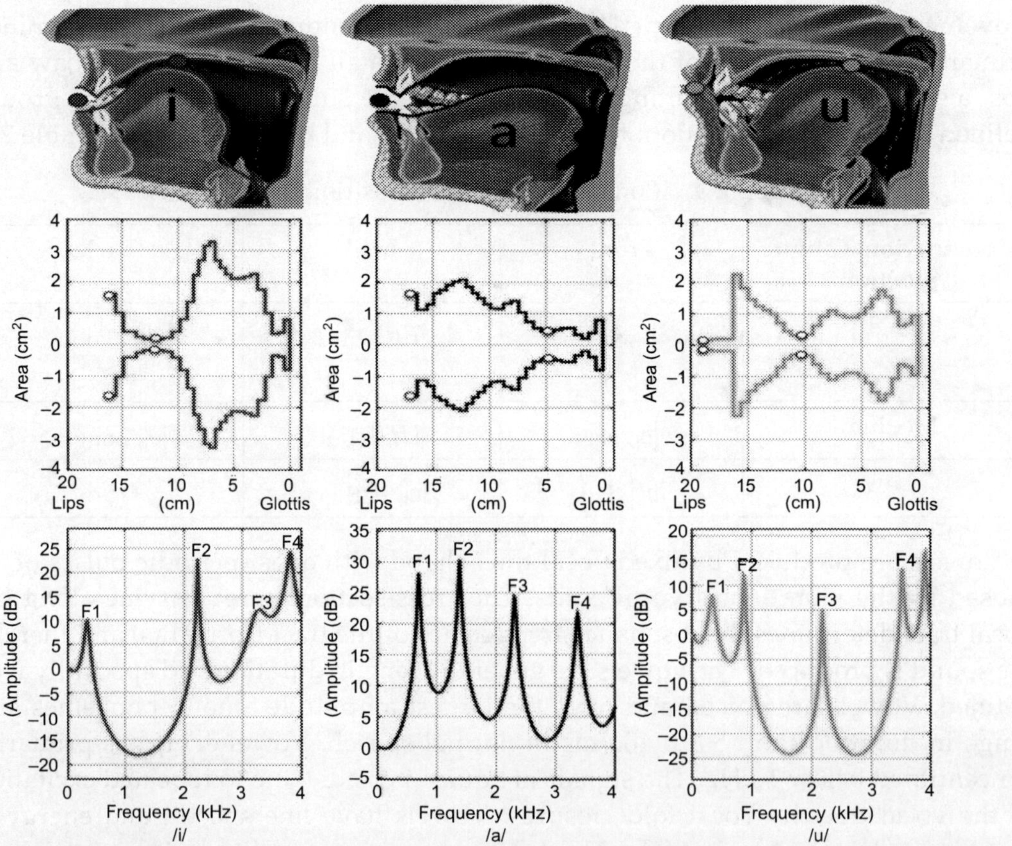

Figure 2.7 Speech signals of the three vowels ($|a|$, $|i|$ and $|u|$), in time and frequency domains

TABLE 2.2 Formant frequencies for different vowels

Symbol for vowel	Vowel	Typical word	F_1	F_2	F_3
IY	$\|i\|$	beet	270	2290	3010
I	$\|I\|$	bit	390	1990	2550
E	$\|i\|$	bet	530	1840	2480
AE	$\|ae\|$	bat	660	1720	2410
AH	$\|i\|$	but	520	1190	2390
A	$\|a\|$	hot	730	1090	2440
OW	$\|i\|$	bought	570	840	2410
U	$\|u\|$	foot	440	1020	2240
OO	$\|u\|$	boot	300	870	2240
ER	$\|i\|$	bird	490	1350	1690

Figure 2.8 shows the plot of the second formant frequency versus the first formant frequency for the vowels. The vowel triangle represents the extremes of formant locations in the F_1 - F_2 plane. At the upper left hand corner of the triangle is the vowel $|i|$ with a low first formant and a high second formant. At the lower left hand corner is the vowel $|u|$ with low first and second formants. The third vertex of the triangle has the vowel is $|a|$ with high first formant and a low second formant.

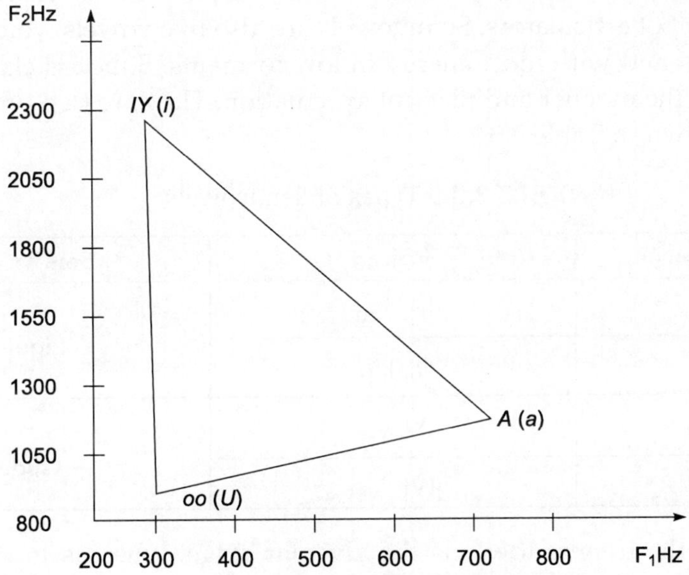

Figure 2.8 The vowel triangle

Figure 2.9 shows the vocal tract profiles for the production of English vowels.

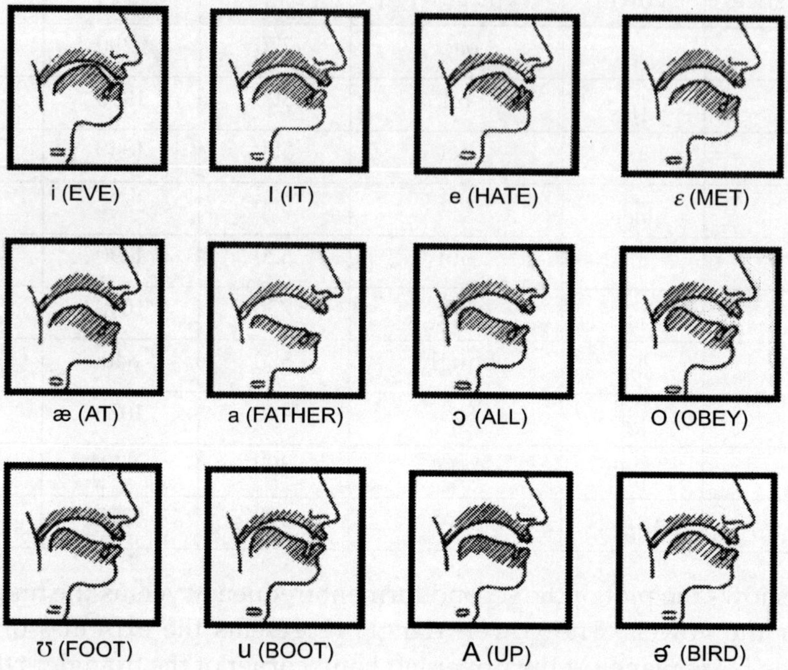

Figure 2.9 Vocal tract profiles for the production of English vowels. (Adapted from Potter, Kopp and Green)

Semivowels: Semivowels are produced in a similar way as vowels but with smooth movement of articulators. Semivowels are also like vowels: voiced, periodic, intense waveform and with most energy in low formants. Subclass classification of semivowels is by the manner and place of articulation. They are two types of liquids and glides as shown in Table 2.3.

TABLE 2.3 Types of Semivowels

Constriction	Voiced	Semivowels
Palate	|r|, run	Liquids
Alveolar	|l|, lap	
Palate	|Y|, yes	Glides
Lips	|W|, wet	

Liquids move the tongue towards the alveolar ridge, whereas in glides the tip of the tongue is raised and the oral cavity is constricted. In liquids, the tongue is shaped in such a way as to form side branches with two openings, one on each side of the contact point with air passing on each side. Hence, in liquids tongue is shaped in such a way, that it has two openings and air passes on both the sides. Hence, liquids differ from glides in the formation of the constriction by the tongue. In glides the tongue tip is raised and the oral cavity is constricted. Glides are very similar to vowels in that they have periodic, intense waveform with most energy at low formants.

Diphthongs: It is the union of two vowel sounds in a single compound sound. Diphthongs are produced by varying the vocal tract smoothly between two vowel configurations. They consist of dynamic vowel sound in which the tongue moves between two vowel positions. American English has 4 diphthongs (Table 2.4.)

TABLE 2.4 Types of diphthongs

Diphthongs	Example
a^y (AY)	buy
a^y (OY)	boy
a^W (AW)	down
e^Y (EY)	bait

Consonants: Consonants are voiced and unvoiced sounds produced by the movement of articulators. Consonants are classified according to the type of excitation and place of articulation of the sound within the vocal tract. The consonants contain a number of subgroups: nasals, fricatives, affricates, stops (plosives) and whispers.

Nasals are voiced sounds. In nasals the velum is lowered and air flows mainly through the nasal cavity. Hence, the oral tract is constricted and the sounds are radiated through the nasal cavity. American English has 3 nasals as shown in Table 2.5. Nasal $|m|$ occurs in the region 750 to 1250 Hz, $|n|$ occurs in the region 1450 to 2200 Hz and above 3 kHz for $|ng|$. Vocal profiles for the nasal consonants are shown in Figure 2.10.

TABLE 2.5 Vocal profiles for the nasal consonants

Costriction	Voiced
Lips	$|m|$ map
Alveolar	$|n|$ no
Back of palate	$|ng|$ Sing

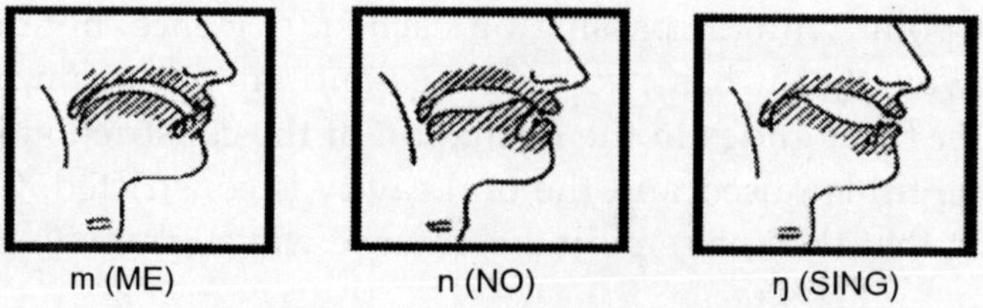

Figure 2.10 Vocal profiles for the nasal consonants. (Adapted from Potter, Kopp and Green)

Stops are formed by the release of a burst of air from a complete constriction. There are two phases the stop (complete constriction) followed by burst (release of air). Stops may be voiced or unvoiced as shown in Table 2.6. The constriction can be located at the lips, just behind the teeth (or at the roof of the mouth, back near the velum) or back of palate.

TABLE 2.6 Types of stops

Constriction	Voiced	Unvoiced
Lips	|*b*|, bat	|*p*| pat
Alveolar	|*d*|, dip	|*t*|, tap
Back of palate	|*g*|, good	|*k*|, cat

Stops have waveforms which are aperiodic, much less intense and usually with most energy at high frequencies. Articulatory profiles for the English stop consonants are shown in Figure 2.11.

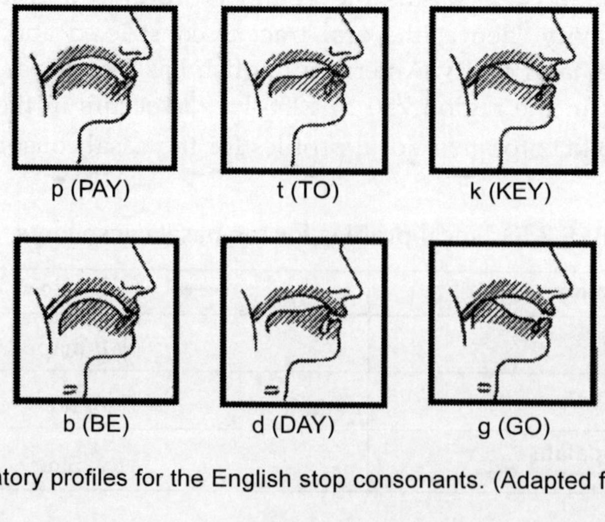

Figure 2.11 Articulatory profiles for the English stop consonants. (Adapted from Potter, Kopp and Green)

Fricatives are produced when turbulence (swirling) of airflow passes through a constriction in the vocal tract. Hence, fricatives employ a narrow constriction in the oral tract or in the glottis. Fricatives include both voiced and unvoiced fricatives as shown in Table 2.7. Fricatives have waveforms with most energy at high frequencies. Vocal tract profiles for the fricative consonants of English are shown in Figure 2.12.

TABLE 2.7 Types of fricatives

Constriction	Voiced	Unvoiced
Teeth/lips	|*v*| vat	|*f*| fit
Teeth	|*Th*| That	|*The*| thaw
Alveolar	|*Z*| zip	|*s*| sap
Palate	|*Zh*| Vision	|*sh*| shape

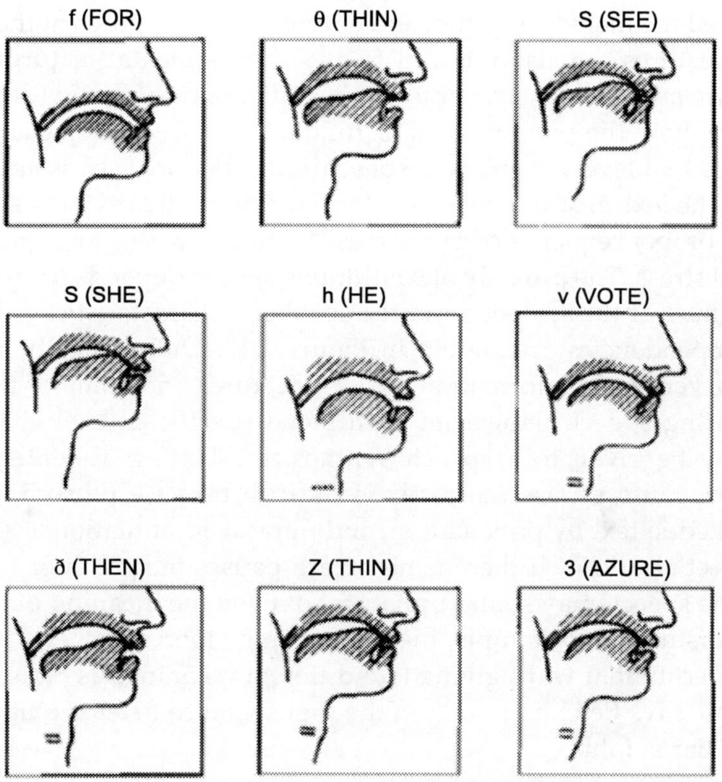

Figure 2.12 Vocal tract profiles for the fricative consonants of English language. The short pairs of lines drawn on the throat represent vocal cord operation. (Adapted from Potter, Kopp and Green)

Affricates are formed by the combination of a plosive (stop) followed by a fricative. It consists of two types of voiced affricates and unvoiced affricates. Voiced affricate $|j|$ can be modelled with the concatenation of the stop $|d|$ and the fricative $|zh|$. For example, "jam" is a combination of $|d|$ followed by $|zh|$. Unvoiced affricate $|ts|$ is a dynamical sound which is modelled with the concatenation of the sound $|t|$ and fricative $|s|$.

Whisper is a consonant similar in the formation to the unvoiced fricative. In whisper, the glottis is open and hence there is no vocal fold vibration. For example, $|h|$, "he" is a whisper.

2.4 PROSODY FEATURES (SUPRASEGMENTALS)

Finding correct intonation, stress, and duration from the written text is probably the most challenging problem. These features together are called prosodic features and may be considered the melody, rhythm, and emphasis of the speech at the perceptual level. Prosodic features are also referred to as suprasegmental features of speech. These features are stress (intensity), intonation (pitch) and quantity (duration). The intonation means how the pitch pattern or fundamental frequency changes during speech. There are two levels of stress in speech: lexical (word) stress and sequential (phrase) stress. The lexical stress pattern is for important identifying a spoken word as the use of the proper sequence of phonemes. A speaker with a foreign accent often misplaces lexical stress. The prosody of continuous speech depends on many aspects, such as the meaning of the sentence and the speaker's characteristics and emotions. The prosodic dependencies are shown in Figure 2.13. Unfortunately, written text usually contains very little information of these features and some of them change dynamically during speech. However, with some specific control characters this information may be given to a speech synthesizer. Timing at sentence level or grouping of words into phrases correctly is difficult because prosodic phrasing is not always marked in text by punctuation, and phrasal accentuation is almost never marked (Santen et al., 1997). If there is no breath pauses in speech or if they are in wrong places, the speech may sound unnatural or even the meaning of the sentence may be misunderstood. For example, the input string "John says Peter is a liar" can be spoken as two different ways giving two different meanings as "John says: Peter is a liar" or "John, says Peter, is a liar". In the first sentence Peter is a liar, and in the second one, the liar is John.

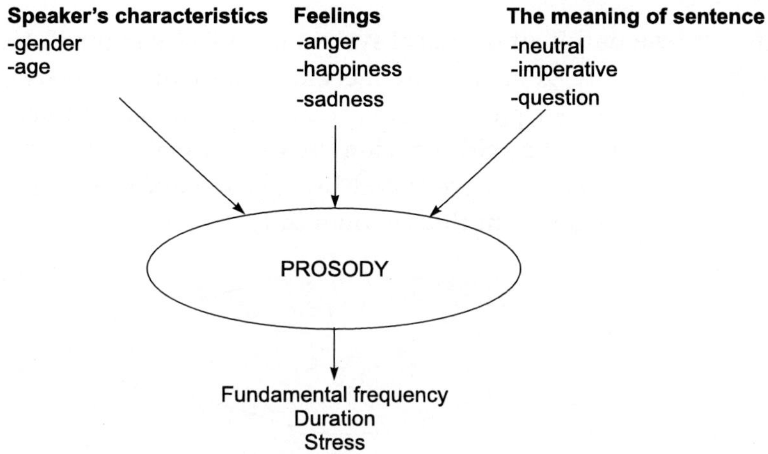

Figure 2.13 Prosodic dependencies

2.5 MODELS FOR SPEECH PRODUCTION

Speech mechanism can be modelled by the excitation generator and variation of impulse response of the vocal tract. The excitation signal generated by the vocal cords or glottis passes via the vocal tract. Based on this assumption, we now give a fundamental speech model for speech production as shown in Figure 2.14.

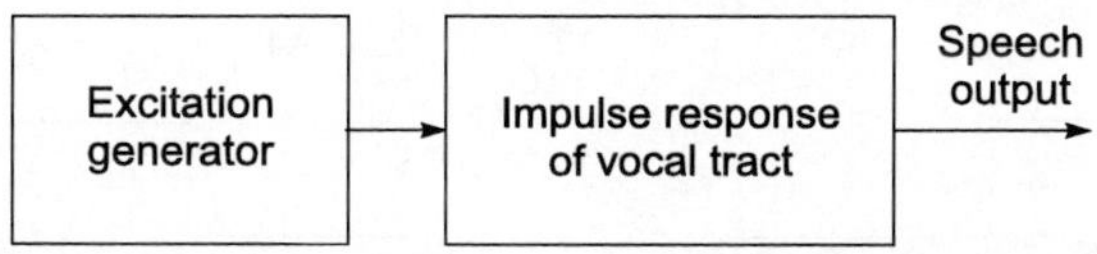

Figure 2.14 Source-system model of speech production

Speech signal consists of a voiced sound or an unvoiced sound. Impulse train generator is used as excitation generator, when a voiced sound is produced. Random signal generator is used as an excitation generator, when unvoiced sound is produced. Vocal tract shape and source characteristics change slowly with time and hence properties of speech signal change relatively with time. Normally, for a speech signal, these parameters are assumed to be constant for a period of 10 to 20 ms. Hence, the parameters over this short segment are described by the linear time invariant (LTI) model. The glottis or vocal cords are represented by the excitation generator, while the vocal tract is represented by the LTI system. If $e(n)$ is the excitation signal and $h(n)$ is the impulse response of the vocal tract, then excitation signal $e(n)$ gets convolved with the impulse response of the vocal tract, $h(n)$ and this convolved signal is speech signal, $s(n)$. Hence, speech signal, $s(n)$, for a short time is given by

$$s(n) = e(n) * h(n) \tag{2.2}$$

The schematic representation of the vocal system is given by Figure 2.15. In voiced sound, the glottal pulses are generated at the glottis and in unvoiced sound; the glottis is open and turbulent air passes from the lungs to the vocal tract. The vocal tract behaves like a lossless tube with varying cross-sectional area and shapes the excitation. The vocal tract can be represented by concatenation of lossless tube of varying cross-sectional area and length as shown in Figure 2.16.

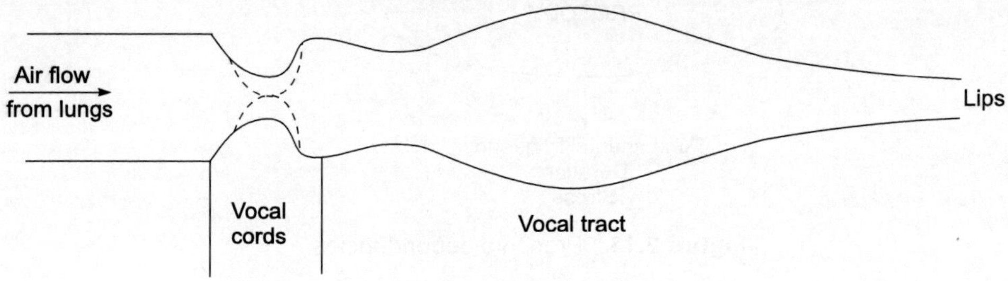

Figure 2.15 Representation of the vocal system

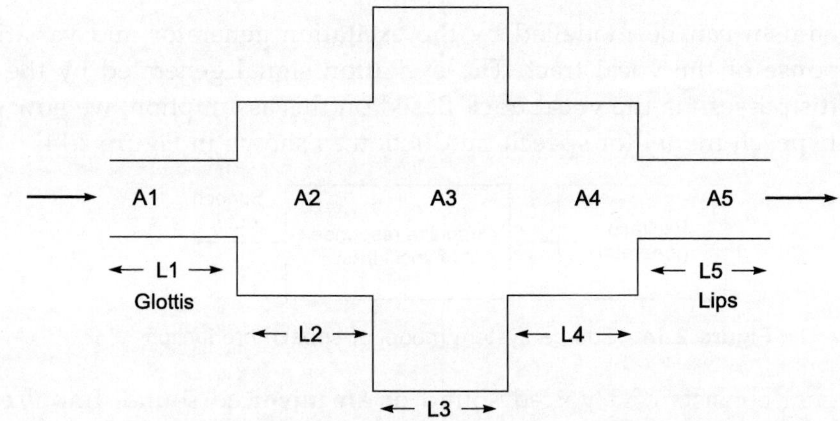

Figure 2.16 Representation of the vocal tract by concatenation of lossless tube

Some vowels are roughly approximated with only two tubes, others require at least three tubes, due to a tongue constriction between the pharyngeal and oral cavities. Lip rounding can form a fourth short tube to model the lips. Hence, the vocal tract is generally modelled as a digital filter (Figure 2.17). Excitation signal is periodic for vowels; its spectrum consists of the fundamental and its harmonics, or in other words, a line spectrum. The line spectrum is filtered to produce speech. So the vocal tract filter is the primary factor for producing speech sounds. When the vocal tract system is excited by an impulse sequence, voiced sound is produced. When the system is excited by random white noise, unvoiced sound is produced. Hence, unvoiced sound sources are modelled by random white noise generator, which includes hardly any periodicity and voiced sounds are modelled by generator of pulses.

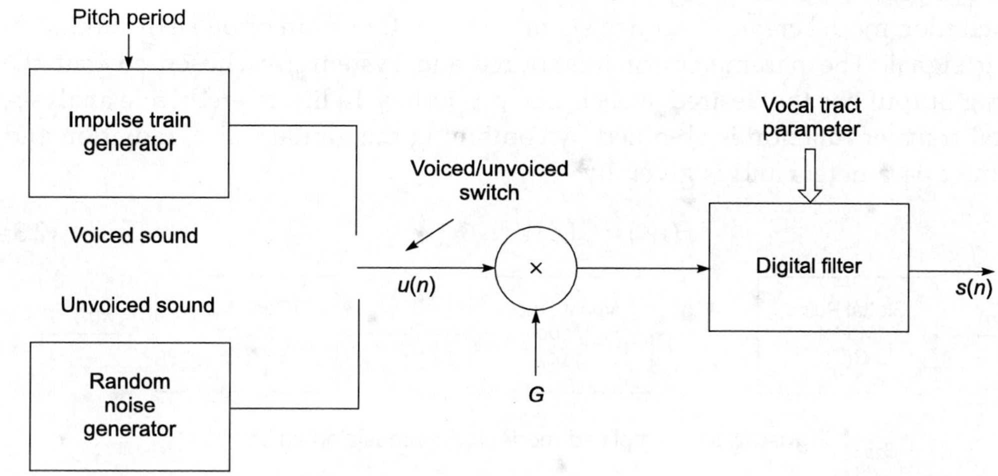

Figure 2.17 Block diagram of simplified model for speech production

2.6 DIGITAL MODEL FOR SPEECH PRODUCTION

Figure 2.18 shows a general block diagram that is representative of numerous models that have been used as the basis for speech processing. These models all have in

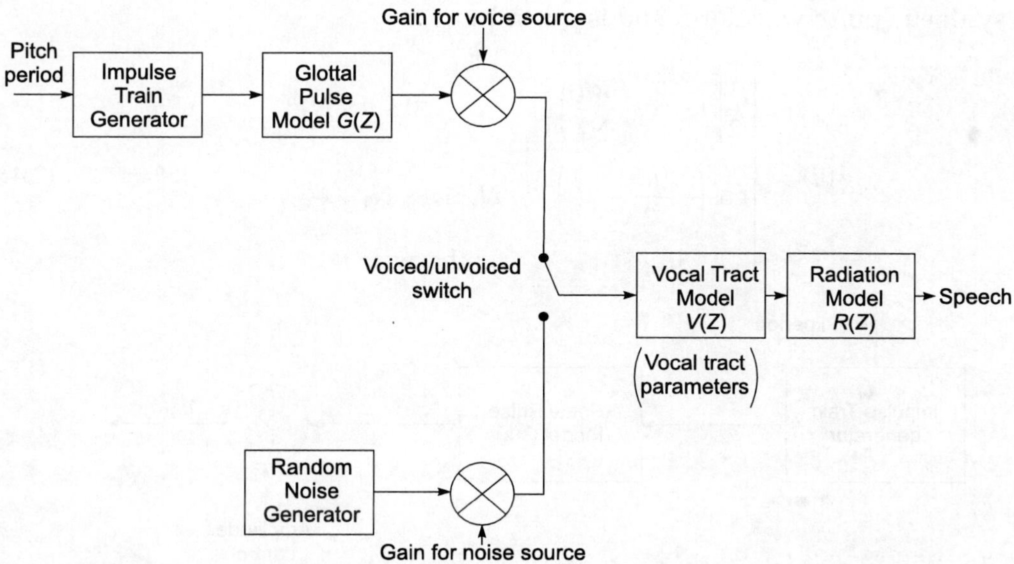

Figure 2.18 General discrete time model for speech production

common that the excitation model separated from the vocal tract and radiation model. Simplified model for synthesizing voiced speech is shown in Figure 2.19. The excitation model creates a source signal that is either train of pulse or randomly varying signal. The parameters of the source and system are chosen so that the resulting output has the desired speech-like properties. In linear predictive analysis, the total transfer function is obtained by combining the glottal pulse, radiation and vocal tract parameters and is given by

$$H(Z) = G(Z)V(Z)R(Z) \tag{2.3}$$

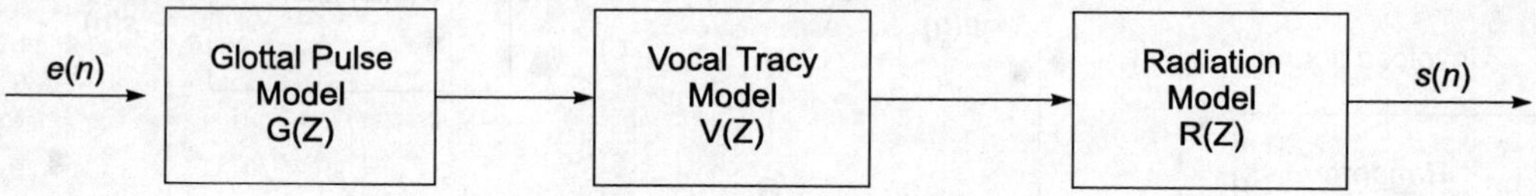

Figure 2.19 Simplified model for synthesizing voice

2.6.1 Excitation or Glottal Pulse Model

When unvoiced sound is produced, the system is excited by random white noise. For unvoiced sounds excitation model is simpler and random number generator provides the excitation signal. When voiced sound is produced, time varying linear system is excited by an impulse sequence. As shown in Figure 2.20, Resenburg developed a glottal pulse model which replaces the natural glottal pulse waveform with synthetic pulse waveform and is given by

$$g(n) = \begin{cases} \dfrac{1}{2}\left[1 - \cos\left(\dfrac{\pi n}{N_1}\right)\right] & 0 \le n \le N_1 \\ \cos\left(\dfrac{\pi(n - N_1)}{2N_2}\right) & N_1 \le n \le N_1 + N_2 \\ 0 & \text{otherwise} \end{cases} \tag{2.4}$$

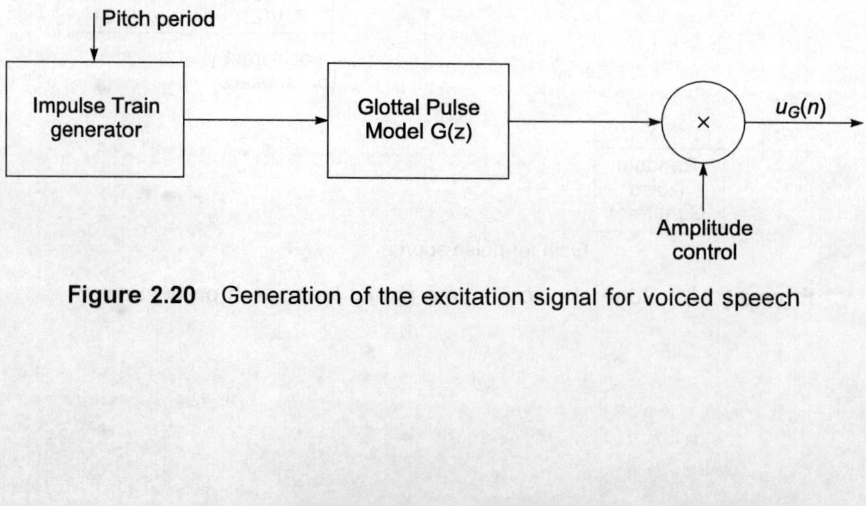

Figure 2.20 Generation of the excitation signal for voiced speech

Figure 2.21 shows a glottal pulse waveform and its Fourier transform magnitude for typical values of N_1 and N_2. The wave shape is very similar in appearance to the pulses shown in Figure 2.4. From its corresponding Fourier transform magnitude the effect of glottal pulse in the frequency domain is to introduce a low-pass filtering effect. Since $g(n)$ has finite length, the Z transform, $G(Z)$ has only zeros.

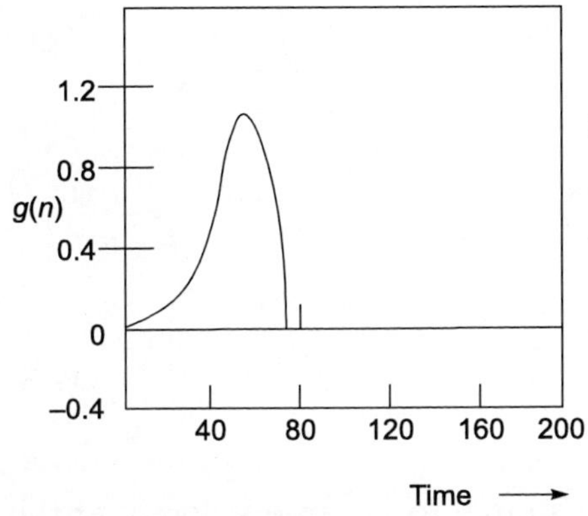

Figure 2.21(a) Rosenberg approximation to glottal pulse

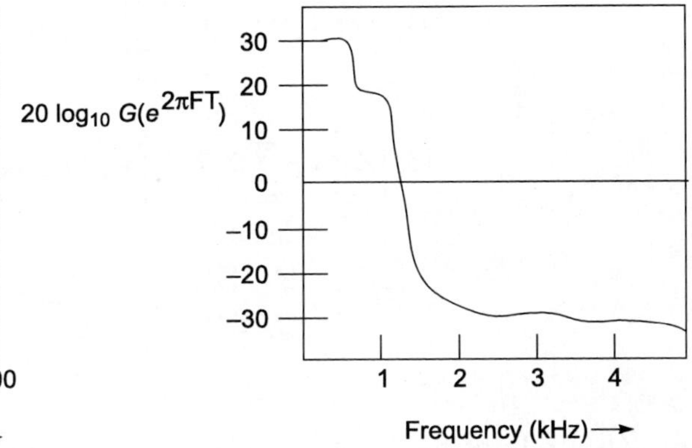

Figure 2.21(b) Corresponding Fourier transform

2.6.2 Vocal Tract Model

Vocal tract can be modelled in a wide variety of ways; important are lossless tube discrete time model and digital filter model.

Lossless tube discrete time model: The vocal tract behaves like a lossless tube with varying cross-sectional area and shapes the excitation (Figure 2.16). The essential feature of this model is depicted in Figure 2.22.

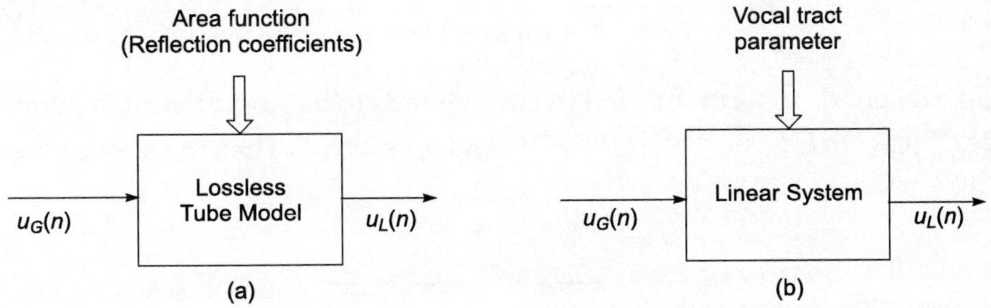

Figure 2.22(a) Representation of the lossless tube model, (b) analog model

Transfer function, $V(Z)$, defines the structure of the model whose parameters vary slowly with time. Parameters of the model are assumed to be constant typically 10-20 msec long. If α_k are vocal tract parameters, then output, $u_L(n)$, is given by

$$u_L(n) = GU_G(n) + \sum_{k=1}^{p} \alpha_k u_L(n-k) \tag{2.5}$$

Taking Z transform on both sides, we obtain

$$U_L(Z) = GU_G(Z) + \sum_{k=1}^{p} \alpha_k Z^{-k} U_L(Z) \tag{2.6}$$

$$U_L(Z)\left[1 - \sum_{k=1}^{p} \alpha_k Z^{-k}\right] = GU_G(Z) \tag{2.7}$$

$$V(Z) = \frac{U_L(Z)}{U_G(Z)} = \frac{G}{\left[1 - \sum_{k=1}^{p} \alpha_k Z^{-k}\right]} \tag{2.8}$$

where, $V(Z)$ relates to the volume velocity at the source to volume velocity at the lips. If the areas of tube model are positive, then all the poles corresponding to $V(Z)$ will be inside the unit circle.

Digital filter model: In this model, samples of speech waveform are assumed to be the output of time varying digital filter. $V(Z)$ is represented as a cascade of the second order system and is given by

$$V(Z) = \prod_{k=1}^{M} V_k(Z) \tag{2.9}$$

$$V_k(Z) = \frac{1 - 2e^{-\alpha_k T}\cos(2\pi F_k T) + e^{-2\alpha_k T}}{1 - 2e^{-\alpha_k T}\cos(2\pi F_k T)Z^{-1} + e^{-2\alpha_k T}Z^{-2}} \tag{2.10}$$

where, F_k corresponds to formants and α_k corresponds to bandwidth (pitch period). If $G_k = 1 - 2e^{-\alpha_k T}\cos(2\pi F_k T) + e^{-2\alpha_k T}, |Z_k| = e^{-\alpha_k T}$ and $\theta_k = 2\pi F_k T$. Then the above equation is given by

$$V_k(Z) = \frac{G_k}{1 - 2|Z_k|\cos(\theta_k)Z^{-1} + |Z_k|^2 Z^{-2}} \tag{2.11}$$

Cascade form of implementation of $V(Z)$ is shown in Figure 2.22.

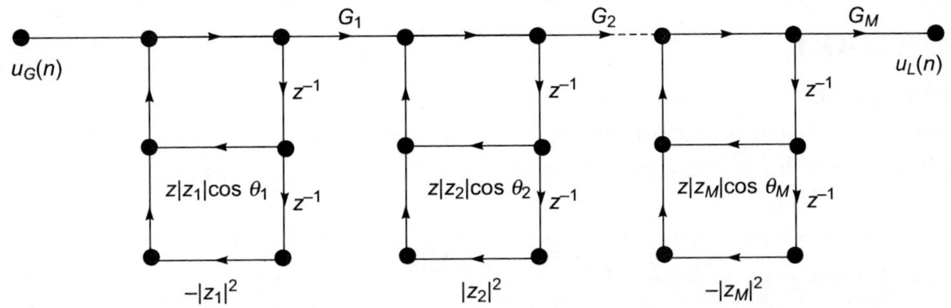

Figure 2.23 Cascade implementation of all-pole transfer function

2.6.3 Radiation Model

Terminal analog model including radiation effects is shown in Figure 2.24. Transfer function $V(Z)$ of the vocal tract relates the volume velocity at the source to volume velocity at the lips.

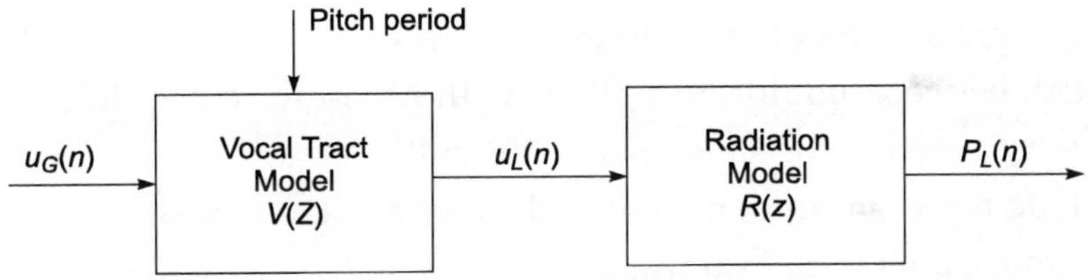

Figure 2.24 Terminal analog model including radiation effects

The function of lips is to convert velocity to pressure. The relation between the pressure and volume velocity is given by

$$P_L(Z) = R(Z)U_L(Z) \tag{2.12}$$

where, $R(Z)$ is the Z transform of lip impedance. Lip impedance is given by

$$Z_L(\Omega) = \frac{j\Omega R_r L_r}{R_r + j\Omega L_r} \tag{2.13}$$

where, $R_r = \dfrac{128}{9\pi^2}$ and $L_r = \dfrac{8a}{3\pi c}$, a is radius of the opening, c is velocity of sound.

Lip impedance can be modelled by a parallel RL circuit, which acts as a high-pass filter with cut-off frequency 6 kHz. Lip impedance behaves like a differentiator; very small at low frequencies and growing at high frequencies.

2.7 SUMMARY

This chapter presents the human vocal system, physiological mechanism of speech production, classification of phonemes of American English, prosody features along with the basic and digital models of speech production mechanism.

EXERCISES

Review Questions

1. Classify phonemes with examples.
2. With the help of a diagram, explain human speech production system.
3. Distinguish between production of voiced and unvoiced speech sounds.
4. Describe the salient features of vowels, semivowels and consonants.
5. Distinguish between voiced and unvoiced fricatives with examples.
6. Explain the production of stops with examples.
7. Distinguish between liquids and glides with necessary examples.
8. Explain why analysis speech signal is difficult.
9. Differentiate between fundamental and formant frequencies.
10. Classify consonants with examples.
11. Explain production and classification of plosives with examples.
12. Describe the features of prosody.
13. With necessary assumptions, describe the LTI model for speech signals.
14. With necessary diagram, explain digital model for speech production.
15. Distinguish between lossless tube discrete time model and digital filter model of the vocal tract.
16. With necessary diagram, explain how radiation model converts velocity to pressure.

3
CHAPTER

Fundamentals of Digital Signal Processing

This chapter provides insights on

- Types of signals and systems
- Convolution and correlation techniques
- Transformation techniques
- Design of digital filters
- Interpolation & decimation techniques

3.1 INTRODUCTION

Digital signal processing (DSP) is one of the most powerful techniques that will shape science and engineering in the 21st century. Digital signal processing is a technique of performing mathematical operations on signals represented as sequences of samples using coding on a digital signal processor (DSP) chip. Digital signal processing is becoming an important modern tool in the fields like speech and audio processing, biomedical, communication, robotics, acoustics, radar, geophysics, and control applications. Speech generation and recognition are used to communicate between humans and machines. After digital recording, speech signals are processed by using DSP algorithms depending on the applications. During playback, processed stored data are converted back into an analog signal. This chapter deals with the basic techniques of digital signal processing, i.e., types of signals, basic DSP algorithms, types of processing systems, and commercial digital signal processors.

3.2 SIGNAL CLASSIFICATION

Signals play an important role in our life and convey information. Examples of signals that we encounter frequently are electrocardiogram (ECG), electromyogram (EMG), electroencephalogram (EEG) and speech signal. Figure 3.1 shows examples for continuous time signal.

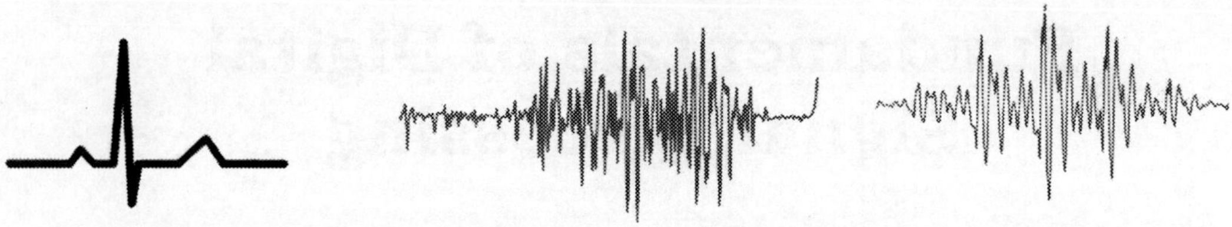

Figure 3.1 Continuous time signals (a) ECG, (b) EMG and (c) speech signal

ECG, EMG, and EEG signals provide information about the condition of the heart, about the muscles, and about the condition of the brain, respectively. All these signals are information bearing signals as a function of time.

A signal is defined as any physical quantity that varies with time, space or any other variable such as frequency, temperature, pressure, distance, position, etc.

Based on modelling, signals are classified into two types.

- Deterministic signal
- Non-deterministic signal

Deterministic signals can be described by mathematical models and hence the characteristics of such a signal can be predicted at any time. Example for this signal is $x(t) = M \sin \omega t$, in this the amplitude of the signal varies sinusoidally, with time and its maximum value is M. Non-deterministic signals cannot be described by mathematical models and are described in terms of probabilities or with the help of their statistical properties. Example for non-deterministic signal is noise in an electrical circuit.

Signals can be classified based on their nature and characteristics in the time domain into the following types.

- Continuous time signal
- Discrete time signal
- Digital signal

Continuous Time Signal

Figure 3.1 shows examples of continuous time signal. Almost all the signals generated from various sources in the nature are continuous time signals. Continuous time signals are also called *analog signals*. In analog signals both amplitude and time are continuous variables. Hence, an analog signal is continuous in amplitude and defined at every instant of time. It is represented by $x(t)$.

Discrete Time Signal

Discrete time signals are obtained by time sampling a continuous time signal as shown in Figure 3.2. Discrete time signals are defined only at certain specific values of time and have discrete amplitude levels. The signal has values only at discrete instants of time. The continuous time variable (t) and discrete variable (n) are related by

$$t = nT_s \qquad n = 0, 1, 2, 3... \tag{3.1}$$

Then analog signal and discrete signal are related by

$$x(t) = x(nT_s) \qquad n = 0, 1, 2, 3... \tag{3.2}$$

This means discrete signal is defined only at n = 0, 1, 2, 3, 4... or $x(t)$ and is represented as samples at 0, T_s, $2T_s$, $3T_s$, ... Time difference between any two adjacent samples is called *sampling time* (T_s) and is determined by the highest frequency (F_{max}) of the input signal. The sampling frequency should be at least twice the maximum frequency of the signal.

$$F_s = \frac{1}{T_s} \geq 2F_{max} \tag{3.3}$$

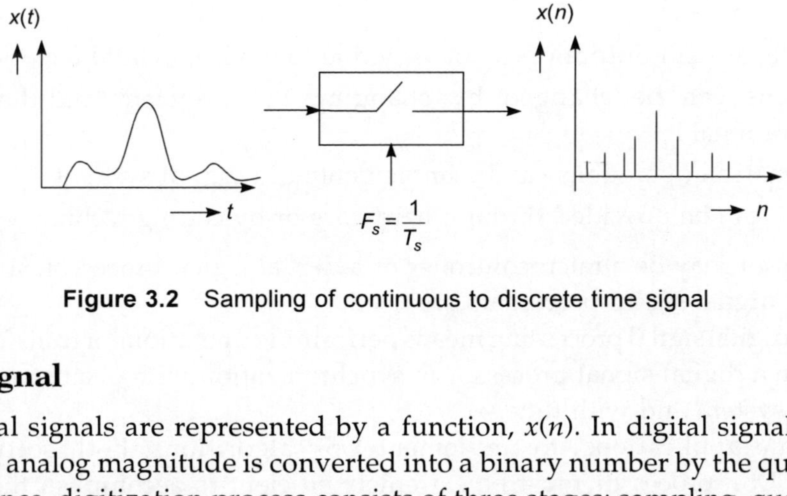

Figure 3.2 Sampling of continuous to discrete time signal

Digital Signal

These digital signals are represented by a function, $x(n)$. In digital signal, sampled value of the analog magnitude is converted into a binary number by the quantization process. Hence, digitization process consists of three stages: sampling, quantization, and encoding. Quantization converts the sampled amplitude into bit streams (a string of 0's and 1's). Encoder represents each quantized sample by a code word. Hence in a digital signal, amplitudes of the signal are represented by 1, 3, 4, 8, 16, 32 or 64 bits.

3.3 DIGITAL SIGNAL PROCESSING SYSTEM

Figure 3.3 represents the block diagram of a typical digital signal processing system. It consists of analog to digital converter (A/D), digital signal processor (DSP) and a digital to analog converter (D/A). Analog signals are converted to digital signal by an analog to digital converter (A/D) consisting of a sampler, a quantizer and an encoder. Signal is sampled in time at a constant rate and analog magnitude is converted into a binary number with an analog to digital converter. Digital signals are processed by using digital signal processor. After processing, digital outputs of the processor are converted into an equivalent analog signal by digital to analog converter.

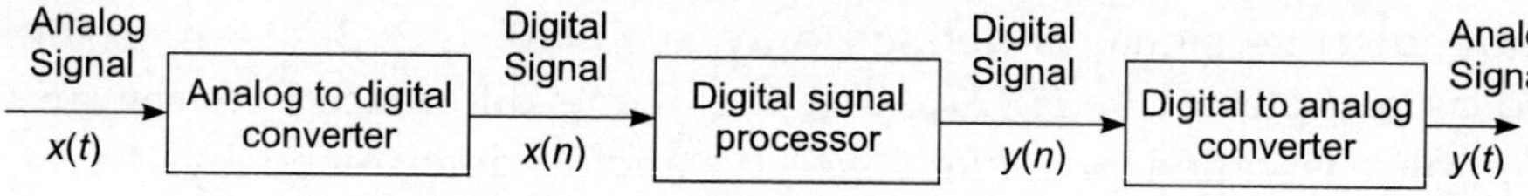

Figure 3.3 Digital Signal Processing System

Following are the advantages of digital signal processing system compared to analog processing system:

- Digital systems are accurate compared to analog systems
- Storage facility
- In digital system output does not vary due to environmental factors
- Functions can be changed by changing the program and hence it is reprogrammable
- Mathematical operations can be implemented in digital systems
- Security can be provided through hardware or by coding techniques

The processor may be a microcontroller or a digital signal processor. In real-time applications, digital signal processors are used.

Real-time digital signal processing means performing operations or transformations of a signal on a digital signal processor in synchronization with events occurring in the physical system and with time.

In real-time applications, to implement DSP algorithms, both software and hardware of DSP processor must be extremely efficient to accomplish this. Digital signal processor is optimized to accomplish these tasks fast enough to maintain real-time operations. DSPs differ from microcontrollers in that they are specially designed by using Sharc or advanced Harvard architecture to perform arithmetic operations and accumulations in a single cycle. They support mathematical functions specifically intended for processing of digital signals using DSP algorithms.

Commercial Digital Signal Processors

Intel announced the first digital signal processor chip Intel 2920 in 1979. In 1990, Intel released digital signal processor chip i960, which is used in DSP applications and in multimedia. In 1982, DSP chips were designed by Texas Instruments TMS 32010 and this chip was capable of 16-bit integer arithmetic operations and can execute 5 million instructions per second (MIPS) and had limited random access memory (RAM), read only memory (ROM) and input/output (I/O) devices. The salient features of some of the DSP processors are summarized in Table 3.1.

TABLE 3.1 Commercial DSP processors

Vendor	DSP Processors	Type of processor	Data Width	Execution Speed (MIPS)
AT&T	DSP16xx	Fixed	16	70.0
	DSP32xx	Floating	32	20.0
Motorola	DSP563xx	Fixed	24	80.0
	DSP96002	Floating	32	20.0
Analog Devices	ADSP-21xx	Fixed	16	33.3
	ADSP-210xx	Floating	32	40.0
Texas Instruments	TMS320C54xx	Fixed	16	50.0
	TMS320C6713	Floating	32	1800.0

3.4 STANDARD DISCRETE TIME SIGNALS

A discrete time signal is one that can be represented by a sequence of numbers. In this section, we will discuss representation of discrete time signals using impulse sequence or unit sample sequence and unit step sequence.

Impulse Sequence or Unit Sample Sequence

The impulse sequence or unit sample sequence is represented by $\delta(n)$. The unit impulse or unit sample sequence is defined by

$$\delta(n)=\begin{cases}1 & n=0\\ 0 & n\neq 0\end{cases} \tag{3.4}$$

In the sequence form it is represented as

$$\delta(n)=\left\{\begin{matrix}\dots\dots\dots.0,0,0,0,1,0,0,0,0,\dots\dots\dots\dots\\ \uparrow\end{matrix}\right\}$$

'↑' (arrow) in the above sequence represents 0th sample. The shifted impulse sequence or shifted unit sample sequence is defined by

$$\delta(n-k)=\begin{cases}1 & n=k\\ 0 & n\neq k\end{cases} \tag{3.5}$$

Figure 3.4(a) shows the representation of unit sample sequence, $\delta(n)$ and Figure 3.4(b) shows the representation of shifted unit sample sequence, $\delta(n-3)$

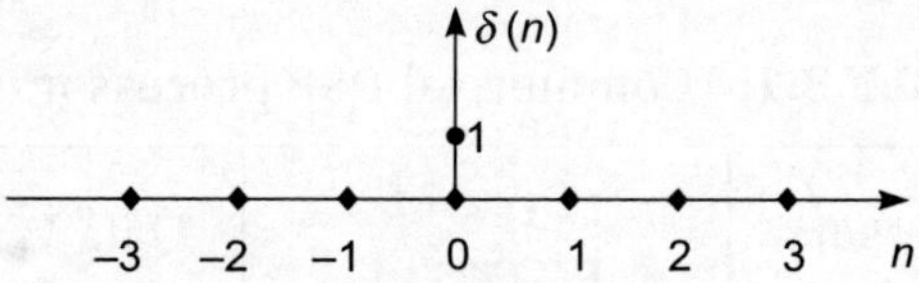

Figure 3.4(a) Unit sample sequence

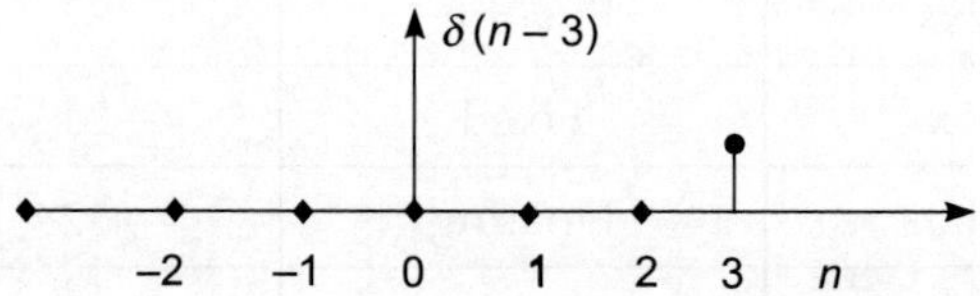

Figure 3.4(b) Delayed unit sample

Example 3.1 If $x(n)=\left\{\begin{matrix}-3, & 6, & 0, & 1, & 2, & -3, & 7\\ & & & & \uparrow & & \end{matrix}\right\}$ then $x(n)$ is represented by unit sample sequence as $x(n)=-3\delta(n+4)+6\delta(n+3)+\delta(n+1)+2\delta(n)-3\delta(n-1)+7\delta(n-2)$

Unit Step Sequence

The unit step sequence is represented by $u(n)$ and is defined by

$$u(n)=\begin{cases}1 & \text{for } n\geq 0\\ 0 & \text{for } n<0\end{cases} \tag{3.6}$$

Shifted unit step sequence, $u(n-k)$ is defined by

$$u(n-k)=\begin{cases}1 & \text{for } n\geq k\\ 0 & \text{for } n<k\end{cases} \tag{3.7}$$

Figure 3.5(a) shows the representation of unit step sequence, $u(n)$ and Figure 3.5(b) shows the representation of delayed unit step sequence, $u(n-4)$.

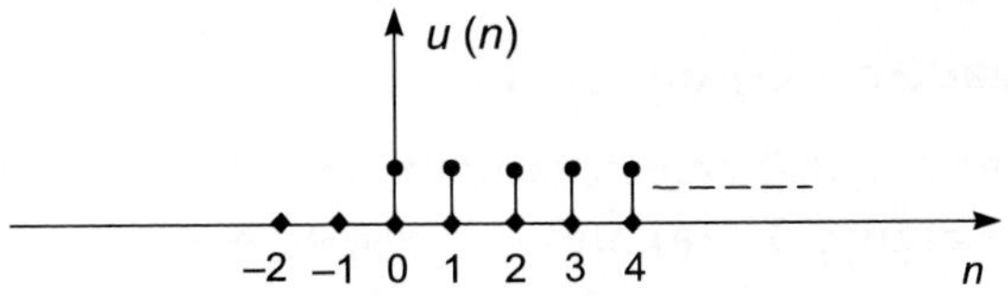

Figure 3.5(a) Unit step sequence

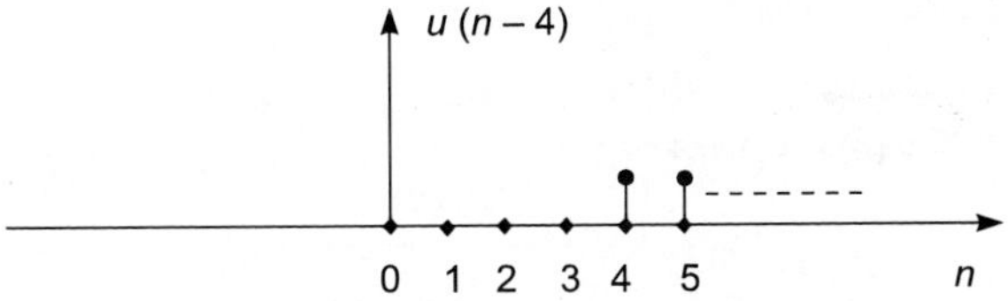

Figure 3.5(b) Delayed unit step sequence

Example 3.2 If $x(n)=\left\{\begin{matrix}2, & 2, & 2, & 2\\ \uparrow & & & \end{matrix}\right\}$, then $x(n)$ is represented by unit step sequence as

$$x(n) = 2[u(n) - u(n-4)]$$

Basic Operation on Discrete Signals

Following basic operations can be carried out on an input signal to obtain output signal.

Multiplication by a Constant

Given sequence $x(n)$ is multiplied by a constant a, results in a new sequence $y(n)=ax(n)$ and a may be a positive or a negative constant. Graphically, representation is shown in Figure 3.6.

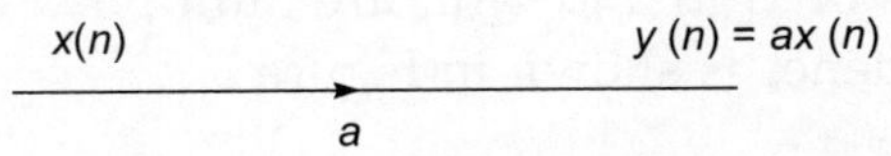

Figure 3.6 Symbol used to represent scaling operation

$$y(n)=ax(n) \tag{3.8}$$

Example 3.3 If $x(n)=\left\{\begin{matrix}4, & 2, & 3\\ \uparrow & & \end{matrix}\right\}$ and $a=2$. Then $y(n)=\left\{\begin{matrix}8, & 4, & 6\\ \uparrow & & \end{matrix}\right\}$

Addition and Subtraction Operations

Two sequences, $x_1(n)$ and $x_2(n)$, can be added or subtracted; results of the new sequences are shown in Figure 3.7 (a) and (b), respectively.

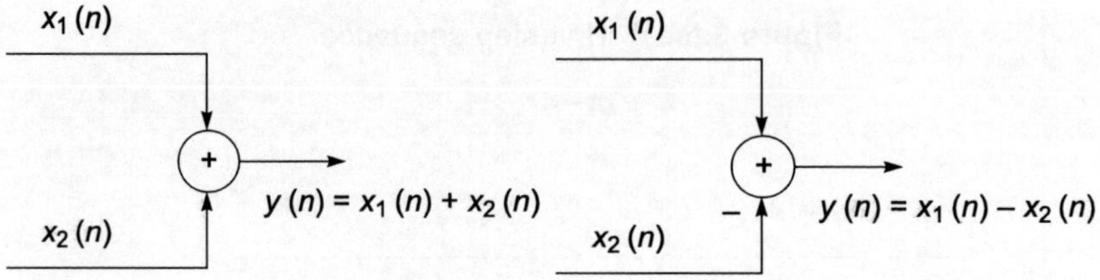

Figure 3.7 Symbol used to represent (a) addition, and (b) subtraction

Two sequences $x_1(n)$ and $x_2(n)$ are added, then $y(n) = x_1(n) + x_2(n)$ (3.9)

Two sequences $x_1(n)$ and $x_2(n)$ are subtracted, then $y(n) = x_1(n) - x_2(n)$ (3.10)

Example 3.4 If $x_1(n) = \left\{\begin{matrix}1, & 2, & 3 \\ \uparrow & & \end{matrix}\right\}$ and $x_2(n) = \left\{\begin{matrix}3, & 5, & 2 \\ \uparrow & & \end{matrix}\right\}$

Then $$y(n) = x_1(n) + x_2(n) = \{1+3, 2+5, 3+2\} = \left\{\begin{matrix}4, & 7, & 5 \\ \uparrow & & \end{matrix}\right\}$$

and $$y(n) = x_1(n) - x_2(n) = \{1-3, 2-5, 3-2\} = \left\{\begin{matrix}-2, & -3, & 1 \\ \uparrow & & \end{matrix}\right\}$$

Multiplication Operation

Two sequences, $x_1(n)$ and $x_2(n)$, can be multiplied as shown in Figure 3.8. Corresponding n^{th} samples of $x_1(n)$ and $x_2(n)$ are multiplied to form the nth sample and the resulting new sequence is shown in Figure 3.8.

$$y(n) = x_1(n) \cdot x_2(n) \tag{3.11}$$

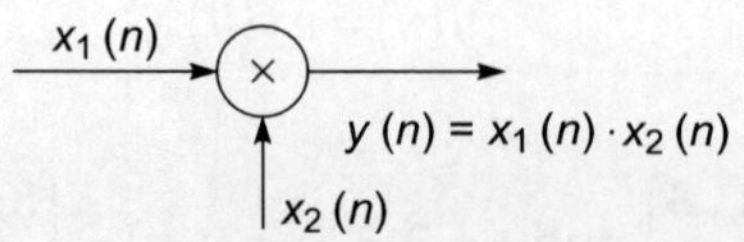

Figure 3.8 Symbol used to represent multiplication

Example 3.5 If $x_1(n) = \left\{ \underset{\uparrow}{-1}, 1, 2 \right\}$ and $x_2(n) = \left\{ \underset{\uparrow}{2}, 3, -1 \right\}$

Then $$y(n) = x_1(n) \cdot x_2(n) = \left\{ \underset{\uparrow}{-1.2}, 1.3, 2, -1 \right\} = \left\{ \underset{\uparrow}{-2}, 3, -2 \right\}$$

Shifting Operation

If the sequence $x(n)$ is shifted left by k units, we obtain new sequence $x(n + k)$ or shifted right by k units, we obtain new sequence $x(n - k)$. The resulting new sequences are shown in Figure 3.9.

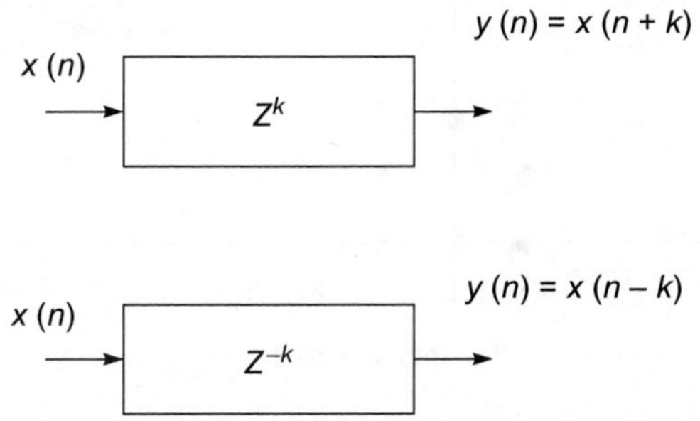

Figure 3.9 Symbol used to represent shifting operation

Example 3.6 If $x(n) = \left\{ \underset{\uparrow}{1}, 2, 1 \right\}$, as shown in Figure 3.10(a), signal $x(n - 2)$ is obtained by shifting $x(n)$ right by 2 units of time as shown in Figure 3.10(b) and signal $x(n + 1)$ is obtained by shifting $x(n)$ left by one unit of time as shown in Figure 3.10(c).

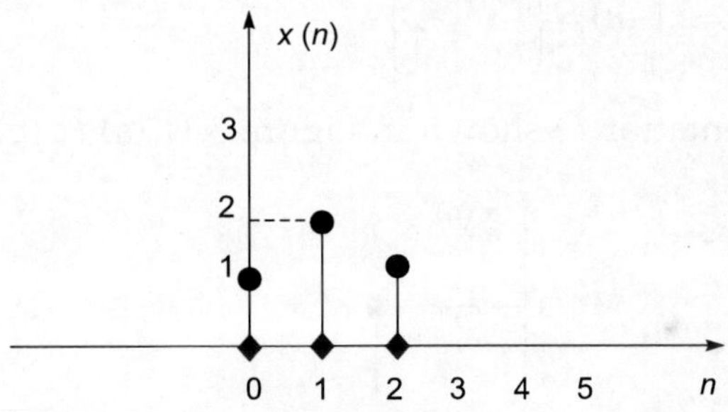

Figure 3.10(a)

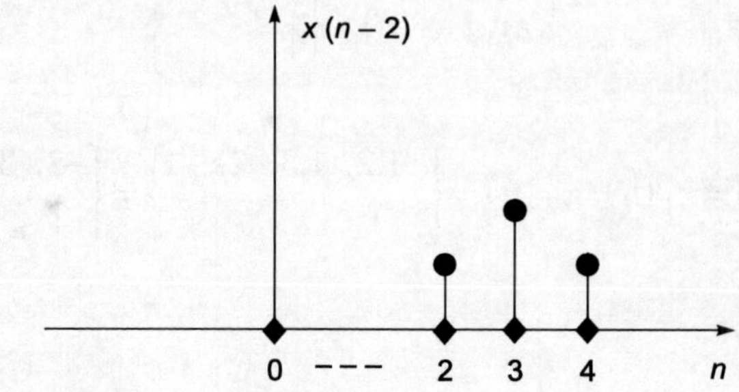

Figure 3.10(b)

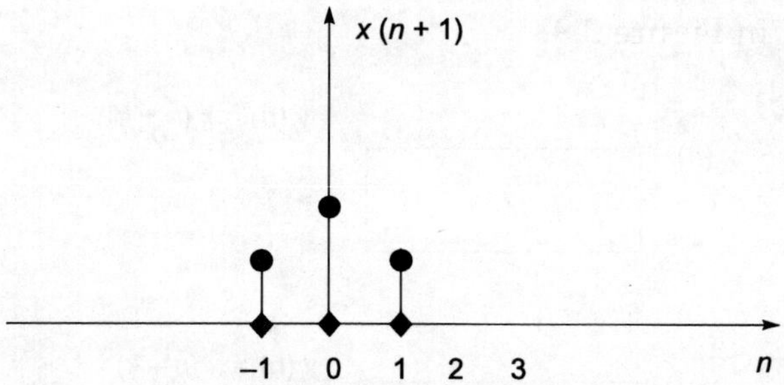

Figure 3.10(c)

Folding or Time Reversal

The operation of folding or time reversal of a digital signal sequence is defined as

$$y(n) = x(-n) \tag{3.12}$$

Example 3.7 Let $x(n) = \left\{ \begin{matrix} 1, & 1, & 3 \\ \uparrow & & \end{matrix} \right\}$, if $x(n)$ is folded at $n = 0$, then

$$y(n) = x(-n) = \left\{ \begin{matrix} 3, & 1, & 1 \\ & & \uparrow \end{matrix} \right\}$$

The graphical representation is shown in Figure 3.11 (a) & (b).

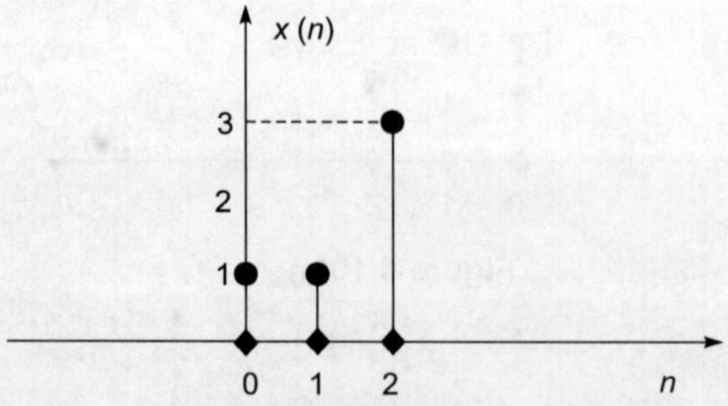

Figure 3.11(a)

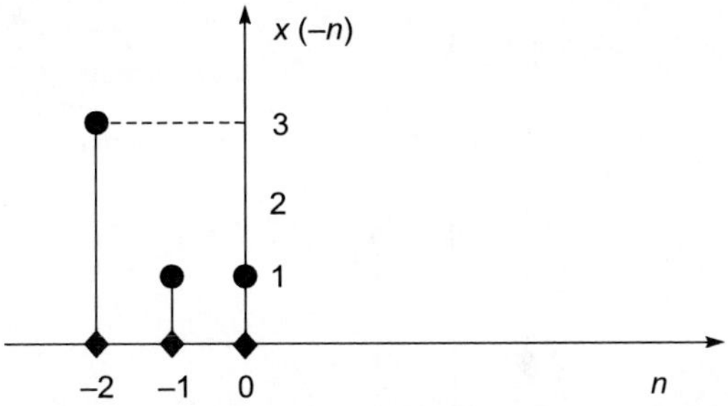

Figure 3.11(b)

Example 3.8 Discrete time signals $x_1(n)$ and $x_2(n)$ are shown in Figure 3.12 (a) & (b). Sketch the following signals.

(a) $y_1(n) = 2x_1(n)$

(b) $y_2(n) = 2\{x_1(n) - x_2(n)\}$

(c) $y_3(n) = x_1(n)\,x_2(n)$

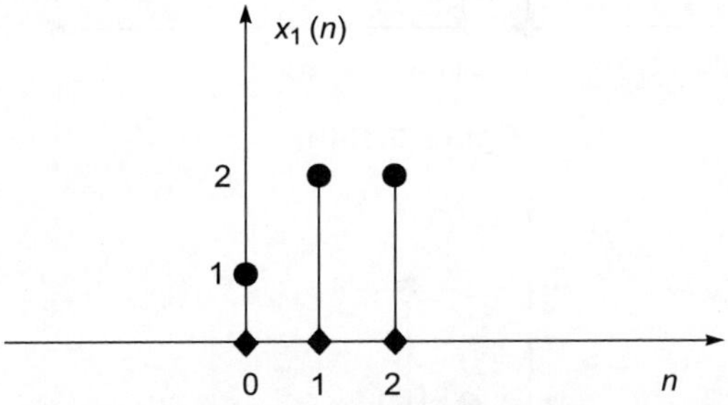

Figure 3.12(a)

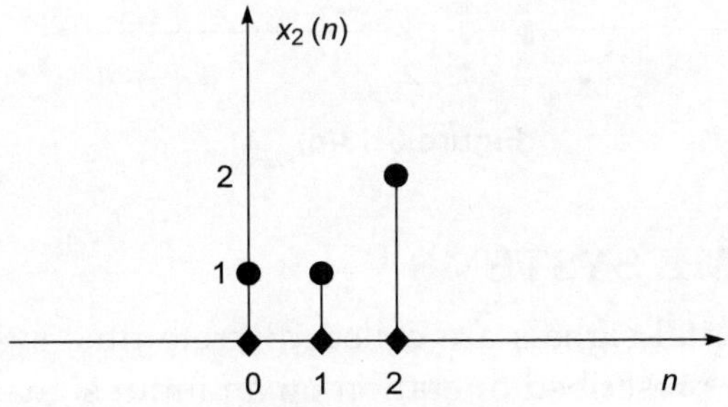

Figure 3.12(b)

Solution

(a) $y_1(n)$ is sketched in Figure 3.13 (a)

(b) $y_2(n)$ is sketched in Figure 3.13 (b)

(c) $y_3(n)$ is sketched in Figure 3.13 (c)

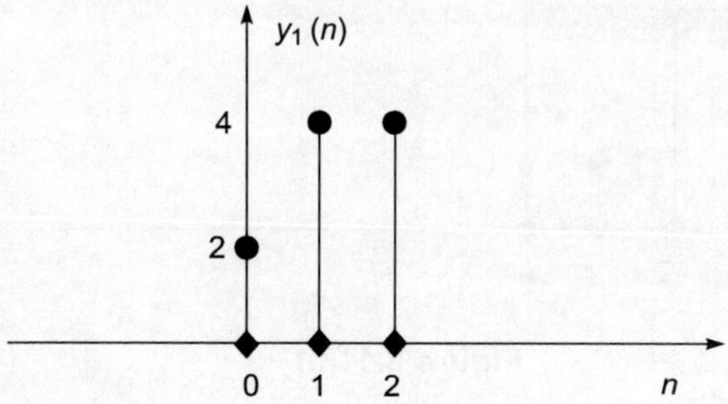

Figure 3.13(a)

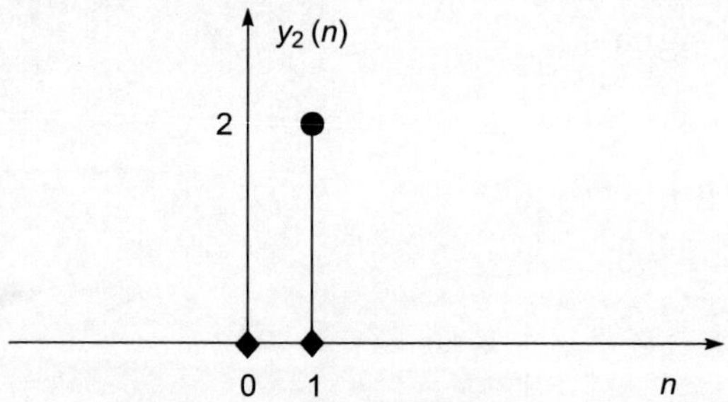

Figure 3.13(b)

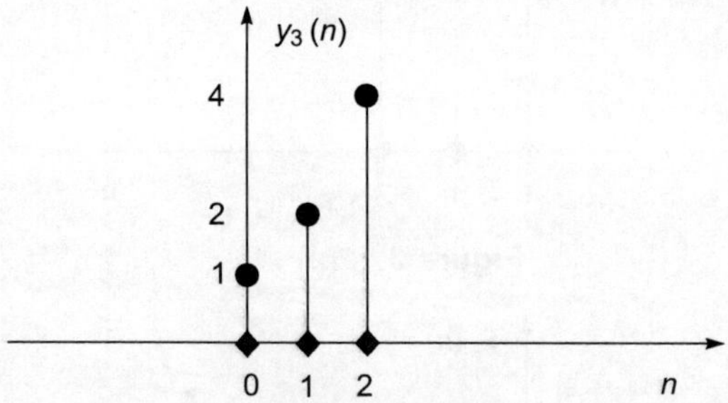

Figure 3.13(c)

3.5 DISCRETE TIME SYSTEMS

Systems that process digital signals are called discrete time systems (DTS). It is a device that performs the prescribed operation on an input signal $x(n)$ and produces another output signal $y(n)$. If T represents transformation operation and $y(n)$, the response to the input is signal $x(n)$, then the input output relationship for a discrete system is as shown in Figure 3.14, and $y(n)$ is given by

$$y(n) = T\left[x(n)\right] \tag{3.13}$$

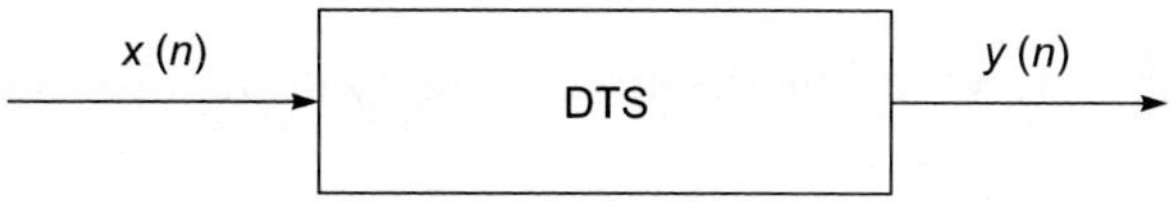

Figure 3.14 Discrete time system

Mathematically, T is defined as an operator that maps an input sequence, $x(n)$, into an output sequence, $y(n)$.

Example 3.9 If discrete time systems are excited by an input sequence $x(n)=\left\{\underset{\uparrow}{1}, -2,\ 2, -3\right\}$. If the system performs scaling operation which multiplies the input sequence, $x(n)$, by 4 (say), then the output sequence, $y(n)$, is given by

$$y(n)=4x(n),\ y(n)=\left\{\underset{\uparrow}{4},\ -8,\ 8,\ -12\right\}$$

Systems are classified according to their characteristics and properties as follows:

- Linear and nonlinear systems
- Time invariant and time variant systems
- Static and dynamic systems
- Causal and non-causal systems
- Stable and unstable systems

3.5.1 Linear and Nonlinear Systems

A system is said to be linear, if it satisfies the principle of superposition. If $x_1(n)$ and $x_2(n)$ are two input sequences, then the system is called linear

$$T\left[cx_1(n)+dx_2(n)\right]=cT\left[x_1(n)\right]+dT\left[x_2(n)\right] \tag{3.14}$$

Figure 3.15 shows the implementation of above equation.

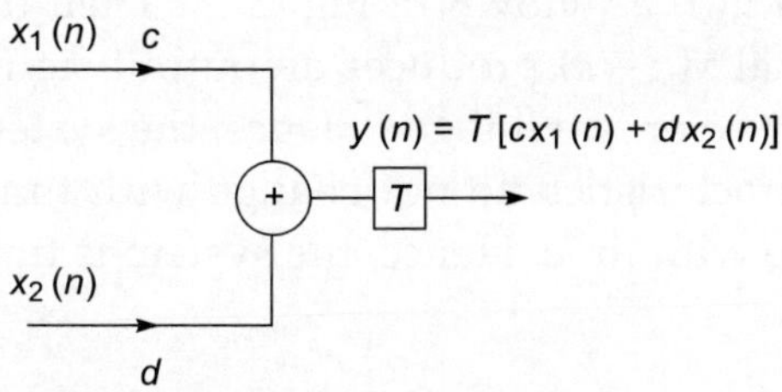

Figure 3.15(a) Representation of LHS of Eq. 3.14

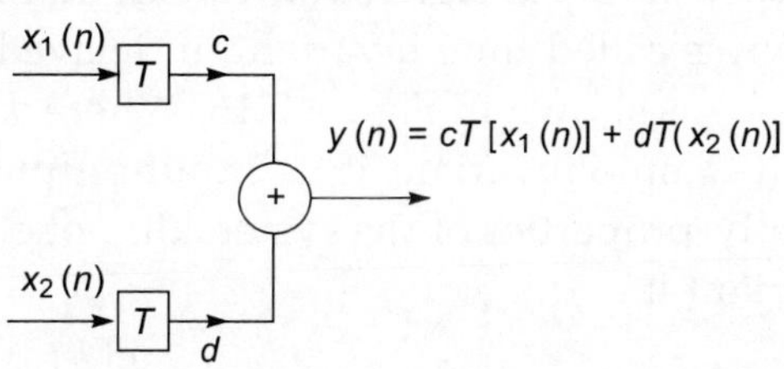

Figure 3.15(b) Representation of RHS of Eq. 3.14

Example 3.10 Determine whether the following systems are linear or nonlinear

1. $y(n) = 5x(n)$
2. $y(n) = 2x^2(n)$

1. Let $x_1(n)$ and $x_2(n)$ be are two inputs, then the corresponding outputs are given by

$$y_1(n) = 5x_1(n) \text{ and } y_2(n) = 5x_2(n)$$

The system is linear if

$$T[ax_1(n) + bx_2(n)] = aT[x_1(n)] + bT[x_2(n)]$$
$$5[ax_1(n) + bx_2(n)] = 5ax_1(n) + 5bx_2(n)$$

LHS = RHS, it obeys the principle of superposition, hence the system is linear.

2. Let $x_1(n)$ and $x_2(n)$ be two inputs, then the corresponding outputs are given by

$$y_1(n) = 2x_1^2(n) \text{ and } y_2(n) = 2x_2^2(n)$$

The system is linear if

$$T[ax_1(n) + bx_2(n)] = aT[x_1(n)] + bT[x_2(n)]$$
$$T[ax_1(n) + bx_2(n)] = 2[ax_1(n) + bx_2(n)]^2$$
$$aT[x_1(n)] + bT[x_2(n)] = 2a[x_1(n)]^2 + 2b[x_2(n)]^2$$

LHS # RHS, it fails to satisfy the superposition theorem, hence the system is non-linear.

3.5.2 Time Invariant and Variant Systems

Suppose an input signal $x(n)$ causes an output $y(n)$ as shown in Fig. 3.16. Then the system is called time invariant, if delayed signal $x(n - k)$ produces an output signal $y(n - k)$, as shown in Figure 3.16. Where, k is a +ve or –ve integer. Hence, the system is called time invariant if its input-output characteristics do not change with time. Usually, properties of the system do not change with time. Hence, the system is time invariant if

$$T[x(n)] = y(n)$$

Then
$$T[x(n-k)] = y(n-k) \tag{3.15}$$

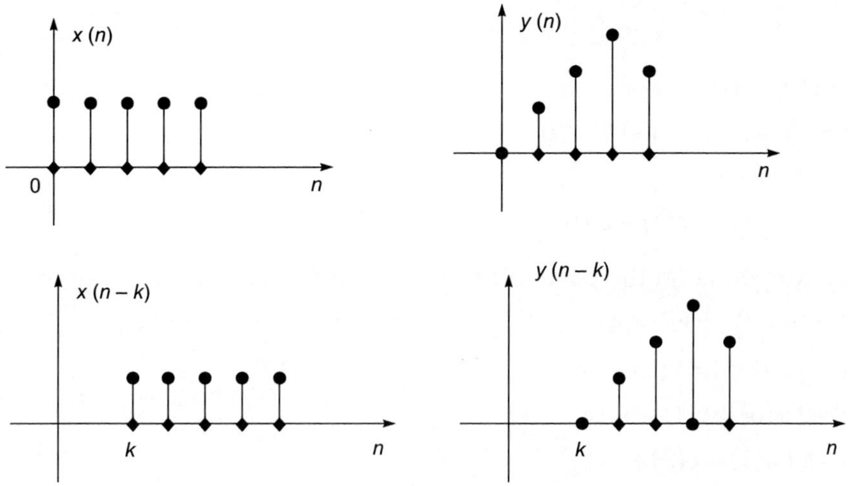

Figure 3.16 Response of a time invariant system to a delayed input signal

A system is called time variant, if it fails to satisfy Eq. (3.15).

Example 3.11 Determine whether the following systems are time invariant or variant

1. $y(n) = nx(n)$
2. $y(n) = x(n) + x(n-1)$

Solution:

1. If $x(n)$ is the input, then output is $y(n) = T[x(n)] = nx(n)$, the system multiplies the input by n. If $x(n - k)$ is input, then output is given by $T[x(n-k)] = nx(n-k) \neq y(n-k)$, hence the system is time variant.
2. If $x(n)$ is the input, then output is $y(n) = T[x(n)] = x(n) + x(n-1)$, the system adds previous input to $x(n)$. If $x(n - k)$ is input, the output is given by $T[x(n-k)] = x(n-k) + x(n-k-1) = y(n-k)$, hence the system is time invariant.

3.5.3 Causal and Non-causal Systems

If output depends only on the past and present values of the input, then such type of discrete time system is called causal system, else the system is called non-causal.

Example 3.12 Determine whether the following systems are causal or non-causal

1. $y(n) = x(n) - x(n-1)$
2. $y(n) = x(n+1) + 3x(n+4)$
3. $y(n) = x(n)^4$

Solution:

1. $y(n) = x(n) - x(n-1)$

   For $n = -1$ $y(-1) = x(-1) - x(0)$

   For $n = 0$ $y(0) = x(0) - x(-1)$

   For $n = +1$ $y(1) = x(1) - x(0)$

The output depends on the past and present input, therefore the system is causal.

2. $y(n) = x(n+1) + 3x(n+4)$

   For $n = -1\, y(-1) = x(0) + 3x(3)$

   For $n = 0\, y(0) = x(1) + 3x(4)$

   For $n = +1\, y(1) = x(2) + 3x(5)$

The output depends on the future input, therefore the system is non-causal

3. $y(n) = x(n^4)$

   For $n = -1 y(-1) = x(1)$

   For $n = 0 y(0) = x(0)$

   For $n = +1 y(1) = x(1)$

For negative values of n, the system depends on future inputs, so the system is non-causal.

3.5.4 Static and Dynamic System

A system is called instantaneous or static or memoryless system, if its response, at $n = n_0$ depends only on the present input, but not on past or future samples of the input. The system is called dynamic, if its response depends on past or future samples of the input.

Example 3.13 Determine whether the following systems are static or dynamic systems

1. $y(n) = 5x(n)$
2. $y(n) = 5x(n) + 3x(n-4)$

Solution:

1. System is static since output depends only on the present input.
2. System is dynamic since output depends not only on the present input but also on the previous values of the input.

3.5.5 Stable and Unstable Systems

A system is called a stable system, if every bounded input sequence produces a bounded output sequence, else the system is called unstable system. For any bounded input, $x(n)$, is defined by

$$|x(n)| \le m_1$$

Then the corresponding output, $y(n)$, is also bounded and is defined by

$$|y(n)| \le m_2$$

In the above conditions m_1 and m_2 are constants. Then we call such type of bounded-input and bounded-output system as a stable system.

3.6 LINEAR TIME INVARIANT SYSTEMS

In the previous section, we studied the properties of a system like linearity, time invariance, causality, stability, static and dynamic. Majority of the discrete time systems are linear and time invariant system. Hence in digital signal processing, we are concerned almost exclusively with a particular class of system known as linear time invariant system. Let us consider a discrete system as shown in Figure 3.17. The output of such a system is given by

$$y(n) = T[x(n)] \tag{3.16}$$

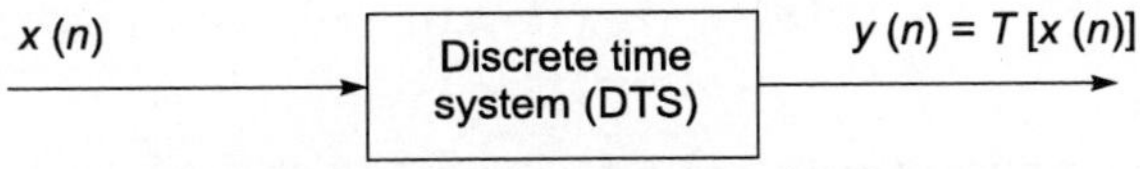

Figure 3.17 Discrete time system

Mathematical expression for output of any LTI system: Let us consider input sequence $x(n)$, where the arrow denotes the 0th sample.

$$x(n) = \left\{ \begin{matrix} \ldots\ldots 0, & 0, & 5, & 2, & 0, & 0, & 5, & -0.5, & 0, & 0, & 0\ldots\ldots \\ & & & \uparrow & & & & & & & \end{matrix} \right\} \tag{3.17}$$

The sequence $x(n)$ can be represented in terms of unit impulse or unit sample sequence and is given by

$$x(n) = 5\delta(n+1) + 2\delta(n) + 5\delta(n-3) - 0.5\delta(n-4) \tag{3.18}$$

Let $x(-1) = 5, x(0) = 2, x(3) = 5, x(4) = -0.5$, then Eq. 3.18 can be written as

$$x(n) = x(-1)\delta(n+1) + x(0)\delta(n) + x(3)\delta(n-3) + x(4)\delta(n-4) \tag{3.19}$$

From the above equation, the sequence $x(n)$ can be expressed in terms of sum of weighted impulses. Now we can generalize Eq. 3.19 for any input and is given by

$$x(n) = \sum_{k=-\infty}^{\infty} x(k)\,\delta(n-k) \tag{3.20}$$

Since the system is linear, it satisfies the principle of superposition theorem. If $x(n)$ is applied as input to LTI system, then we can write response, $y(n)$, of the system as

$$y(n) = T\{x(n)\} = T\left[\sum_{k=-\infty}^{\infty} x(k)\delta(n-k)\right] = \sum_{k=-\infty}^{\infty} x(k)T[\delta(n-k)] \tag{3.21}$$

For input $x(n) = \delta(n)$ and if the response of the system is $h(n)$, then output $y(n)$ of a discrete system is given by

$$y(n) = T[\delta(n)] = h(n) \tag{3.22}$$

Since the system is time invariant, output $y(n)$ of a discrete system for input $x(n) = \delta(n-k)$ is given by

$$y(n) = T[\delta(n-k)] = h(n-k) \tag{3.23}$$

Substituting Eq. 3.23 in Eq. 3.21, we obtain general expression for $y(n)$ and is given by

$$y(n) = \sum_{k=-\infty}^{\infty} x(k)h(n-k) \tag{3.24}$$

The above equation is commonly called convolution sum and is represented by the following notation

$$y(n) = x(n) * h(n)$$

where, $*$ denotes the convolution operation.

Hence, for a linear time invariant system, if the input sequence is $x(n)$ and response of the system is $h(n)$, then the output $y(n)$ is given by

$$y(n) = x(n) * h(n) = \sum_{k=-\infty}^{\infty} x(k)h(n-k) \tag{3.25}$$

3.6.1 Properties of Convolution

Now let us discuss properties of convolution operation:

- Commutative property of convolution

$$x(n) * h(n) = h(n) * x(n) \tag{3.26}$$

- Distributive property of convolution with respect to addition

$$x(n)*\left[h_1(n)+h_2(n)\right]=x(n)*h_1(n)+x(n)*h_2(n) \tag{3.27}$$

- Associative property of convolution

$$x(n)*\left[y(n)*h(n)\right]=\left[x(n)*y(n)\right]*h(n) \tag{3.28}$$

Commutative Property of Convolution

The convolution operation is commutative if, $x(n)*h(n)=h(n)*x(n)$

Proof: From the definition of convolution, we have

$$x(n)*h(n)=\sum_{k=-\infty}^{\infty}x(k)h(n-k)$$

Let $n-k=m$, as $k\to\infty, m\to-\infty$, and as $k\to-\infty, m\to\infty$

$$x(n)*h(n)=\sum_{m=-\infty}^{\infty}x(n-m)h(m)=\sum_{m=-\infty}^{\infty}h(m)x(n-m)=h(n)*x(n)$$

$$x(n)*h(n)=h(n)*x(n)$$

Hence, it is commutative.

Distributive Property of Convolution with Respect to Addition

A convolution operation is distributive with respect to addition if,

$$x(n)*\left[h_1(n)+h_2(n)\right]=x(n)*h_1(n)+x(n)*h_2(n)$$

Proof: From Eq. 3.25, we have

$$x(n)*\left[h_1(n)+h_2(n)\right]=\sum_{k=-\infty}^{\infty}x(k)\left[h_1(n-k)+h_2(n-k)\right]$$

$$x(n)*\left[h_1(n)+h_2(n)\right]=\sum_{k=-\infty}^{\infty}x(k)h_1(n-k)+\sum_{k=-\infty}^{\infty}x(k)h_2(n-k)$$

$$x(n)*\left[h_1(n)+h_2(n)\right]=x(n)*h_1(n)+x(n)*h_2(n)$$

Hence, it is distributive with respect to addition.

Associative Property of Convolution

The convolution operation is associative if $x(n)*\left[y(n)*h(n)\right]=\left[x(n)*y(n)\right]*h(n)$

Proof: From Eq. 3.25, we have

$$x(n)*\left[y(n)*h(n)\right]=\sum_{k=-\infty}^{\infty}x(k)\left[y(n-k)*h(n-k)\right]$$

Since convolution is commutative

$$y(n-k)*h(n-k)=h(n-k)*y(n-k)$$

Hence,

$$x(n)*\left[y(n)*h(n)\right]=\sum_{k=-\infty}^{\infty}x(k)\left[h(n-k)*y(n-k)\right]$$

$$x(n)*\left[y(n)*h(n)\right]=\sum_{k=-\infty}^{\infty}x(k)\sum_{m=-\infty}^{\infty}h(p)y[(n-k)-m]$$

$$x(n)*\left[y(n)*h(n)\right]=\sum_{m=-\infty}^{\infty}h(m)\sum_{k=-\infty}^{\infty}x(k)y[(n-m)-k]$$

$$x(n)*\left[y(n)*h(n)\right]=\sum_{m=-\infty}^{\infty}h(m)\left[x(n-m)*y(n-m)\right]$$

$$x(n)*\left[y(n)*h(n)\right]=h(n)*\left[x(n)*y(n)\right]$$

$$x(n)*\left[y(n)*h(n)\right]=\left[x(n)*y(n)\right]*h(n)$$

Hence, it is associative.

3.6.2 Connection of LTI Systems

Following are the methods to connect two LTI systems:

- Cascade system
- Parallel system

Cascade System

Cascade connection of two LTI systems is shown in Figure 3.18. In this LTI system output of the first system is applied as input to the second system. Let $w(n)$ be the output of the first system and $y(n)$ the output of the second system and $h(n)$ is response of the complete system.

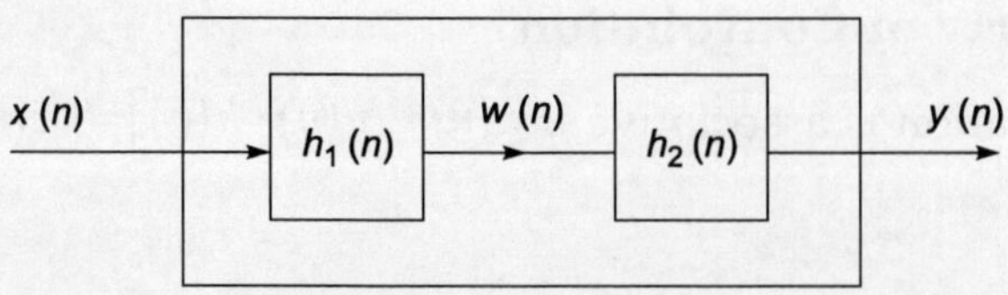

Figure 3.18 Cascade connection of two systems

Then

$$w(n) = x(n) * h_1(n)$$

Hence, we have

$$y(n) = w(n) * h_2(n) = [x(n) * h_1(n)] * h_2(n)$$

Using the associative property of convolution, the above equation can be written as

$$y(n) = x(n) * [h_1(n) * h_2(n)] = x(n) * h(n)$$

Hence, the impulse response of cascade of two LTI systems is given by

$$h(n) = h_1(n) * h_2(n) \tag{3.29}$$

Since convolution is commutative, impulse response of cascade of two LTI systems is also given by

$$h(n) = h_2(n) * h_1(n) \tag{3.30}$$

Parallel System

Parallel connection of two LTI systems is shown in Figure 3.19. Let $y_1(n)$ be the output of the first system and $y_2(n)$ be the output of the second system and let $y(n)$ be the output and $h(n)$ be the response of the complete system, then output $y(n)$ is given by

$$y(n) = y_1(n) + y_2(n) = \sum_{k=-\infty}^{\infty} h_1(k)x(n-k) + \sum_{k=-\infty}^{\infty} h_2(k)x(n-k)$$

$$y(n) = \sum_{k=-\infty}^{\infty} [h_1(k) + h_2(k)]x(n-k)$$

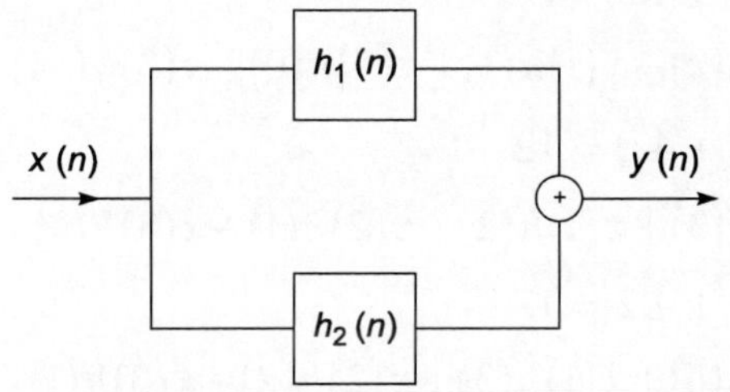

Figure 3.19 Parallel connection of two systems

Since convolution is distributive with respect to addition, hence output is given by

$$y(n) = [h_1(n) + h_2(n)] * x(n)$$

Hence, the impulse response of parallel connection of two LTI systems is given by

$$h(n) = h_1(n) + h_2(n) \tag{3.31}$$

Example 3.14 The input sequence, $x(n) = \left\{ \begin{matrix} 1, & 2, & 3, & 4 \\ \uparrow & & & \end{matrix} \right\}$, is applied to DSP system, with unit sample response, $h(n) = \left\{ \begin{matrix} 2, & 3 \\ \uparrow & \end{matrix} \right\}$. What is the resulting output sequence.

Solution: Given $x(n) = \left\{ \begin{matrix} 1, & 2, & 3, & 4 \\ \uparrow & & & \end{matrix} \right\}$ and $h(n) = \left\{ \begin{matrix} 2, & 3 \\ \uparrow & \end{matrix} \right\}$

Length of $y(n) = 4 + 2 - 1 = 5$

Lower limits of $y(n) = 0 + 0 = 0$

Upper limits of $y(n) = 3 + 1 = 4$

Method 1: By the convolution definition, we have

$$y(n) = x(n) * h(n) = \sum_{k=n_1}^{n_2} x(k)h(n-k)$$

$$y(n) = \sum_{k=0}^{3} x(k)h(n-k) = x(0)h(n) + x(1)h(n-1) + x(2)h(n-2) + x(3)h(n-3)$$

For $n = 0$, $y(0) = x(0)h(0) + x(1)h(-1) + x(2)h(-2) + x(3)h(-3)$

$= 1.2 + 2.0 + 3.0 + 4.0 = 2$

For $n = 1$, $y(0) = x(0)h(1) + x(1)h(0) + x(2)h(-1) + x(3)h(-2)$

$= 1.3 + 2.2 + 3.0 + 4.0 = 7$

For $n = 2$, $y(2) = x(0)h(2) + x(1)h(1) + x(2)h(0) + x(3)h(-1)$

$= 1.0 + 2.3 + 3.2 + 4.0 = 12$

For $n = 3$, $y(3) = x(0)h(3) + x(1)h(2) + x(2)h(1) + x(3)h(0)$

$= 1.0 + 2.0 + 3.3 + 4.2 = 17$

For $n = 4$, $y(4) = x(0)h(4) + x(1)h(3) + x(2)h(2) + x(3)h(1)$

$= 1.0 + 2.0 + 3.0 + 4.3 = 12$

$$y(n) = \left\{ \begin{matrix} 2, & 7, & 12, & 17, & 12 \\ \uparrow & & & & \end{matrix} \right\}$$

Method 2: By the convolution definition, we have

$$y(n) = h(n) * x(n) = \sum_{k=p_1}^{p_2} h(k)x(n-k)$$

$$y(n) = h(n) * x(n) = \sum_{k=0}^{1} h(k)x(n-k) = h(0)x(n) + h(1)x(n-1)$$

For $n = 0,\ y(0) = h(0)x(0) + h(1)x(-1) = 2(1) + 3(0) = 2$

For $n = 1,\ y(1) = h(0)x(1) + h(1)x(0) = 2(2) + 3(1) = 7$

For $n = 2,\ y(2) = h(0)x(2) + h(1)x(1) = 2(3) + 3(2) = 12$

For $n = 3,\ y(3) = h(0)x(3) + h(1)x(2) = 2(4) + 3(3) = 17$

For $n = 4,\ y(4) = h(0)x(4) + h(1)x(3) = 2(0) + 3(4) = 12$

$$y(n) = \begin{Bmatrix} 2, & 7, & 12, & 17, & 12 \\ \uparrow & & & & \end{Bmatrix}$$

In the above example, we find $y(n)$ using method 1 and method 2. Method 1 requires 24 numbers of multiplications and method 2 requires 18 numbers of multiplications. Method 1 or 2 is selected depending on the length of the sequence. In this example, length of $h(n)$ is less than length of $x(n)$, so method 2 can be used to compute $y(n)$.

Method 3: This method can be used for verification of convolution result as shown in Figure 3.20.

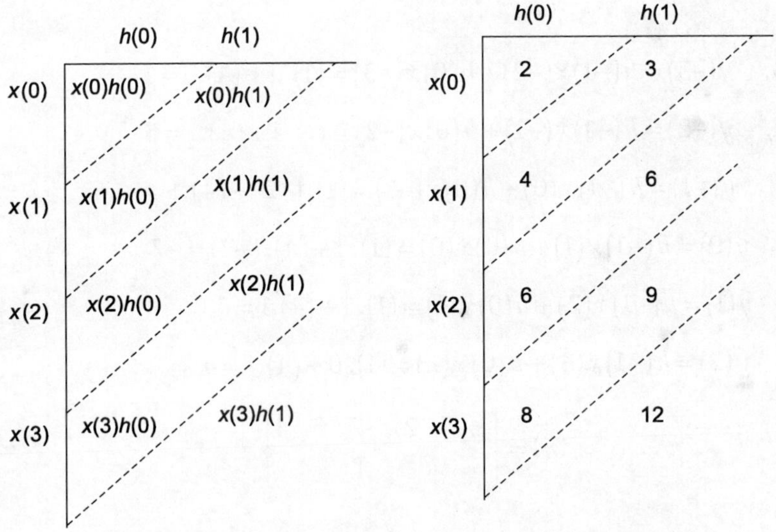

Figure 3.20 Computation of convolution using tabulation

$$y(0)=2, \quad y(1)=4+3=7, \quad y(2)=6+6=12, \quad y(3)=8+9=17 \text{ and } y(4)=12$$

Hence, $y(n)=\left\{\begin{matrix} 2, & 7, & 12, & 17, & 12 \\ \uparrow & & & & \end{matrix}\right\}$

Example 3.15 The input sequence, $x(n)=\left\{\begin{matrix} 1, & 1, & -2, & 1, & 1 \\ & & \uparrow & & \end{matrix}\right\}$, is applied to a DSP system, with unit sample response, $h(n)=\left\{\begin{matrix} 1, & 4 \\ & \uparrow \end{matrix}\right\}$. What is the resulting output sequence of the processor?

Solution:

Length of $y(n)=5+2-1=6$

Lower limits of $y(n)=(-2)+(-1)=-3$

Upper limits of $y(n)=2+0=2$

Method 1: In this example, length of $h(n)$ is 2 and is less than the length of $x(n)$, hence the output, $y(n)$, is computed by using Eq. 3.26.

$$y(n)=h(n)*x(n)=\sum_{k=p_1}^{p_2} h(k)x(n-k)$$

$$y(n)=h(n)*x(n)=\sum_{k=-1}^{0} h(k)x(n-k)=h(-1)x(n+1)+h(0)x(n)$$

For $n=-3$, $y(-3)=h(-1)x(-2)+h(0)x(-3)=(1).1+(4).0=1$

For $n=-2$, $y(-2)=h(-1)x(-1)+h(0)x(-2)=(1).1+(4).1=5$

For $n=-1$, $y(-1)=h(-1)x(0)+h(0)x(-1)=(1).(-2)+(4).1=2$

For $n=0$, $y(0)=h(-1)x(1)+h(0)x(0)=(1).1+(4).(-2)=-7$

For $n=1$, $y(1)=h(-1)x(2)+h(0)x(1)=(1).1+(4).1=5$

For $n=2$, $y(2)=h(-1)x(3)+h(0)x(2)=(1).0+(4).1=4$

$$y(n)=\left\{\begin{matrix} 1, & 5, & 2, & -7, & 5, & 4 \\ & & & \uparrow & & \end{matrix}\right\}$$

Method 2: The result can be verified by using method 2 as shown in Figure 3.21.

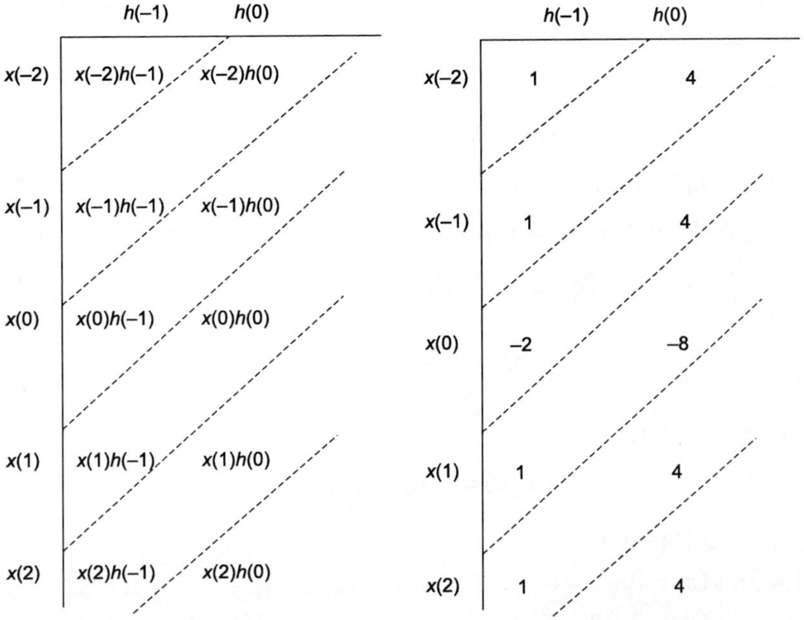

Figure 3.21 Computation of convolution using tabulation

$$y(-3)=1,\quad y(-2)=4+1=5,\quad y(-1)=-2+4=2,\quad y(0)=1-8=-7$$

$$y(1)=1+4=5 \text{ and } y(2)=4$$

Hence, $y(n)=\left\{\begin{matrix}1, & 5, & 2, & -7, & 5, & 4\\ & & & \uparrow & & \end{matrix}\right\}$

Example 3.16 The input sequence, $x(n)=\left\{\begin{matrix}3, & 1, & 2\\ \uparrow & & \end{matrix}\right\}$, is applied to a DSP system, having two subsystems with unit sample response $h_1(n)=\left\{\begin{matrix}5, & 8, & 1\\ \uparrow & & \end{matrix}\right\}$ and $h_2(n)=\left\{\begin{matrix}1, & 1\\ \uparrow & \end{matrix}\right\}$ connected in cascade. What is the overall impulse response and resulting output sequence of the processor?

Solution: If they are connected in cascade, overall impulse response is given by

$$h(n)=h_1(n)*h_2(n)=\sum_{k=-\infty}^{\infty}h_1(k)h_2(n-k)$$

$$h(n)=\sum_{k=0}^{2}h_1(k)h_2(n-k)=h_1(0)h_2(n)+h_1(1)h_2(n-1)+h_1(2)h_2(n-2)$$

Length of $h(n) = 3+2-1 = 4$

Lower limits of $h(n) = 0+0 = 0$

Upper limits of $h(n) = 2+1 = 3$

For $n = 0$, $h(0) = h_1(0)h_2(0) + h_1(1)h_2(-1) + h_1(2)h_2(-2) = 5.1 + 8.0 + 1.0 = 5$

For $n = 1$, $h(1) = h_1(0)h_2(1) + h_1(1)h_2(0) + h_1(2)h_2(-1) = 5.1 + 8.1 + 1.0 = 13$

For $n = 2$, $h(2) = h_1(0)h_2(2) + h_1(1)h_2(1) + h_1(2)h_2(0) = 5.0 + 8.1 + 1.1 = 9$

For $n = 3$, $h(3) = h_1(0)h_2(3) + h_1(1)h_2(2) + h_1(2)h_2(1) = 5.0 + 8.0 + 1.1 = 1$

$$h(n) = \left\{ \begin{matrix} 5, & 13, & 9, & 1 \\ \uparrow & & & \end{matrix} \right\}$$

Output of the processor is given by

$$y(n) = x(n) * h(n)$$

Length of $y(n) = 3+4-1 = 6$

Lower limits of $y(n) = 0+0 = 0$

Upper limits of $y(n) = 2+3 = 5$

$$y(n) = \sum_{k=0}^{2} x(k)h(n-k) = x(0)h(n) + x(1)h(n-1) + x(2)h(n-2)$$

For $n = 0$, $y(0) = x(0)h(0) + x(1)h(-1) + x(2)h(-2) = 3.5 + 1.0 + 2.0 = 15$

For $n = 1$, $y(1) = x(0)h(1) + x(1)h(0) + x(2)h(-1) = 3.13 + 1.5 + 2.0 = 44$

For $n = 2$, $y(2) = x(0)h(2) + x(1)h(1) + x(2)h(0) = 3.9 + 1.13 + 2.5 = 50$

For $n = 3$, $y(3) = x(0)h(3) + x(1)h(2) + x(2)h(1) = 3.1 + 1.9 + 2.13 = 38$

For $n = 4$, $y(4) = x(0)h(4) + x(1)h(3) + x(2)h(2) = 3.0 + 1.1 + 2.9 = 19$

For $n = 5$, $y(5) = x(0)h(5) + x(1)h(4) + x(2)h(3) = 3.0 + 1.0 + 2.1 = 2$

$$y(n) = \left\{ \begin{matrix} 15, & 44, & 50, & 38, & 19, & 2 \\ \uparrow & & & & & \end{matrix} \right\}$$

Example 3.17 The input sequence $x(n) = \left\{ \begin{matrix} 1, & 2, & 4, & 8 \\ \uparrow & & & \end{matrix} \right\}$ is applied to a DSP system, having two subsystems with unit sample response, $h_1(n) = \left\{ \begin{matrix} 1, & 1, & 1 \\ \uparrow & & \end{matrix} \right\}$ and $h_2(n) = \left\{ \begin{matrix} 2, & 2, & 2 \\ \uparrow & & \end{matrix} \right\}$, connected in parallel. What is the overall impulse response and the resulting output sequence of the processor?

Solution: If they are connected in parallel, overall impulse response is given by

$$h(n) = h_1(n) + h_2(n) = \left\{ \underset{\uparrow}{3}, 3, 3 \right\}$$

Output of the processor is given by

$$y(n) = x(n) * h(n) = \sum_{k=0}^{3} x(k)h(n-k)$$

Length of $y(n) = 4 + 3 - 1 = 6$
Lower limits of $y(n) = 0 + 0 = 0$
Upper limits of $y(n) = 3 + 2 = 5$

$$y(n) = x(0)h(n) + x(1)h(n-1) + x(2)h(n-2) + x(3)h(n-3)$$

For $n = 0$, $y(0) = x(0)h(0) + x(1)h(-1) + x(2)h(-2) + x(3)h(-3)$
$= 1.3 + 2.0 + 4.0 + 8.0 = 3$

For $n = 1$, $y(1) = x(0)h(1) + x(1)h(0) + x(2)h(-1) + x(3)h(-2)$
$= 1.3 + 2.3 + 4.0 + 8.0 = 9$

For $n = 2$, $y(2) = x(0)h(2) + x(1)h(1) + x(2)h(0) + x(3)h(-1)$
$= 1.3 + 2.3 + 4.3 + 8.0 = 21$

For $n = 3$, $y(3) = x(0)h(3) + x(1)h(2) + x(2)h(1) + x(3)h(0)$
$= 1.0 + 2.3 + 4.3 + 8.3 = 42$

For $n = 4$, $y(4) = x(0)h(4) + x(1)h(3) + x(2)h(2) + x(3)h(1)$
$= 1.0 + 2.0 + 4.3 + 8.3 = 36$

For $n = 5$, $y(5) = x(0)h(5) + x(1)h(4) + x(2)h(3) + x(3)h(2)$
$= 1.0 + 2.0 + 4.0 + 8.3 = 24$

$$y(n) = \left\{ \underset{\uparrow}{3}, 9, 21, 42, 36, 24 \right\}$$

Example 3.18 Determine the overall impulse response of two LTIs having impulse response

$$h_1(n) = \left(\frac{1}{2}\right)^n u(n) \text{ and } h_2(n) = \left(\frac{1}{4}\right)^n u(n)$$

(a) If they are connected in cascade
(b) If they are connected in parallel

Solution:

(a) If they are connected in cascade, overall impulse response is given by

$$h(n) = h_1(n) * h_2(n) = \sum_{k=-\infty}^{\infty} h_1(k)h_2(n-k) = \sum_{k=-\infty}^{\infty} h_1(k)h_2(n-k)$$

$$h(n) = h_1(n) * h_2(n) = \sum_{k=-\infty}^{\infty} \left(\frac{1}{2}\right)^k u(k)\left(\frac{1}{4}\right)^{n-k} u(n-k)$$

$$\text{But } u(k)u(n-k) = \begin{cases} 1 & \text{for 0 to } n \\ 0 & \text{otherwise} \end{cases}$$

$$h(n) = \left(\frac{1}{4}\right)^n \sum_{k=0}^{n} \left(\frac{1}{2}\right)^k \left(\frac{1}{4}\right)^{-k} = \left(\frac{1}{4}\right)^n \sum_{k=0}^{n} (2)^k = \left(\frac{1}{4}\right)^n \left[\frac{1-2^{n+1}}{1-2}\right]$$

$$= \left(\frac{1}{4}\right)^n [2^{n+1} - 1]u(n)$$

(b) If they are connected in parallel, overall impulse response is given by

$$h(n) = h_1(n) + h_2(n) = \left[\left(\frac{1}{2}\right)^n + \left(\frac{1}{4}\right)^n\right]u(n)$$

3.7 DECONVOLUTION

If system impulse response $h(n)$ and input $x(n)$ are given, the output of an LTI system is computed by taking the convolution of $x(n)$ and $h(n)$ (discussed in section 3.5). Instead, if $x(n)$ and $y(n)$ are given, $h(n)$ is computed by deconvolution process. Deconvolution helps in computing response of the system, i.e., system identification. As discussed in section 3.5, from Eq. 3.26 the output $y(n)$ of a causal discrete LTI system is given by

$$y(n) = h(n) * x(n) = \sum_{k=0}^{n} h(k)x(n-k) = \sum_{k=0}^{n-1} h(k)x(n-k) + h(n)x(0)$$

Hence, $h(n)$ is given by

$$h(n) = \frac{1}{x(0)}\left[y(n) - \sum_{k=0}^{n-1} h(k)x(n-k)\right] \tag{3.32}$$

Hence, response of the system, $h(n)$, depends not only on the input and output values, but also on the previous values of $h(n)$. Hence, deconvolution algorithm is implemented by recursion.

Example 3.19 For input sequence $x(n)=\begin{Bmatrix}1, & 2, & 3, & 4\\ \uparrow & & & \end{Bmatrix}$ is applied to a discrete time processor with output $y(n)=\begin{Bmatrix}2, & 7, & 12, & 17, & 12\\ \uparrow & & & & \end{Bmatrix}$. What is the system response, $h(n)$?

Solution:

Length of $h(n)$ = Length of $y(n)$ – Length of $x(n)+1=6-4+1=3$

Lower limits of $h(n)$ = Lower limits of $y(n)$ – Lower limits of $x(n)=0-0=0$

Upper limits of $h(n)$ = Upper limits of $y(n)$ – Upper limits of $x(n)=4-3=1$

From Eq. 3.26, we have

$$y(n)=x(n)*h(n)=h(n)*x(n)=\sum_{k=0}^{1}h(k)x(n-k)$$

For n = 0

$$y(0)=\sum_{k=0}^{1}h(n)x(-k)=h(0)x(0)+h(1)x(-1)=h(0)x(0)+h(1)\cdot 0$$

$$h(0)=\frac{y(0)}{x(0)}=\frac{2}{1}=2$$

For n = 1

$$y(1)=\sum_{k=0}^{1}h(k)x(1-k)=h(0)x(1)+h(1)x(0)=h(0)x(1)+h(1)\cdot x(0)$$

$$h(1)=\frac{1}{x(0)}\left[y(1)-h(0)x(1)\right]=\frac{1}{1}[7-2.2]=3$$

System response, $h(n)=\begin{Bmatrix}2, & 3\\ \uparrow & \end{Bmatrix}$

Example 3.20 For input sequence $y(n)=\begin{Bmatrix}1, & 1, & -2, & 1, & 1\\ & & \uparrow & & \end{Bmatrix}$ is applied to a discrete time processor with output $y(n)=\begin{Bmatrix}1, & 5, & 2, & -7, & 5, & 4\\ & & & \uparrow & & \end{Bmatrix}$. What is the system response, $h(n)$?

Solution:

Length of $h(n)$ = Length of $h(n)$ – Length of $x(n)+1=6-5+1=2$

Lower limits of $h(n)$ = Lower limits of $y(n)$ – Lower limits of $x(n)=-3-(-2)=-1$

Upper limits of $h(n)$ = Upper limits of $y(n)$ – Upper limits of $x(n)=2-2=0$

From Eq. 3.26, we have

$$y(n)=x(n)*h(n)=h(n)*x(n)=\sum_{k=-1}^{0}h(k)x(n-k)$$

For $n = -3$

$$y(-3)=\sum_{k=-1}^{0}h(k)x(-3-k)=h(-1)x(-2)+h(0)x(-3)=h(-1)x(-2)+h(0).0$$

$$h(-1)=\frac{y(-3)}{x(-2)}=\frac{1}{1}=1$$

For $n = -2$

$$y(-2)=\sum_{k=-1}^{0}h(k)x(-2-k)=h(-1)x(-1)+h(0)x(-2)=h(-1)x(-1)+h(0)x(-2)$$

$$h(0)=\frac{1}{x(-2)}\left[y(-2)-h(-1)x(-1)\right]=\frac{1}{1}[5-(1)1]=4$$

System response, $h(n)=\left\{\begin{matrix}1, & 4\\ & \uparrow\end{matrix}\right\}$

3.8 CORRELATIONS OF DISCRETE TIME SIGNALS

Correlation provides information about the similarity between two sequences. Mathematical operations of correlation closely resemble convolution. In both convolution and correlation two signals are involved. As discussed in section 3.6 convolution produces an output by convolving the given input signal, $x(n)$, with the impulse response, $h(n)$, of an LTI system and output, $y(n)$, is given by Eq. 3.25.

$$y(n)=\sum_{k=-\infty}^{\infty}x(k)h(n-k) \tag{3.33}$$

In correlation there are two types.

- Cross-correlation
- Autocorrelation

3.8.1 Cross-correlation of Discrete Time Signals

Cross-correlation, $r_{xy}(k)$, is used to measure the similarity between two sequences, $x(n)$ and $y(n)$. Cross-correlation of two sequences, $x(n)$ and $y(n)$, is defined by

$$r_{xy}(k) = \sum_{n=-\infty}^{\infty} x(n)y(n-k) \qquad k = 0 \pm 1 \pm 2 \ldots \tag{3.34}$$

or, equivalently as

$$r_{xy}(k) = \sum_{n=-\infty}^{\infty} x(n+k)y(n) \qquad k = 0 \pm 1 \pm 2 \ldots \tag{3.35}$$

In Eq. (3.35) the sequence, $y(n)$, is fixed and for positive values of k, the sequence, $x(n)$, is shifted left by k units in time; for negative values of k, the sequence, $x(n)$, is shifted right by k units in time. Here, index k takes only integer value. In Eq. (3.34), the sequence, $x(n)$, is fixed and for positive values of k the sequence, $y(n)$, is shifted right by k units in time; for negative values of k the sequence, $y(n)$, is shifted left by k units in time. Hence, computation of both the equations yields identical cross-correlation. If we reverse the sequences $x(n)$ and $y(n)$, we obtain cross-correlation sequence, $r_{yx}(k)$

$$r_{yx}(k) = \sum_{n=-\infty}^{\infty} y(n)x(n-k) \qquad k = 0 \pm 1 \pm 2 \ldots \tag{3.36}$$

or, equivalently as

$$r_{yx}(k) = \sum_{n=-\infty}^{\infty} y(n+k)x(n) \qquad k = 0 \pm 1 \pm 2 \ldots \tag{3.37}$$

3.8.2 Autocorrelation of Discrete Time Signals

Autocorrelation is used to compare the similarity between two samples of same sequence. If $y(n) = x(n)$, then we get correlation of a signal $x(n)$ which itself is called autocorrelation. Autocorrelation, $r_{xx}(k)$, of sequences, $x(n)$, is given by

$$r_{xx}(k) = \sum_{n=-\infty}^{\infty} x(n)x(n-k) \qquad k = 0 \pm 1 \pm 2 \ldots \tag{3.38}$$

or, equivalently as

$$r_{xx}(k) = \sum_{n=-\infty}^{\infty} x(n+k)x(n) \qquad l = 0 \pm 1 \pm 2 \ldots \tag{3.39}$$

3.8.3 Properties of Correlation Sequences

1. $r_{xy}(-k) = r_{yx}(k)$ (3.40)
2. $r_{xy}(n) = x(n) * y(-n)$

Proof:

1. Correlation of sequences, $x(n)$ and $y(n)$, is given by

$$r_{xy}(k) = \sum_{n=-\infty}^{\infty} x(n)y(n-k)$$

Substituting, $k = -k$ in the above equation and from Eq. 3.37, we get

$$r_{xy}(-k) = \sum_{n=-\infty}^{\infty} y(n+k)x(n) = r_{yx}(k)$$

Hence, $r_{yx}(k) = r_{xy}(-k)$ or $r_{yx}(k)$ is simply a folded version of $r_{xy}(k)$.

2. Linear convolution of two sequences, $x(n)$ and $y(n)$, is given by

$$x(n) * y(n) = \sum_{k=-\infty}^{\infty} x(k)y(n-k)$$

If the sequence is $y(n)$ folded, the folded sequence becomes $y(-n)$, the convolution of $x(n)$ and $y(-n)$ is given by

$$x(n) * y(-n) = \sum_{l=-\infty}^{\infty} x(k)y(-(n-k)) = \sum_{n=-\infty}^{\infty} x(k)y(k-n) \tag{3.41}$$

From Eq. (3.36) cross-correlation of two sequences, $x(n)$ and $y(n)$, is given by

$$r_{xy}(k) = \sum_{n=-\infty}^{\infty} x(n)y(n-k)$$

Since n and k are index, change n by l and change k by n

$$r_{xy}(n) = \sum_{k=-\infty}^{\infty} x(l)y(l-n)$$

change l by k

$$r_{xy}(n) = \sum_{l=-\infty}^{\infty} x(k)y(k-n) \tag{3.42}$$

From Eq. (3.41) and Eq. (3.42)

$$r_{xy}(n) = x(n) * y(-n) = \sum_{l=-\infty}^{\infty} x(k)y(k-n) \tag{3.43}$$

Hence, the cross-correlation of sequences, $x(n)$ and $y(n)$, is equivalent to convolution of $x(n)$ and the flipped signal, $y(n)$.

Example 3.21 Determine the correlation of sequence $x(n)=\left\{\begin{matrix}2, & 1, & 2, & 3\\ \uparrow & & & \end{matrix}\right\}$ and $y(n)=\left\{\begin{matrix}2, & 1, & 2, & 3\\ \uparrow & & & \end{matrix}\right\}$

From Eq. (3.41) we have

$$r_{xy}(n)=x(n)*y(-n)=\sum_{k=-\infty}^{\infty}x(k)y(k-n)$$

Length of $r_{xy}(n)=$ Length of $x(n)$ + Length of $y(n)-1=4+4-1=7=8-1=7$

Lower limits of $r_{xy}(n)=$ Lower limits of $x(n)$ + Lower limits of $y(-n)=0+(-3)=-3$

Upper limits of $r_{xy}(n)=$ Upper limits of $x(n)$ + Upper limits of $y(-n)=3+0=3$

Hence, n varies from –3 to +3.

From Eq. 3.43 and the substituting limits, we have

At $\quad n=-3\ r_{xy}(-3)=\sum_{k=0}^{3}x(k)y(k+3)=x(0)y(3)+x(1)y(4)+x(2)y(5)+x(3)y(6)$

$$r_{xy}(-3)=2.3+1.0+2.0+3.0=6$$

At $\quad n=-2,\ r_{xy}(-2)=\sum_{k=0}^{3}x(k)y(k+2)=x(0)y(2)+x(1)y(3)+x(2)y(4)+x(3)y(5)$

$$r_{xy}(-2)=2.2+1.3+2.0+3.0=7$$

At $\quad n=-1,\ r_{xy}(-1)=\sum_{k=0}^{3}x(k)y(k+1)=x(0)y(1)+x(1)y(2)+x(2)y(3)+x(3)y(4)$

$$r_{xy}(-1)=2.1+1.2+2.3+3.0=10$$

At $\quad n=0,\ r_{xy}(0)=\sum_{k=0}^{3}x(k)y(k)=x(0)y(0)+x(1)y(1)+x(2)y(2)+x(3)y(3)$

$$r_{xy}(0)=2.2+1.1+2.2+3.3=18$$

At $\quad n=1,\ r_{xy}(1)=\sum_{k=0}^{3}x(k)y(k-1)=x(0)y(-1)+x(1)y(0)+x(2)y(1)+x(3)y(2)$

$$r_{xy}(1)=2.0+1.2+2.1+3.2=2+2+6=10$$

At $\quad n=2, r_{xy}(2)=\sum_{k=0}^{3} x(k)y(k-2)=x(0)y(-2)+x(1)y(-1)+x(2)y(0)+x(3)y(1)$

$$r_{xy}(2)=2.0+1.0+2.2+3.1=7$$

At $\quad n=3, r_{xy}(3)=\sum_{k=0}^{3} x(k)y(k-3)=x(0)y(-3)+x(1)y(-2)+x(2)y(-1)+x(3)y(0)$

$$r_{xy}(3)=2.0+1.0+2.0+3.2=6$$

Hence,
$$r_{xy}(n)=\left\{\begin{matrix}6, & 7, & 10, & 18, & 10, & 7, & 6\\ & & & \uparrow & & & \end{matrix}\right\}$$

Example 3.22 Determine the correlation of sequences, $x(n)=\left\{\begin{matrix}1, & 2, & 3, & 1\\ \uparrow & & & \end{matrix}\right\}$ and $y(n)=\left\{\begin{matrix}1, & 1, & 1, & 1\\ \uparrow & & & \end{matrix}\right\}$

From Eq. 3.43, we have

$$r_{xy}(n)=x(n)*y(-n)=\sum_{k=-\infty}^{\infty} x(k)y(k-n)$$

Length of $r_{xy}(n)=$ Length of $x(n)+$ Length of $y(-n)-1=4+4-1=7$

Lower limits of $r_{xy}(n)=$ Lower limits of $x(n)$ + Lower limits of $y(-n)=0+(-3)=-3$

Upper limits of $r_{xy}(n)$ = Upper limits of $x(n)$ + Upper limits of $y(-n)=3+0=3$

Hence, n varies from –3 to +3.

From Eq. 3.43, and the substituting limits, we have

At $\quad n=-3, r_{xy}(-3)=\sum_{k=0}^{3} x(k)y(k+3)=x(0)y(3)+x(1)y(4)+x(2)y(5)+x(3)y(6)$

$$r_{xy}(-3)=1.1+2.0+3.0+1.0=1$$

At $\quad n=-2, r_{xy}(-2)=\sum_{n=0}^{3} x(k)y(k+2)=x(0)y(2)+x(1)y(3)+x(2)y(4)+x(3)y(5)$

$$r_{xy}(-2)=1.1+2.1+3.0+1.0=3$$

At $\quad n=-1, r_{xy}(-1)=\sum_{k=0}^{3} x(k)y(k+1)=x(0)y(1)+x(1)y(2)+x(2)y(3)+x(3)y(4)$

$$r_{xy}(-1)=1.1+2.1+3.1+1.0=6$$

At $n=0, r_{xy}(0)=\sum_{k=0}^{3} x(k)y(k)=x(0)y(0)+x(1)y(1)+x(2)y(2)+x(3)y(3)$

$$r_{xy}(0)=1.1+1.2+1.3+1.1=7$$

At $n=1, r_{xy}(1)=\sum_{k=0}^{3} x(k)y(k-1)=x(0)y(-1)+x(1)y(0)+x(2)y(1)+x(3)y(2)$

$$r_{xy}(1)=1.0+2.1+3.1+1.1=6$$

At $n=2, r_{xy}(2)=\sum_{k=0}^{3} x(k)y(k-2)=x(0)y(-2)+x(1)y(-1)+x(2)y(0)+x(3)y(1)$

$$r_{xy}(2)=1.0+2.0+3.1+1.1=4$$

At $n=3, r_{xy}(3)=\sum_{k=0}^{3} x(k)y(k-3)=x(0)y(-3)+x(1)y(-2)+x(2)y(-1)+x(3)y(0)$

$$r_{xy}(3)=1.0+2.0+3.0+1.1=1$$

Hence, $$r_{xy}(n)=\left\{\begin{matrix}1, & 3, & 6, & 7, & 6, & 4, & 1\\ & & & \uparrow & & & \end{matrix}\right\}$$

Example 3.23 Determine the correlation of sequences $x(n)=\left\{\begin{matrix}-1, & 2, & 2\\ & \uparrow & \end{matrix}\right\}$ and $y(n)=\left\{\begin{matrix}5, & 1, & 2\\ & \uparrow & \end{matrix}\right\}$

Solution: From Eq. 3.43, and the substituting limits, we have

$$r_{xy}(n)=x(n)*y(-n)=\sum_{k=-\infty}^{\infty} x(k)y(k-n)$$

Length of $r_{xy}(n)=$ Length of $x(n)$ + Length of $y(-n)-1=3+3-1=5$

Lower limits of $r_{xy}(n)=$ Lower limits of $x(n)$ + Lower limits of $y(-n)=-1+(-1)=-2$

Upper limits of $r_{xy}(n)$ = Upper limits of $x(n)$ + Upper limits of $y(-n)=1+1=2$

Hence, (n) varies from -2 to $+2$.

From Eq. 3.41, we have

At $n=-2, r_{xy}(-2)=\sum_{k=-1}^{1} x(k)y(k+2)=x(-1)y(1)+x(0)y(2)+x(1)y(3)$

$$r_{xy}(-2)=-1.2+2.0+2.0=-2$$

At $n = -1,\ r_{xy}(-1) = \sum_{k=-1}^{1} x(k)y(k+1) = x(-1)y(0) + x(0)y(1) + x(1)y(2)$

$$r_{xy}(-1) = -1.1 + 2.2 + 2.0 = 3$$

At $n = 0,\ r_{xy}(0) = \sum_{k=-1}^{1} x(k)y(k) = x(-1)y(-1) + x(0)y(0) + x(1)y(1)$

$$r_{xy}(0) = -1.5 + 2.1 + 2.2 = 1$$

At $n = 1,\ r_{xy}(1) = \sum_{k=-1}^{1} x(k)y(k-1) = x(-1)y(-2) + x(0)y(-1) + x(1)y(0)$

$$r_{xy}(1) = -1.0 + 2.5 + 2.1 = 12$$

At $n = 2,\ r_{xy}(2) = \sum_{k=-1}^{1} x(k)y(k-2) = x(-1)y(-3) + x(0)y(-2) + x(1)y(-1)$

$$r_{xy}(2) = -1.0 + 2.0 + 2.5 = 10$$

Hence, $$r_{xy}(n) = \left\{ \begin{matrix} -2, & 3, & 1, & 12, & 10 \\ & & \uparrow & & \end{matrix} \right\}$$

3.9 CORRELATION COEFFICIENT

Correlation coefficient, (r), measures the strength and direction of a linear relationship between two variables. Correlation coefficients take values between –1 and +1. Correlation coefficient of –1 means the variables are perfectly negatively related; while +1 means a perfect positive relation, and coefficient of 0 means the variable is not related. A negative value implies a negative association between variables (i.e., high values of one variable are associated with low values of the other) and positive value implies a positive association between variables (i.e., high values of one variable are associated with high values of the other). Following points are the accepted guidelines for interpreting the correlation coefficient.

1. 0 means no linear relationship.
2. +1 means a perfect positive linear relationship
3. –1 means a perfect negative linear relationship.
4. Values between 0 and 0.3 (0 and –0.3) mean a weak positive (negative) linear relationship via a shaky linear rule.
5. Values between 0.3 and 0.7 (–0.3 and –0.7) mean a moderate positive (negative) linear relationship via a fuzzy-firm linear rule.

6. Values between 0.7 and 1.0 (–0.7 and –1.0) mean a strong positive (negative) linear relationship via a firm linear rule.

For N pairs of data, $x(n)$ and $y(n)$, the cross-correlation coefficient is given by

$$r_{xy} = \frac{\sum_{k=0}^{N-1}\{x(k)-\bar{x}\}\{y(k)-\bar{y}\}}{\sqrt{\sum_{k=0}^{N-1}\{x(k)-\bar{x}\}^2 \sum_{k=0}^{N-1}\{y(k)-\bar{y}\}^2}} \tag{3.44}$$

where, $\bar{x}$ and $\bar{y}$ are mean values of $x(n)$ and $y(n)$ signals respectively.

If finite length sequences, $x(n)$ and $y(n)$, are to be analyzed, then we must define cross-correlation coefficient of the two signals by

$$r_{xy}(l) = \frac{\sum_{k=0}^{N-1}\{x(k)-\bar{x}\}\{y(k+l)-\bar{y}\}}{\sqrt{\sum_{k=0}^{N-1}\{x(k)-\bar{x}\}^2 \sum_{k=0}^{N-1}\{y(k)-\bar{y}\}^2}}$$

where, $\bar{x}$ and $\bar{y}$ are mean values of $x(n)$ and $y(n)$ signals, respectively.

In the case of correlation coefficient of a signal within itself is called autocorrelation coefficient and is given by

$$r_{xx}(l) = \frac{\sum_{k=0}^{N-1}\{x(k)-\bar{x}\}\{x(k+l)-\bar{x}\}}{\sum_{k=0}^{N-1}\{x(k)-\bar{x}\}^2}$$

where, $\bar{x}$ and $\bar{y}$ are mean of $x(n)$ and $y(n)$ signal.

Example 3.24 Compute correlation coefficient of the sequence $x(n)=\left\{\begin{matrix}4, & 2, & 3\\ \uparrow & & \end{matrix}\right\}$ and $y(n)=\left\{\begin{matrix}4, & 2, & 3\\ \uparrow & & \end{matrix}\right\}$

Solution: Correlation coefficient is given by

$$r_{xy} = \frac{\sum_{k=0}^{N-1}\{x(k)-\bar{x}\}\{y(k)-\bar{y}\}}{\sqrt{\sum_{k=0}^{N-1}\{x(k)-\bar{x}\}^2 \sum_{k=0}^{N-1}\{y(k)-\bar{y}\}^2}}$$

For $x(n)=\left\{\begin{matrix}1, & 2, & 3\\ \uparrow & & \end{matrix}\right\}$ and $y(n)=\left\{\begin{matrix}1, & 2, & 3\\ \uparrow & & \end{matrix}\right\}$ correlation coefficient is given by

$$r_{xy}=\frac{\sum_{k=0}^{2}\{x(k)-\overline{x}\}\{y(k)-\overline{y}\}}{\sqrt{\sum_{k=0}^{2}\{x(k)-\overline{x}\}^2\sum_{k=0}^{2}\{y(k)-\overline{y}\}^2}}$$

where, $\overline{x}$ and $\overline{y}$ are mean of $x(n)$ and $y(n)$ signal.

$$\overline{x}=\frac{4+2+3}{3}=\frac{9}{3}=3$$

and

$$\overline{y}=\frac{4+2+3}{3}=\frac{9}{3}=3$$

$$\text{Num}=[x(0)-\overline{x}][y(0)-\overline{y}]+[x(1)-\overline{x}][y(1)-\overline{y}]+[x(2)-\overline{x}][y(2)-\overline{y}]$$

$$\text{Num}=[4-3][4-3]+[2-3][2-3]+[3-3][3-3]=1+1+0=2$$

$$\text{Den}=\sqrt{\{[x(0)-\overline{x}]^2+[x(1)-\overline{x}]^2+[x(2)-\overline{x}]^2\}\{[y(0)-\overline{y}]^2+[y(1)-\overline{y}]^2+[y(2)-\overline{y}]^2\}}$$

$$=\sqrt{(1+1+0)(1+1+0)}=2$$

Hence, correlation coefficient $r_{xy}=\frac{2}{2}=1$

Example 3.25 Compute correlation coefficient of the sequence $x(n)=\left\{\begin{matrix}4, & 2, & 3\\ \uparrow & & \end{matrix}\right\}$ and $y(n)=\left\{\begin{matrix}-4, & -2, & -3\\ \uparrow & & \end{matrix}\right\}$

Solution: Correlation coefficient is given by

$$r_{xy}=\frac{\sum_{k=0}^{N-1}\{x(k)-\overline{x}\}\{y(k)-\overline{y}\}}{\sqrt{\sum_{k=0}^{N-1}\{x(k)-\overline{x}\}^2\sum_{k=1}^{N}\{y(k)-\overline{y}\}^2}}$$

For $x(n)=\left\{\begin{matrix}4, & 2, & 3\\ \uparrow & & \end{matrix}\right\}$ and $y(n)=\left\{\begin{matrix}-4, & -2, & -3\\ \uparrow & & \end{matrix}\right\}$ correlation coefficient is given by

$$r_{xy}=\frac{\sum_{k=0}^{2}\{x(k)-\overline{x}\}\{y(k)-\overline{y}\}}{\sqrt{\sum_{k=0}^{2}\{x(k)-\overline{x}\}^2\sum_{k=0}^{2}\{y(k)-\overline{y}\}^2}}$$

where, $\overline{x}$ and $\overline{y}$ are mean of $x(n)$ and $y(n)$ signal.

$$\overline{x}=\frac{4+2+3}{3}=\frac{9}{3}=3$$

and

$$\overline{y}=\frac{-4-2-3}{3}=-3$$

$$\text{Num}=[x(0)-\overline{x}][y(0)-\overline{y}]+[x(1)-\overline{x}][y(1)-\overline{y}]+[x(2)-\overline{x}][y(2)-\overline{y}]$$

$$\text{Num}=[4-3][-4+3]+[2-3][-2+3]+[3-3][-3+3]=-1-1-0=-2$$

$$\text{Den}=\sqrt{\{[x(0)-\overline{x}]^2+[x(1)-\overline{x}]^2+[x(2)-\overline{x}]^2\}\{[y(0)-\overline{y}]^2+[y(1)-\overline{y}]^2+[y(2)-\overline{y}]^2\}}$$

$$\text{Den}=\sqrt{(1+1+0)(1+1+0)}=2$$

Hence, correlation coefficient, $r_{xy}=\frac{-2}{2}=-1$

3.10 Z TRANSFORMS

As discussed in section 3.5, output of an LTI system is computed by taking the convolution of input and the response of the system. It is a basic operation in signal processing and is computationally very cumbersome. *Z* transforms provide a straightforward way of computing the output of an LTI system. For example, the *Z* transform of the sequence, $x(n)=\left\{\begin{matrix}3, & 1, & 2, & -1\\ \uparrow & & & \end{matrix}\right\}$, is given by X(Z) = (3 + Z^{-1} + $2Z^{-2}$ – Z^{-3}) and the *Z* transform of the sequence, $h(n)=\left\{\begin{matrix}2, & 3, & 1\\ \uparrow & & \end{matrix}\right\}$, is given by $(Z)=\left(3+2Z^{-1}+Z^{-2}\right)$.

The product of $X(Z)$ and $H(Z)$ is given by

$$Y(Z) = X(Z)H(Z) = \left(3 + Z^{-1} + 2Z^{-2} - Z^{-3}\right)\left(2 + 3Z^{-1} + Z^{-2}\right)$$

$$\begin{array}{l} 6 + 2Z^{-1} + 4Z^{-2} - 2Z^{-3} \\ \quad 9Z^{-1} + 3Z^{-2} + 6Z^{-3} - 3Z^{-4} \\ \qquad 3Z^{-2} + Z^{-3} + 2Z^{-4} - Z^{-5} \\ \hline 6 + 11Z^{-1} + 10Z^{-2} + 5Z^{-3} - Z^{-4} - Z^{-5} \end{array}$$

The inverse Z transform of $Y(Z)$ is given by $y(n) = \left\{ \underset{\uparrow}{6}, 11, 10, 5, -1, -1 \right\}$. Hence, result is obtained by multiplication of $X(Z)$ and $H(Z)$ rather than by the convolution of $x(n)$ and $h(n)$.

The Z transform of a discrete signal, $x(n)$, is defined as

$$X(Z) = \sum_{n=-\infty}^{\infty} x(n) z^{-n} \tag{3.45}$$

For convenience the Z transform of a signal, $x(n)$, will be denoted by

$$X(Z) = Z\{x(n)\}$$

The relationship between $x(n)$ and $X(Z)$ will be indicated by

$$x(n) \leftrightarrow X(Z)$$

3.10.1 Pole-Zero Models

Pole-zero models help to get information about the stability and steady state frequency response of a system. The values of Z for which denominator polynomial is zero are known as poles and the values of Z for which numerator polynomial is zero are known as zeros. If all the poles lie inside the unit circle then the system is stable, else indicates unstable behaviour.

Example 3.26 Find the positions of poles and zeros of

(a) $H(Z) = \dfrac{Z}{(Z + \alpha)}$ (b) $H(Z) = \dfrac{Z(Z-1)}{(Z^2 - Z + 0.5)}$

Solution: In (a) the system contains one pole at $Z = -\alpha$ and one zero at Z = 0. Pole-zero plot is shown in Figure 3.20 (a). Poles are indicated by × and zeros are indicated by ○ in 3.20 (b). System contains two poles at $Z = 0.5 + j0.5$ and at $Z = 0.5 - j0.5$ and it contains two zeros at $Z = 0$ and at $Z = 1$. Pole-zero plot is as shown in Fig. 3.22(a) and (b). Poles are indicated by × and zeros are indicated by ○.

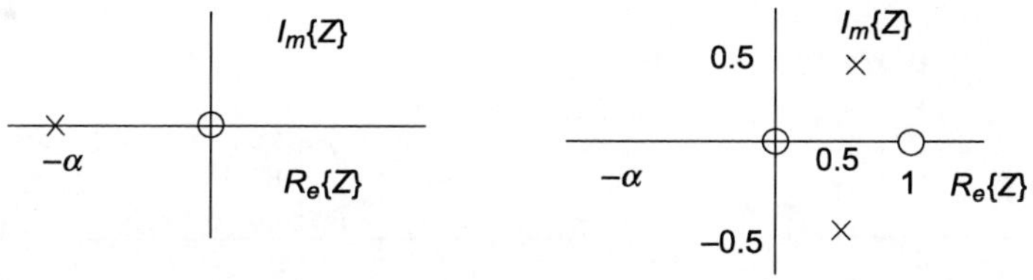

Figure 3.22 (a) and (b)- Pole zero plot

3.10.2 Relationship between Fourier Transform and Z Transform

By substituting Z in polar coordinates in Eq. 3.45, i.e., $Z = re^{j\omega}$, we get

$$X(re^{j\omega}) = \sum_{n=-\infty}^{\infty} x(n)(re^{j\omega})^{-n}$$

If we evaluate the above equation on a unit circle, i.e., if **r = 1,** then the above equation is given by

$$X(e^{j\omega}) = \sum_{n=-\infty}^{\infty} x(n)e^{-j\omega n}$$

Hence, if Z transform of the sequence is evaluated on a unit circle, the Z transform of the sequence, $x(n)$, is equal to Fourier transform $X(e^{j\omega})$ of the sequence $x(n)$ (covered in the next section).

3.11 FOURIER TRANSFORM AND DISCRETE FOURIER TRANSFORM

Let us consider a discrete signal, $x(n)$. Its Fourier transform (FT) of $x(n)$ is given by

$$X(e^{j\omega}) = \sum_{n=-\infty}^{\infty} x(n)e^{-j\omega n} \tag{3.46}$$

The function, $X(e^{j\omega})$ defined in the above equation, is called Fourier transform (FT) of $x(n)$ inverse Fourier transform of $X(e^{j\omega})$ is given by

$$x(n) = \frac{1}{2\pi}\int_{-\pi}^{\pi} X(e^{j\omega})e^{j\omega n}\,d\omega \tag{3.47}$$

$X(e^{j\omega})$ represents the frequency contents of $x(n)$ and it extracts frequency information of $x(n)$. As shown in Figure 3.23, Fourier transform, $X(e^{j\omega})$, computed from discrete time samples, $x(n)$, is a continuous function of the frequency variable. This implies that FT is not suitable for processing discrete time signals in a digital computer.

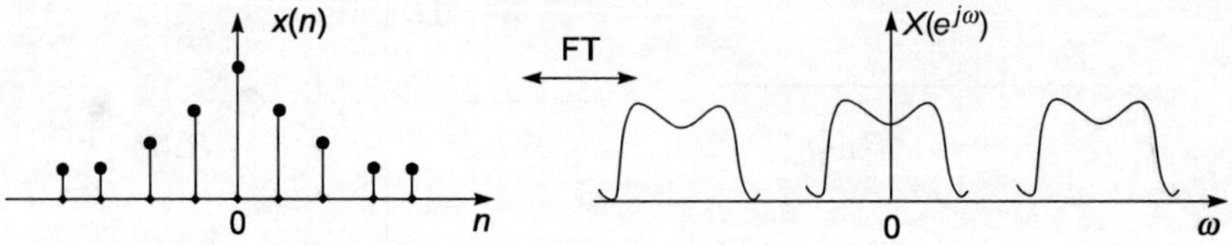

Figure 3.23 Spectrum of discrete signal

Discrete Fourier transform (DFT) is developed to convert discrete time signal, $x(n)$, to discrete spectrum by sampling the continuous spectrum, $X(e^{j\omega})$, of the signal. Hence, Fourier transform calculated at discrete points is called **discrete Fourier transform** (DFT). Discrete Fourier transform is a powerful computation tool for performing frequency analysis of discrete time signals. In this section, we will study the definition, expressions for direct and inverse discrete Fourier transforms.

From Eq. 3.46 Fourier transform of a sequence, $x(n)$, is given by

$$X\left(e^{j\omega}\right) = \sum_{n=-\infty}^{\infty} x(n)e^{-j\omega n}$$

Here $x(n)$ is the discrete signal and $X(e^{j\omega})$ is continuous function of ω and the range of ω is $-\pi$ to π or 0 to 2π. Discrete Fourier transform is an algorithm, which computes $X(e^{j\omega})$ only at discrete values of ω and it is denoted by $X(k)$. Hence, $X(k)$ is obtained by sampling $X(e^{j\omega})$ at N equally spaced points. The spectral spacing is $\frac{2\pi}{N}$ and $X(k)$ is computed at $k = 0, 1, \ldots N-1$, i.e., at N discrete points and $X(k)$ is given by

$$X(k) = \sum_{n=0}^{N-1} x(n)e^{-j\left(\frac{2\pi}{N}\right)kn} \quad k = 0, 1, \ldots N-1 \tag{3.48}$$

Let us define $W_N = e^{-j\left(\frac{2\pi}{N}\right)}$, this is called twiddle factor. Then DFT $X(k)$ is given by

$$X(k) = \sum_{n=0}^{N-1} x(n)W_N^{kn} \; k = 0, 1, \ldots N-1 \tag{3.49}$$

Similarly, the sequence, $x(n)$, can be obtained from $X(k)$ by taking inverse discrete Fourier transform (IDFT) and is given by

$$x(n)=\frac{1}{N}\sum_{k=0}^{N-1}X(k)W_N^{-kn} \quad n=0,1,\ldots N-1 \tag{3.50}$$

Proof: $x(n)$ can be obtained from $X(k)$ and is given by

$$x(n)=\frac{1}{N}\sum_{k=0}^{N-1}X(k)W_N^{-kn} \quad n=0,1,\ldots N-1$$

Substituting $X(k)$ value from Eq. 3.49, we obtain

$$x(n)=\frac{1}{N}\sum_{k=0}^{N-1}\left[\sum_{p=0}^{N-1}x(p)W_N^{kp}\right]W_N^{-kn}=\frac{1}{N}\sum_{p=0}^{N-1}x(p)\sum_{k=0}^{N-1}W_N^{-(n-p)k}$$

It can be shown that

$$\begin{cases}\displaystyle\sum_{k=0}^{N-1}W_N^{-(n-p)k}=\sum_{k=0}^{N-1}e^{j\left(\frac{2\pi}{N}\right)(n-p)k}=N & \text{for } n=p\\ \displaystyle\sum_{k=0}^{N-1}W_N^{-(n-p)k}=\sum_{k=0}^{N-1}e^{j\left(\frac{2\pi}{N}\right)(n-p)k}=0 & \text{for } n\neq p\end{cases}$$

$$\therefore \qquad \text{RHS}=\frac{1}{N}\sum_{p=0}^{N-1}x(p)N\delta(n-p)=x(p)\big|_{p=n}=x(n)$$

and hence, $$x(n)=\frac{1}{N}\sum_{k=0}^{N-1}X(k)W_N^{-kn} \qquad n=0,1,\ldots N-1$$

In the above DFT and IDFT equations require N^2 complex multiplications. Equation 3.49 is known as analysis equation and Eq. 3.50 is known as synthesis equation. $X(k)$ is a complex quantity and can represented as

$$X(k)=|X(k)|\angle X(k) \text{ for } k=0,1,2\ldots N-1$$

The plot of $|X(k)|$ versus k is called magnitude spectrum and the plot of $\angle X(k)$ versus k is called angle or phase spectrum.

Example 3.27 If $x(n)=\{\delta(n),\delta(n-2)\}$. Find DFT of the periodic signal $x(n)$ with N = 4.

Solution: DFT of $x(n)$ is given by

$$X(k) = \sum_{n=0}^{N-1} x(n) W_N^{kn} \quad \text{for } k = 0, 1, \ldots N-1$$

$$X(k) = \sum_{n=0}^{3} x(n) W_N^{kn} = x(0)W_N^0 + x(1)W_N^k + x(2)W_N^{2k} + x(3)W_N^{3k} \quad \text{for } k = 0, 1, 2, 3$$

Substituting, $x(0) = 1, x(1) = 0, x(2) = 1, x(3) = 0$, we get

$$X(k) = 1 + W_N^{2k} = 1 + e^{\frac{-j4\pi k}{4}} = 1 + e^{-j\pi k} \quad \text{for } k = 0, 1, 2, 3$$

$$X(0) = 1 + 1 = 2$$

$$X(1) = 1 + e^{-j\pi} = 0$$

$$X(2) = 1 + e^{-j2\pi} = 2$$

$$X(3) = 1 + e^{-j3\pi} = 0$$

$$X(k) = \left\{ \begin{matrix} 2, & 0, & 2, & 0 \\ \uparrow & & & \end{matrix} \right\}$$

Example 3.28 If $X(k) = \left\{ \begin{matrix} 2, & 0, & 2, & 0 \\ \uparrow & & & \end{matrix} \right\}$, find the periodic signal $x(n)$ with $N = 4$.

Solution: Inverse discrete Fourier transform (IDFT) is given by

$$x(n) = \frac{1}{N} \sum_{k=0}^{N-1} X(k) W_N^{-kn} \quad \text{for } n = 0, 1, \ldots N-1$$

We know that $N = 4$

$$x(n) = \frac{1}{4} \sum_{k=0}^{3} X(k) W_N^{-kn} \quad \text{for } n = 0, 1, 2, 3$$

$$x(n) = \frac{1}{4} \left[X(0) + X(1) e^{\frac{j\pi n}{2}} + X(2) e^{j\pi n} + X(3) e^{\frac{j3\pi n}{2}} \right]$$

Evaluating $x(n)$ for $n = 0, 1, 2, 3$, we get

$$x(0) = \frac{1}{4}[X(0) + X(1) + X(2) + X(3)] = \frac{1}{4}[2 + 0 + 2 + 0] = 1$$

$$x(1)=\frac{1}{4}\left[X(0)+X(1)e^{\frac{j\pi}{2}}+X(2)e^{j\pi}+X(3)e^{\frac{j3\pi}{2}}\right]=\frac{1}{4}[2+0+(-2)+0]=0$$

$$x(2)=\frac{1}{4}\left[X(0)+X(1)e^{j\pi}+X(2)e^{j\pi 2}+X(3)e^{j3\pi}\right]=\frac{1}{4}[2+0+2+0]=1$$

$$x(3)=\frac{1}{4}\left[X(0)+X(1)e^{\frac{j\pi 3}{2}}+X(2)e^{j\pi 3}+X(3)e^{\frac{j9\pi}{2}}\right]=\frac{1}{4}[2+0+(-2)+0]=0$$

Hence, $$x(n)=\left\{\begin{matrix}1, 0, 1, 0\\ \uparrow\end{matrix}\right\}$$

Example 3.29 If $x(n)=\{4\delta(n), 2\delta(n-2), 3\delta(n-3)\}$. Find DFT of the periodic signal $x(n)$ with $N = 4$.

Solution: DFT of $x(n)$ is given by

$$X(k)=\sum_{n=0}^{N-1}x(n)W_N^{kn} \quad \text{for} \quad k=0,1,\ldots N-1$$

$$X(k)=\sum_{n=0}^{3}x(n)W_N^{kn}=x(0)W_N^{0}+x(1)W_N^{k}+x(2)W_N^{2k}+x(3)W_N^{3k} \quad \text{for } k=0,1,2,3$$

Substituting, $x(0)=4, x(1)=0, x(2)=2, x(3)=3$, we get

$$X(k)=4+2W_4^{2k}+3W_4^{3k}=4+2e^{-j\pi k}+3e^{\frac{-j3\pi k}{2}} \quad \text{for } k=0,1,2,3$$

$$X(0)=4+2+3=9$$

$$X(1)=1+e^{-j\pi}+3e^{\frac{-j3\pi}{2}}=2+j3$$

$$X(2)=1+2e^{-j2\pi}+3e^{-j3\pi}=3$$

$$X(3)=1+2e^{-j3\pi}+3e^{\frac{-j9\pi}{2}}=2-j3$$

$$X(k)=\left\{\begin{matrix}9, & 2+j3, & 3, & 2-j3\\ \uparrow & & & \end{matrix}\right\}$$

Example 3.30 If $X(k)=\left\{\begin{matrix}9, & 2+j3, & 3, & 2-j3\\ \uparrow & & & \end{matrix}\right\}$. Find the periodic signal $x(n)$ with $N=4$.

Solution: Inverse discrete Fourier transform (IDFT) is given by

$$x(n) = \frac{1}{N}\sum_{k=0}^{N-1} X(k)W_N^{-kn} \quad \text{for } n = 0, 1, \ldots N-1$$

We know that $N = 4$

$$x(n) = \frac{1}{4}\sum_{k=0}^{3} X(k)W_N^{-kn} \quad \text{for } n = 0, 1, 2, 3$$

$$x(n) = \frac{1}{4}\left[X(0) + X(1)e^{\frac{j\pi n}{2}} + X(2)e^{j\pi n} + X(3)e^{\frac{j3\pi n}{2}}\right]$$

Evaluating $x(n)$ for $n = 0, 1, 2, 3$, we get

$$x(0) = \frac{1}{4}[X(0) + X(1) + X(2) + X(3)] = \frac{1}{4}[9 + (2 + j3) + 3 + (2 - j3)] = 4$$

$$x(1) = \frac{1}{4}\left[X(0) + X(1)e^{\frac{j\pi}{2}} + X(2)e^{j\pi} + X(3)e^{\frac{j3\pi}{2}}\right]$$

$$= \frac{1}{4}\left[9 + (2 + j3)e^{\frac{j\pi}{2}} + 3e^{j\pi} + (2 - j3)e^{\frac{j3\pi}{2}}\right] = 0$$

$$x(2) = \frac{1}{4}[X(0) + X(1)e^{j\pi} + X(2)e^{j\pi 2} + X(3)e^{j3\pi}] = \frac{1}{4}$$

$$= \frac{1}{4}\left[9 + (2 + j3)e^{j\pi} + 3e^{j2\pi} + (2 - j3)e^{\frac{j3\pi}{2}}\right] = 2$$

$$x(3) = \frac{1}{4}\left[X(0) + X(1)e^{\frac{j\pi 3}{2}} + X(2)e^{j\pi 3} + X(3)e^{\frac{j9\pi}{2}}\right]$$

$$= \frac{1}{4}\left[9 + (2 + j3)e^{\frac{j\pi 3}{2}} + 3e^{j\pi 3} + (2 - j3)e^{\frac{j9\pi}{2}}\right] = 3$$

Hence, $x(n) = \left\{\begin{matrix} 4, & 0, & 2, & 3 \\ \uparrow & & & \end{matrix}\right\}$

Computation of DFT involves a large number of multiplications and additions. For N point DFT requires N^2 number of multiplications and $N(N-1)$ number of additions. If $N = 512$, then it requires $(512)^2$ multiplications and (512) (511) additions. Approximately, 0.5 million complex multiplications and additions are required.

3.12 RADIX-2 FAST FOURIER TRANSFORM ALGORITHM

In 1965, Cooley and Tukey developed a method of computing the DFT and the algorithm requires $\frac{N}{2}\log_2 N$ computations, when N is a power of 2. Algorithm developed by Cooley and Tukey, which can achieve this, is called fast Fourier transform (FFT). For $N=512$, direct DFT algorithm requires $N^2=(512)^2$ number of multiplications. FFT requires $\frac{N}{2}\log_2 N=\frac{512}{2}\log_2 2^9=2304$ number of multiplications and reduces the number of multiplications by a factor of 113.78. In this section, we will discuss Radix-2 FFT algorithm. In this algorithm, N point DFT is successively decomposed into smallest DFT of size $N=2$. In Radix-2, two types algorithm are used to achieve this and they are:

- Decimation in time FFT algorithm
- Decimation in frequency FFT algorithm

Decimation in time FFT algorithm is based on by dividing input sequence and this process is continued until two-point DFT is obtained. Decimation in frequency FFT algorithm is based on the dividing output sequence and this process is continued until two-point DFT is obtained. Both the algorithms require $\frac{N}{2}\log_2 N$ number of complex multiplications to compute DFT. In DIT algorithm, the input is a bit reversed and output is in natural order but in DIF algorithm, the input is natural order and output is bit-reversed order. For $N = 8$, both the algorithms require 12 number of multiplications as shown in Fig. 3.24.

Example 3.31 If $x(n)=\left\{\begin{matrix}2, & 1, & 0, & 1, & 3, & 1, & 5, & 4\\ \uparrow & & & & & & & \end{matrix}\right\}$, find $X(k)$ by using decimation in time FFT algorithm.

Solution: $N=8$, number of stages required $=\log_2 N=\log_2 8=3$. Signal flow diagram for $N = 8$ is shown in Figure 3.24.

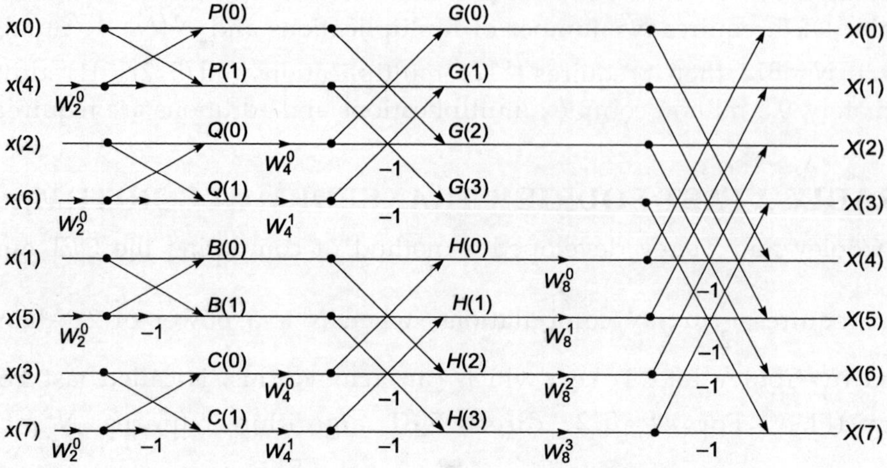

Figure 3.24 Signal flow graph of decimation in time FFT algorithm for N = 8

$$W_8^0 = W_4^0 = W_2^0 = 1, W_8^2 = W_4^1 = -j, W_8^1 = 0.707 - j0.707 \text{ and } W_8^3 = -0.707 - j0.707$$

The first stage outputs are given by

$P(0) = x(0) + x(4) = 2 + 3 = 5$

$P(1) = x(0) - x(4) = 2 - 3 = -1$

$Q(0) = x(2) + x(6) = 0 + 5 = 5$

$Q(1) = x(2) - x(6) = 0 - 5 = -5$

$B(0) = x(1) + x(5) = 1 + 1 = 2$

$B(1) = x(1) - x(5) = 1 - 1 = 0$

$C(0) = x(3) + x(7) = 4 + 1 = 5$

$C(1) = x(3) - x(7) = 1 - 4 = -3$

The second stage outputs are given by

$G(0) = P(0) + W_4^0 Q(0) = 5 + (1)5 = 10$

$G(1) = P(1) + W_4^1 Q(1) = -1 + (-j)(-5) = -1 + j5$

$G(2) = P(0) - W_4^0 Q(0) = 5 - (1)(5) = 0$

$G(3) = P(1) - W_4^1 Q(1) = -1 - (-j)(-5) = -1 - j5$

$H(0) = B(0) + W_4^0 C(0) = 2 + (1)5 = 7$

$H(1) = B(1) + W_4^1 C(1) = 0 + (-j)(-3) = j3$

$H(2) = B(0) - W_4^0 C(0) = 2 - (1)5 = -3$

$H(3) = B(1) - W_4^1 C(1) = 0 - (-j)(-3) = -j3$

The third stage outputs are given by

$$X(0) = G(0) + W_8^0 H(0) = 10 + (1)7 = 17$$

$$X(1) = G(1) + W_8^1 H(1) = (-1 + j5) + (0.707 - j0.707)(j3) = 1.12 + j7.12$$

$$X(2) = G(2) + W_8^2 H(2) = 0 + (-j)(-3) = j3$$

$$X(3) = G(3) + W_8^3 H(3) = (-1 - j5) + (-0.707 - j0.707)(-j3) = -3.121 - j2.87$$

$$X(4) = G(0) - W_8^0 H(0) = 10 - (1)7 = 3$$

$$X(5) = G(1) - W_8^1 H(1) = (-1 + j5) - (0.707 - j0.707)(j3) = -3.12 + j2.87$$

$$X(6) = G(2) - W_8^2 H(2) = 0 - (-j)(-3) = -j3$$

$$X(7) = G(3) - W_8^3 H(3) = (-1 - j5) - (-0.707 - j0.707)(-j3) = 1.12 - j7.12$$

Hence, $X(k)$ is given by

$$X(k) = \left\{ \begin{matrix} 17, & 1.12 + j7.12, & j3, & -3.121 - j2.87, & 3, & -3.12 + j2.87, & -j3, & 1.12 - j7.12 \\ \uparrow & & & & & & & \end{matrix} \right\}$$

Example 3.32 If $x(n) = \left\{ \begin{matrix} 1, & 2, & 3, & 4, & 0, & 1, & 2, & 3 \\ \uparrow & & & & & & & \end{matrix} \right\}$, find $X(k)$ by using decimation in time FFT algorithm.

Solution: N = 8, number of stages required = $\log_2 N = \log_2 8 = 3$. Signal flow diagram for N = 8 is shown in Figure 3.25.

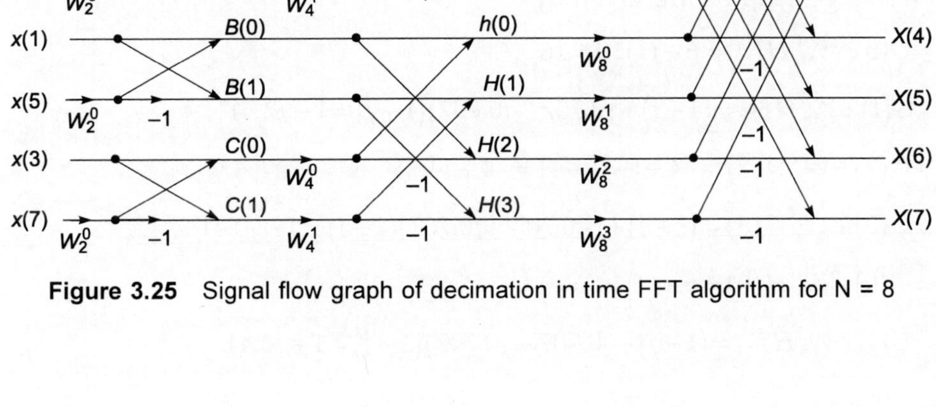

Figure 3.25 Signal flow graph of decimation in time FFT algorithm for N = 8

$$W_8^0 = W_4^0 = W_2^0 = 1, W_8^2 = W_4^1 = -j, W_8^1 = 0.707 - j0.707 \text{ and } W_8^3 = -0.707 - j0.707$$

The first stage outputs are given by

$$P(0) = x(0) + x(4) = 1 + 0 = 1$$

$$P(1) = x(0) - x(4) = 1 - 0 = 1$$

$$Q(0) = x(2) + x(6) = 3 + 2 = 5$$

$$Q(1) = x(2) - x(6) = 3 - 2 = 1$$

$$B(0) = x(1) + x(5) = 2 + 1 = 3$$

$$B(1) = x(1) - x(5) = 2 - 1 = 1$$

$$C(0) = x(3) + x(7) = 4 + 3 = 7$$

$$C(1) = x(3) - x(7) = 4 - 3 = 1$$

The second stage outputs are given by

$$G(0) = P(0) + W_4^0 Q(0) = 1 + (1)5 = 6$$

$$G(1) = P(1) + W_4^1 Q(1) = 1 + (-j)(1) = 1 - j$$

$$G(2) = P(0) - W_4^0 Q(0) = 1 - (1)(5) = -4$$

$$G(3) = P(1) - W_4^1 Q(1) = 1 - (-j)(1) = 1 + j$$

$$H(0) = B(0) + W_4^0 C(0) = 3 + (1)7 = 10$$

$$H(1) = B(1) + W_4^1 C(1) = 1 + (-j)(1) = 1 - j$$

$$H(2) = B(0) - W_4^0 C(0) = 3 - (1)7 = -4$$

$$H(3) = B(1) - W_4^1 C(1) = 1 - (-j)(1) = 1 + j$$

The third stage outputs are given by

$$X(0) = G(0) + W_8^0 H(0) = 6 + (1)10 = 16$$

$$X(1) = G(1) + W_8^1 H(1) = (1 - j) + (0.707 - j0.707)(1 - j) = 1 - j2.41$$

$$X(2) = G(2) + W_8^2 H(2) = -4 + (-j)(-4) = -4 + j4$$

$$X(3) = G(3) + W_8^3 H(3) = (1 + j) + (-0.707 - j0.707)(1 + j) = 1 - j0.41$$

$$X(4) = G(0) - W_8^0 H(0) = 6 - (1)10 = -4$$

$$X(5) = G(1) - W_8^1 H(1) = (1 - j) - (0.707 - j0.707)(1 - j) = 1 + j0.41$$

$$X(6) = G(2) - W_8^2 H(2) = -4 - (-j)(-4) = -4 - j4$$

$$X(7) = G(3) - W_8^3 H(3) = (1+j) - (-0.707 - j0.707)(1+j) = 1 + j2.41$$

Hence, $X(k)$ is given by

$$X(k) = \left\{ \begin{matrix} 16, & 1-j2.41, & -4+j4, & 1-j0.41, & -4, & 1+j0.41, & -4-j4, & 1+j2.41 \\ \uparrow & & & & & & & \end{matrix} \right\}$$

Example 3.33 If $x(n) = \left\{ \begin{matrix} 2, & 1, & 0, & 1, & 3, & 1, & 5, & 4 \\ \uparrow & & & & & & & \end{matrix} \right\}$, find $X(k)$ by using decimation in frequency FFT algorithm.

Solution: $N = 8$, number of stages required $= \log_2 N = \log_2 8 = 3$. Signal flow diagram for $N = 8$ is shown in Figure 3.26.

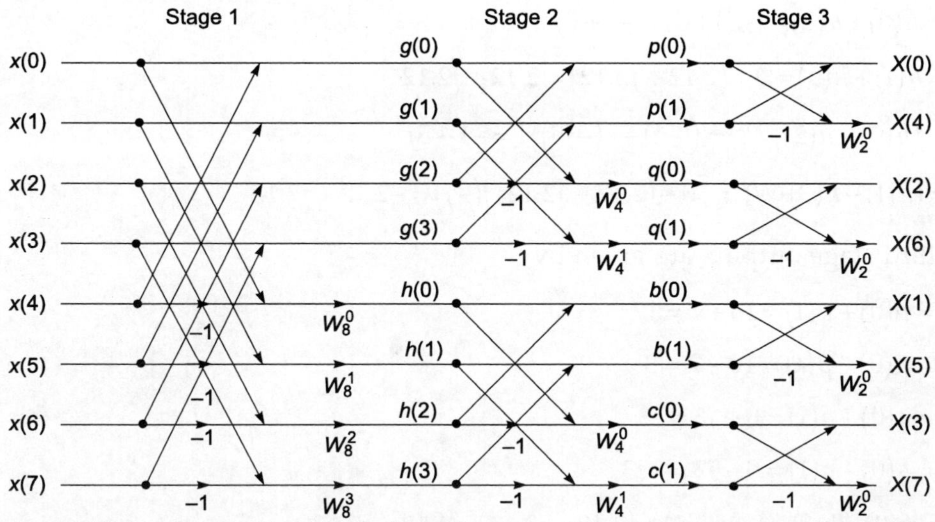

Figure 3.26 Flow graph of Decimation frequency FFT algorithm for N = 8

$$W_8^0 = W_4^0 = W_2^0 = 1, W_8^2 = W_4^1 = -j, W_8^1 = 0.707 - j0.707 \text{ and } W_8^3 = -0.707 - j0.707$$

The first stage outputs are given by

$$g(0) = x(0) + x(4) = 2 + 3 = 5$$

$$g(1) = x(1) + x(5) = 1 + 1 = 2$$

$$g(2) = x(2) + x(6) = 0 + 5 = 5$$

$$g(3) = x(3) + x(7) = 1 + 4 = 5$$

$$h(0)=(x(0)-x(4))W_8^0=(2-3)1=-1$$

$$h(1)=(x(1)-x(5))W_8^1=(1-1)(0.707-j0.707)=0$$

$$h(2)=(x(2)-x(6))W_8^2=(0-5)(-j)=j5$$

$$h(3)=(x(3)-x(7))W_8^3=(1-4)(-0.707-j0.707)=2.12+j2.12$$

The second stage outputs are given by

$$p(0)=g(0)+g(2)=5+5=10$$

$$p(1)=g(1)+g(3)=2+5=7$$

$$q(0)=(g(0)-g(2))W_4^0=(5-5)1=0$$

$$q(1)=(g(1)-g(3))W_4^1=(2-5)(-j)=j3$$

$$b(0)=h(0)+h(2)=(-1)+5j=-1+5j$$

$$b(1)=h(1)+h(3)=0+(2.12+j2.12)=2.12+j2.12$$

$$c(0)=(h(0)-h(2))W_4^0=((-1)—(-j5))1=-1-5j$$

$$c(1)=(h(1)-h(3))W_4^1=[0-(2.12+j2.12)](-j)=-2.12+j2.12$$

The third stage outputs are given by

$$X(0)=p(0)+p(1)=10+7=17$$

$$X(4)=p(0)-p(1)=10-7=3$$

$$X(2)=q(0)+q(1)=0+j3=j3$$

$$X(6)=q(0)-q(1)=0-j3=-j3$$

$$X(1)=b(0)+b(1)=(-1+j5)+(2.12+j2.12)=1.12+j7.12$$

$$X(5)=b(0)-b(1)=(-1+j5)-(2.12+j2.12)=-3.12+j2.88$$

$$X(3)=c(0)+c(1)=(-1-j5)+(-2.12+j2.12)=-3.12-j2.88$$

$$X(7)=c(0)-c(1)=(-1-j5)-(-2.12+j2.12)=1.12-j7.12$$

Hence, $X(k)$ is given by

$$X(k)=\left\{\begin{matrix}17, & 1.12+j7.12, & j3, & -3.121-j2.87, & 3, & -3.12+j2.87,+j2.87, & -j3, & 1.12-j7.12\\ \uparrow & & & & & & & \end{matrix}\right\}$$

Example 3.34 If $x(n)=\left\{\begin{matrix}1, & 2, & 3, & 4, & 0, & 1, & 2, & 3\\ \uparrow & & & & & & & \end{matrix}\right\}$, find $X(k)$ by using decimation in frequency FFT algorithm.

Solution: N = 8, number of stages required $=\log_2 N=\log_2 8=3$. Signal flow diagram for N = 8 is shown in Fig. 3.27.

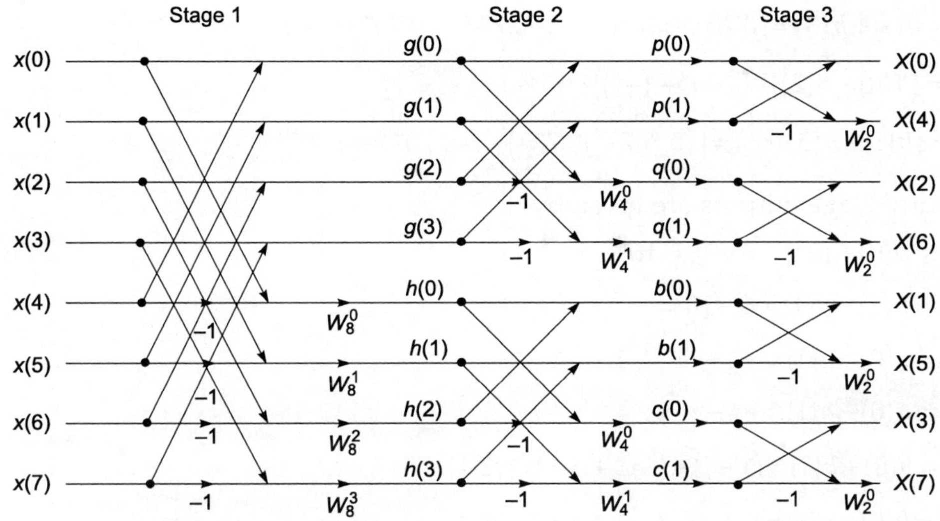

Figure 3.27 Flow graph of decimation frequency FFT algorithm for N = 8

$$W_8^0=W_4^0=W_2^0=1, W_8^2=W_4^1=-j, W_8^1=0.707-j0.707 \text{ and } W_8^3=-0.707-j0.707$$

The first stage outputs are given by

$g(0)=x(0)+x(4)=1+0=1$

$g(1)=x(1)+x(5)=2+1=3$

$g(2)=x(2)+x(6)=3+2=5$

$g(3)=x(3)+x(7)=4+3=7$

$h(0)=(x(0)-x(4))W_8^0=(1-0)1=1$

$h(1)=(x(1)-x(5))W_8^1=(2-1)(0.707-j0.707)=(0.707-j0.707)$

$h(2)=(x(2)-x(6))W_8^2=(3-2)(-j)=-j$

$h(3)=(x(3)-x(7))W_8^3=(4-3)(-0.707-j0.707)=(-0.707-j0.707)$

The second stage outputs are given by

$p(0)=g(0)+g(2)=1+5=6$

$$p(1) = g(1) + g(3) = 3 + 7 = 10$$

$$q(0) = (g(0) - g(2))W_4^0 = (1-5)1 = -4$$

$$q(1) = (g(1) - g(3))W_4^1 = (3-7)(-j) = j4$$

$$b(0) = h(0) + h(2) = 1 - j$$

$$b(1) = h(1) + h(3) = (0.707 - j0.707) + (-0.707 - j0.707) = -j1.41$$

$$c(0) = (h(0) - h(2))W_4^0 = (1 - (-j))1 = 1 + j$$

$$c(1) = (h(1) - h(3))W_4^1 = [(0.707 - j0.707) - (-0.707 - j0.707)](-j) = -j1.41$$

The third stage outputs are given by

$$X(0) = p(0) + p(1) = 6 + 10 = 16$$

$$X(4) = p(0) - p(1) = 6 - 10 = -4$$

$$X(2) = q(0) + q(1) = -4 + j4 = -4 + j4$$

$$X(6) = q(0) - q(1) = -4 - j4 = -4 - j4$$

$$X(1) = b(0) + b(1) = (1 - j) + (-j1.41) = 1 - j2.41$$

$$X(5) = b(0) - b(1) = (1 - j) - (-j1.41) = 1 + j0.41$$

$$X(3) = c(0) + c(1) = (1 + j) + (-j1.41) = 1 - j0.41$$

$$X(7) = c(0) - c(1) = (1 + j) - (-j1.41) = 1 + j2.41$$

Hence, $X(k)$ is given by

$$X(k) = \left\{ \begin{matrix} 16, & 1-j2.41, & -4+j4, & 1-j0.41, & -4, & 1+j0.41, & -4-j4, & 1+j2.41 \\ \uparrow & & & & & & & \end{matrix} \right\}$$

3.13 RADIX-2 INVERSE FAST FOURIER TRANSFORM

Similarly, the sequence, $x(n)$, can be obtained from $X(k)$ by taking inverse fast Fourier transform (IFFT) and is given by

$$x(n) = \frac{1}{N} \sum_{k=0}^{N-1} X(k) W_N^{-nk} \tag{3.51}$$

where, $$W_N = e^{-j\left(\frac{2\pi}{N}\right)}$$

By taking complex conjugate (replacing j by $-j$) on both the sides we get, Eq. 3.51 with W_N^{nk} and conjugate functions of $x(n)$ and $X(k)$.

$$x^*(n) = \frac{1}{N}\sum_{k=0}^{N-1} X^*(k)W_N^{nk} \qquad \text{or} \qquad Nx^*(n) = \sum_{k=0}^{N-1} X^*(k)W_N^{nk}$$

Hence,

$$x(n) = \frac{1}{N}\left[\sum_{k=0}^{N-1} X^*(k)W_N^{nk}\right]^* = \frac{1}{N}\left[Nx^*(n)\right]^* \tag{3.52}$$

In decimation in time IFFT algorithm, the input is a bit reversed and the output is in natural order and in decimation in frequency IFFT algorithm the input is natural order and output is in bit reversed.

Example 3.35 If

$$X(k) = \left\{ \begin{array}{l} 17,\ 1.12+j7.12,\ j3,\ -3.121-j2.87,\ 3,\ -3.12+j2.87,\ -j3,\ 1.12-j7.12 \\ \uparrow \end{array} \right\}$$

Find $x(n)$ by using decimation in time IFFT algorithm.

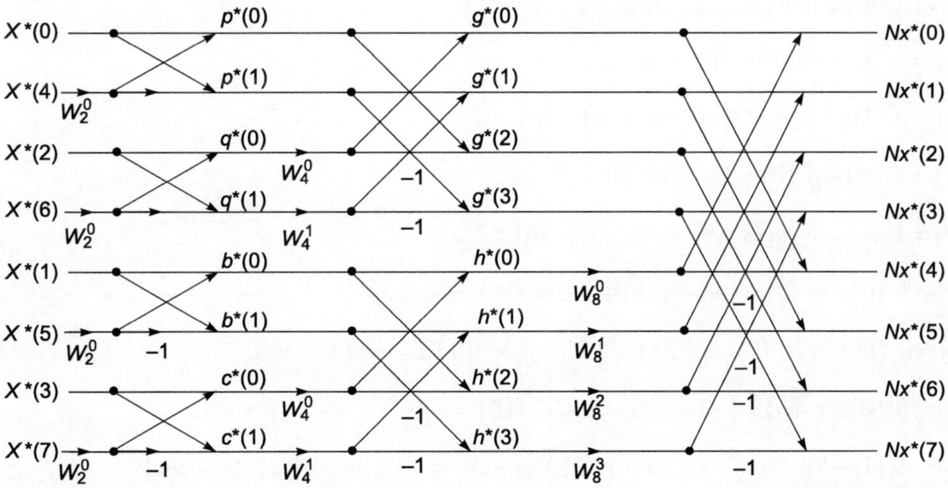

Figure 3.28 Signal flow graph of decimation in time IFFT algorithm for N = 8

Solution: In DIT IFFT algorithm, the input is a bit reversed and the output is in natural order.

$$X^*(k) = \left\{ \begin{array}{l} 17,\ 1.12-j7.12,\ -j3,\ -3.121+j2.87,\ 3,\ -3.12-j2.87,\ j3,\ 1.12+j7.12 \\ \uparrow \end{array} \right\}$$

$N = 8$, number of stages required $= \log_2 N = \log_2 8 = 3$. Signal flow diagram for $N = 8$ is shown in Figure 3.28.

$$W_8^0 = W_4^0 = W_2^0 = 1, W_8^2 = W_4^1 = -j, W_8^1 = 0.707 - j0.707 \text{ and } W_8^3 = -0.707 - j0.707$$

The output of first stage is given by

$$p^*(0) = X^*(0) + X^*(4) = 17 + 3 = 20$$

$$p^*(1) = X^*(0) - X^*(4) = 17 - 3 = 14$$

$$q^*(0) = X^*(2) + X^*(6) = (-j3) + (j3) = 0$$

$$q^*(1) = X^*(2) - X^*(6) = (-j3) - (j3) = -j6$$

$$b^*(0) = X^*(1) + X^*(5) = (1.12 - j7.12) + (-3.12 - j2.87) = -2 - j10$$

$$b^*(1) = X^*(1) - X^*(5) = (1.12 - j7.12) - (-3.12 - j2.87) = 4.24 - j4.25$$

$$c^*(0) = X^*(3) + X^*(7) = (-3.12 + j2.87) + (1.12 + j7.12) = -2 + j10$$

$$c^*(1) = X^*(3) - X^*(7) = (-3.12 + j2.87) - (1.12 + j7.12) = -4.24 - j4.25$$

The output of the second stage is given by

$$g^*(0) = p^*(0) + q^*(0) = 20 + 0 = 20$$

$$g^*(1) = p^*(1) + W_4^1 q^*(1) = (14) + (-j)(-j6) = 8$$

$$g^*(2) = p^*(0) - q^*(0) = 20 - 0 = 20$$

$$g^*(3) = p^*(1) - W_4^1 q^*(1) = (14) - (-j)(-j6) = 20$$

$$h^*(0) = b^*(0) + c^*(0) = (-2 - j10) + (-2 + j10) = -4$$

$$h^*(1) = b^*(1) + W_4^1 c^*(1) = (4.24 - j4.25) + (-j)(-4.24 - j4.25) = 0$$

$$h^*(2) = b^*(0) - c^*(0) = (-2 - j10) - (-2 + j10) = -j20$$

$$h^*(3) = b^*(1) - W_4^1 c^*(1) = (4.24 - j4.25) - (-j)(-4.24 - j4.25) = 8.49 - j8.49$$

The output of third stage is given by

$$Nx^*(0) = g^*(0) + W_8^0 h^*(0) = 20 + (1)(-4) = 16$$

$$Nx^*(1) = g^*(1) + W_8^1 h^*(1) = 8 + (0.707 - j0.707)(0) = 8$$

$$Nx^*(2) = g^*(2) + W_8^2 h^*(2) = 20 + (-j)(-j20) = 0$$

$$Nx^*(3) = g^*(3) + W_8^3 h^*(3) = 20 + (-0.707 - j0.707)(8.49 - j8.49) = 8$$

$$Nx^*(4) = g^*(0) - W_8^0 h^*(0) = (20) - (1)(-4) = 24$$

$$Nx^*(5) = g^*(1) - W_8^1 h^*(1) = 8 - (0.707 - j0.707)(0) = 8$$

$$Nx^*(6) = g^*(2) - W_8^2 h^*(2) = 20 - (-j)(-j20) = 40$$

$$Nx^*(7) = g^*(3) - W_8^3 h^*(3) = 20 - (-0.707 - j0.707)(8.49 - j8.49) = 32$$

$$x(n) = \frac{1}{N}[Nx^*(n)]$$

Hence, $x(n)$ is given by

$$x(n) = x(n) = \left\{\begin{matrix} 2, & 1, & 0, & 1, & 3, & 1, & 5, & 4 \\ \uparrow & & & & & & & \end{matrix}\right\}$$

Example 3.36 If

$$X(k) = \left\{\begin{matrix} 16, & 1-j2.41, & -4+j4, & 1-j0.41, & -4, & 1+j0.41, & -4-j4, & 1+j2.41 \\ \uparrow & & & & & & & \end{matrix}\right\}$$

Find $x(n)$ by using decimation in time IFFT algorithm.

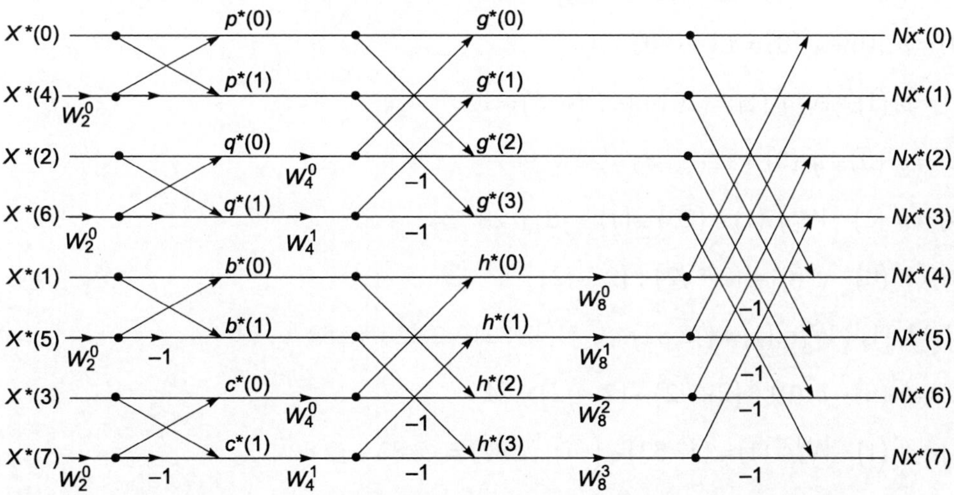

Figure 3.29 Signal flow graph of decimation in time IFFT algorithm for N = 8

Solution: In DIT IFFT algorithm, the input is bit reversed and output is in natural order.

$$X(k) = \left\{\begin{matrix} 16, & 1+j2.41, & -4-j4, & 1+j0.41, & -4, & 1-j0.41, & -4+j4, & 1-j2.41 \\ \uparrow & & & & & & & \end{matrix}\right\}$$

$N = 8$, number of stages required $= \log_2 N = \log_2 8 = 3$. Signal flow diagram for $N = 8$ is shown in Fig. 3.29.

$$W_8^0 = W_4^0 = W_2^0 = 1, W_8^2 = W_4^1 = -j, W_8^1 = 0.707 - j0.707 \text{ and } W_8^3 = -0.707 - j0.707$$

The output of first stage is given by

$$p^*(0) = X^*(0) + X^*(4) = 16 + (-4) = 12$$

$$p^*(1) = X^*(0) - X^*(4) = 16 - (-4) = 20$$

$$q^*(0) = X^*(2) + X^*(6) = (-4 - j4) + (-4 + j4) = -8$$

$$q^*(1) = X^*(2) - X^*(6) = (-4 - j4) - (-4 + j4) = -j8$$

$$b^*(0) = X^*(1) + X^*(5) = (1 + j2.41) + (1 - j0.41) = 2 + j2$$

$$b^*(1) = X^*(1) - X^*(5) = (1 + j2.41) - (1 - j0.41) = j2.82$$

$$c^*(0) = X^*(3) + X^*(7) = (1 + j0.41) + (1 - j2.41) = 2 - j2$$

$$c^*(1) = X^*(3) - X^*(7) = (1 + j0.41) - (1 - j2.41) = j2.82$$

The output of the second stage is given by

$$g^*(0) = p^*(0) + q^*(0) = 12 + (-8) = 4$$

$$g^*(1) = p^*(1) + W_4^1 q^*(1) = (20) + (-j)(-j8) = 12$$

$$g^*(2) = p^*(0) - q^*(0) = 12 - (-8) = 20$$

$$g^*(3) = p^*(1) - W_4^1 q^*(1) = (20) + (j)(-j8) = 28$$

$$h^*(0) = b^*(0) + c^*(0) = (2 + j2) + (2 - j2) = 4$$

$$h^*(1) = b^*(1) + W_4^1 c^*(1) = (j2.82) + (-j)(j2.82) = 2.82 + j2.82$$

$$h^*(2) = b^*(0) - c^*(0) = (2 + j2) - (2 - j2) = j4$$

$$h^*(3) = b^*(1) - W_4^1 c^*(1) = (j2.82) - (-j)(j2.82) = -2.82 + j2.82$$

The output of third stage is given by

$$Nx^*(0) = g^*(0) + W_8^0 h^*(0) = 4 + (1)(4) = 8$$

$$Nx^*(1) = g^*(1) + W_8^1 h^*(1) = 12 + (0.707 - j0.707)(2.82 + j2.82) = 16$$

$$Nx^*(2) = g^*(2) + W_8^2 h^*(2) = 20 + (-j)(j4) = 24$$

$$Nx^*(3) = g^*(3) + W_8^3 h^*(3) = 28 + (-0.707 - j0.707)(-2.82 + j2.82) = 32$$

$$Nx^*(4) = g^*(0) - W_8^0 h^*(0) = (4) - (1)(4) = 0$$

$$Nx^*(5) = g^*(1) - W_8^1 h^*(1) = 12 - (0.707 - j0.707)(2.82 + j2.82) = 8$$

$$Nx^*(6) = g^*(2) - W_8^2 h^*(2) = 20 - (-j)(j4) = 16$$

$$Nx^*(7) = g^*(3) - W_8^3 h^*(3) = 28 - (-0.707 - j0.707)(-2.82 + j2.82) = 24$$

$$x(n) = \frac{1}{N}[Nx^*(n)]$$

Hence, $x(n)$ is given by $x(n) = \left\{ \begin{matrix} 1, & 2, & 3, & 4, & 0, & 1, & 2, & 3 \\ \uparrow & & & & & & & \end{matrix} \right\}$

Example 3.37 If

$$X(k) = \left\{ \begin{matrix} 17, & 1.12 + j7.12, & j3, & -3.121 - j2.87, & 3, & -3.12 + j2.87, & -j3, & 1.12 - j7.12 \\ \uparrow & & & & & & & \end{matrix} \right\},$$

find $x(n)$ by using decimation in frequency IFFT algorithm.

Solution: In DIF IFFT algorithm, the input is natural order and output is bit-reversed order. For N = 8, number of stages required $= \log_2 N = \log_2 8 = 3$. Signal flow diagram is shown in Figure 3.30.

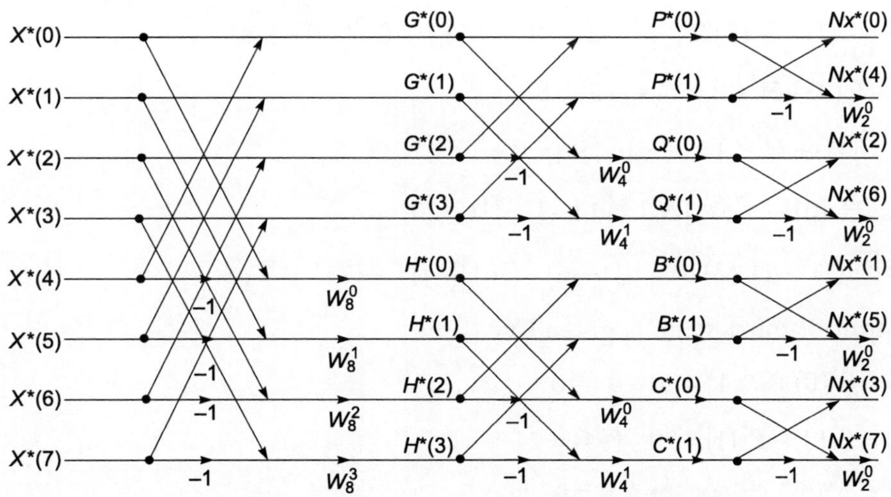

Figure 3.30 Flow graph of decimation in frequency IFFT algorithm for N = 8

$$W_8^0 = W_4^0 = W_2^0 = 1, W_8^2 = W_4^1 = -j, W_8^1 = 0.707 - j0.707 \text{ and } W_8^3 = -0.707 - j0.707$$

$$X^*(k) = \left\{ \begin{matrix} 17, & 1.12 - j7.12, & -j3, & -3.121 + j2.87, & 3, & -3.12 - j2.87, & j3, & 1.12 + j7.12 \\ \uparrow & & & & & & & \end{matrix} \right\}$$

The output of first stage is given by

$G^*(0) = X^*(0) + X^*(4) = 17 + 3 = 20$

$G^*(1) = X^*(1) + X^*(5) = (1.12 - j7.12) + (-3.12 - j2.87) = -2 - j10$

$G^*(2) = X^*(2) + X^*(6) = (-j3) + (j3) = 0$

$G^*(3) = X^*(3) + X^*(7) = (-3.121 + j2.87) + (1.12 + j7.12) = -2 + j10$

$H^*(0) = [X^*(0) - X^*(4)]W_8^0 = (17 - 3)1 = 14$

$H^*(1) = [X^*(1) - X^*(5)]W_8^1 = [(1.12 + j7.12) - (-3.121 - j2.87)](0.707 - j0.707) = -j6$

$H^*(2) = \left[X^*(2) - X^*(6)\right]W_8^2 = [(-j3) - (j3)](-j) = -6$

$H^*(3) = [X^*(3) - X^*(7)]W_8^3 = [(-3.121 + j2.87) - (1.12 + j7.12)](-0.707 - j0.707) = j6$

The output of second stage is given by

$P^*(0) = G^*(0) + G^*(2) = 20 + 0 = 20$

$P^*(1) = G^*(1) + G^*(3) = [(-2 - j10) + (-2 + j10)] = -4$

$Q^*(0) = \left[G^*(0) - G^*(2)\right]W_4^0 = [20 - 0] = 20$

$Q^*(1) = [G^*(1) - G^*(3)]W_4^1 = [(-2 - j10) - (-2 + j10)](-j) = -20$

$B^*(0) = H^*(0) + H^*(2) = 14 + (-6) = 8$

$B^*(1) = H^*(1) + H^*(3) = (-j6) + (j6) = 0$

$C^*(0) = \left[H^*(0) - H^*(2)\right]W_4^0 = [14 - (-6)]1 = 20$

$C^*(1) = \left[H^*(1) - H^*(3)\right]W_4^1 = [(-j6) - (j6)](-j) = -12$

The output of third stage is given by

$Nx^*(0) = P^*(0) + P^*(1) = 20 - 4 = 16$

$Nx^*(4) = P^*(0) - P^*(1) = 20 - (-4) = 24$

$Nx^*(2) = Q^*(0) + Q^*(1) = 20 + (-20) = 0$

$Nx^*(6) = Q^*(0) - Q^*(1) = 20 - (-20) = 40$

$Nx^*(1) = B^*(0) + B^*(1) = 8 + 0 = 8$

$Nx^*(5) = B^*(0) - B^*(1) = 8 - 0 = 8$

$Nx^*(3) = C^*(0) + C^*(1) = 20 + (-12) = 8$

$$Nx^*(7) = C^*(0) - C^*(1) = 20 - (-12) = 32$$

$$x(n) = \frac{1}{N}[Nx^*(n)]$$

Hence, $x(n)$ is given by $x(n) = \left\{ \begin{matrix} 2, & 1, & 0, & 1, & 3, & 1, & 5, & 4 \\ \uparrow & & & & & & & \end{matrix} \right\}$

Example 3.38 If

$$X(k) = \left\{ \begin{matrix} 16, & 1-j2.41, & -4+j4, & 1-j0.41, & -4, & 1+j0.41, & -4-j4, & 1+j2.41 \\ \uparrow & & & & & & & \end{matrix} \right\}$$

Find $x(n)$ by using decimation in frequency IFFT algorithm.

Solution: In DIF algorithm, the input is natural order and output is a bit-reversed order. For $N = 8$, number of stages required $= \log_2 N = \log_2 8 = 3$. Signal flow diagram is shown in Figure 3.31.

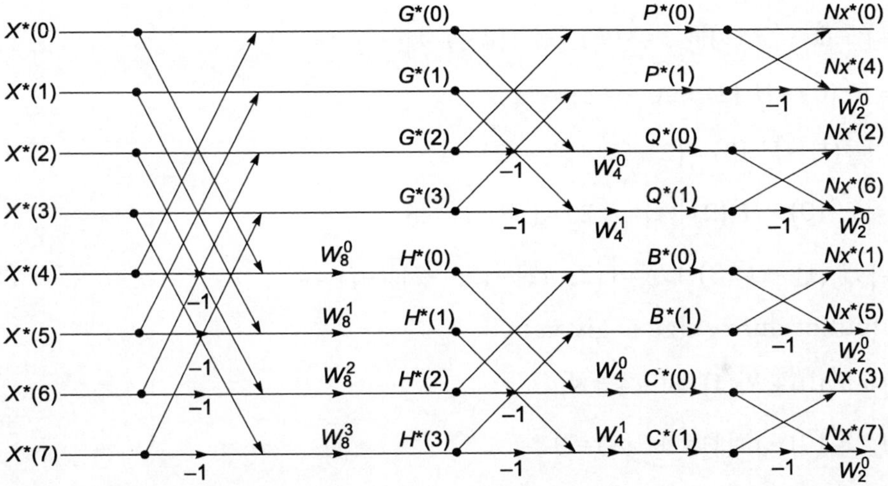

Figure 3.31 Flow graph of decimation in frequency IFFT algorithm for N = 8

$$W_8^0 = W_4^0 = W_2^0 = 1, W_8^2 = W_4^1 = -j, W_8^1 = 0.707 - j0.707 \text{ and } W_8^3 = -0.707 - j0.707$$

$$X(k) = \left\{ \begin{matrix} 16, & 1+j2.41, & -4-j4, & 1+j0.41, & -4, & 1-j0.41, & -4+j4, & 1-j2.41 \\ \uparrow & & & & & & & \end{matrix} \right\}$$

Output of first stage is given by

$$G^*(0) = X^*(0) + X^*(4) = 16 - 4 = 12$$

$$G^*(1) = X^*(1) + X^*(5) = (1 + j2.41) + (1 - j0.41) = 2 + j2$$

$$G^*(2) = X^*(2) + X^*(6) = (-4 - j4) + (-4 + j4) = -8$$

$$G^*(3) = X^*(3) + X^*(7) = (1 + j0.41) + (1 - j2.41) = 2 - j2$$

$$H^*(0) = [X^*(0) - X^*(4)]W_8^0 = (16 - (-4))1 = 20$$

$$H^*(1) = [X^*(1) - X^*(5)]W_8^1 = [(1 + j2.41) - (1 - j0.41)](0.707 - j0.707) = 2 + j2$$

$$H^*(2) = [X^*(2) - X^*(6)]W_8^2 = [(-4 - j4) - (-4 + j4)](-j) = -8$$

$$H^*(3) = [X^*(3) - X^*(7)]W_8^3 = [(1 + j0.41) - (1 - j2.41)](-0.707 - j0.707) = 2 - j2$$

The output of second stage is given by

$$P^*(0) = G^*(0) + G^*(2) = 12 + (-8) = 4$$

$$P^*(1) = G^*(1) + G^*(3) = [(2 + j2) + (2 - j2)] = 4$$

$$Q^*(0) = [G^*(0) - G^*(2)]W_4^0 = [12 - (-8)] = 20$$

$$Q^*(1) = [G^*(1) - G^*(3)]W_4^1 = [(2 + j2) - (2 - j2)](-j) = 4$$

$$B^*(0) = H^*(0) + H^*(2) = 20 + (-8) = 12$$

$$B^*(1) = H^*(1) + H^*(3) = (2 + j2) + (2 - j2) = 4$$

$$C^*(0) = [H^*(0) - H^*(2)]W_4^0 = [20 - (-8)]1 = 28$$

$$C^*(1) = [H^*(1) - H^*(3)]W_4^1 = [(2 + j2) - (2 - j2)](-j) = 4$$

The output of third stage is given by

$$Nx^*(0) = P^*(0) + P^*(1) = 4 + 4 = 8$$

$$Nx^*(4) = P^*(0) - P^*(1) = 4 - (4) = 0$$

$$Nx^*(2) = Q^*(0) + Q^*(1) = 20 + (4) = 24$$

$$Nx^*(6) = Q^*(0) - Q^*(1) = 20 - (4) = 16$$

$$Nx^*(1) = B^*(0) + B^*(1) = 12 + 4 = 16$$

$$Nx^*(5) = B^*(0) - B^*(1) = 12 - 4 = 8$$

$$Nx^*(3) = C^*(0) + C^*(1) = 28 + 4 = 32$$

$$Nx^*(7) = C^*(0) - C^*(1) = 28 - 4 = 24$$

$$x(n) = \frac{1}{N}[Nx^*(n)]$$

Hence, $x(n)$ is given by

$$x(n)=\left\{\begin{matrix}1, & 2, & 3, & 4, & 0, & 1, & 2, & 3\\ \uparrow & & & & & & & \end{matrix}\right\}$$

3.14 DIGITAL FILTERS

Digital filter is a linear time-invariant system that rejects unwanted frequencies of the input signal and allows the desired frequency components of a signal. Digital filters perform the same function as analog filters, except it operates with digital samples of the signal and with digital samples of the filter characteristics. Digital filters are classified into two types: finite impulse response system (FIR), and infinite impulse response system (IIR). If the impulse response sequence of the filter is of finite duration, then the filter is called FIR filter and if the impulse response sequence is of infinite duration, then it is called an IIR filter.

FIR filters are of non-recursive type and output depends on past and present samples of the input.

$$y(n)=\sum_{i=0}^{N} a_i x(n-i) \tag{3.53}$$

Whereas IIR system is recursive type and output depends on past and present samples of the input and also on past output samples.

$$y(n)=\sum_{i=0}^{N} a_i x(n-i)-\sum_{k=1}^{M} b_k y(n-k) \tag{3.54}$$

3.14.1 Classification of Digital Filters

Filter is essentially a system that passes certain desired frequency components in an input sequence without any distortion and blocks other frequency components. Filters are of four types. They are:

- Low-pass filter
- High-pass filter
- Band-pass filter
- Band-reject filter

Low-pass filter allows the signal between $(0<\omega<\omega_c)$ to pass and signal above $(\omega>\omega_c)$ are blocked. Magnitude responses of ideal and practical low-pass filter are shown in Figure 3.32 (a).

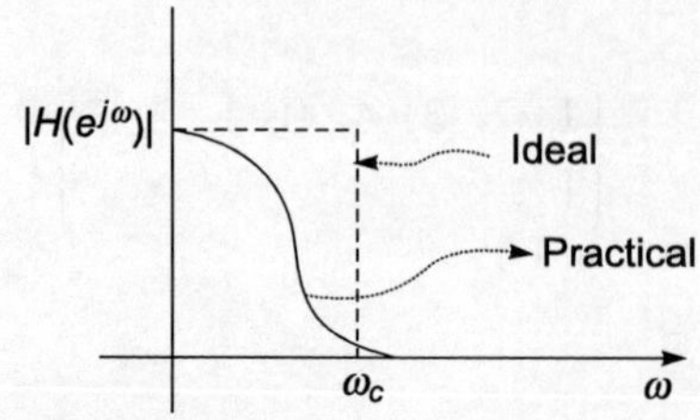

Figure 3.32(a) Low-pass filter

High-pass filter rejects the signal between $0 < \omega < \omega_c$ and allows the signal above $\omega > \omega_c$. Magnitude responses of ideal and practical high-pass filter are shown in Figure 3.32 (b).

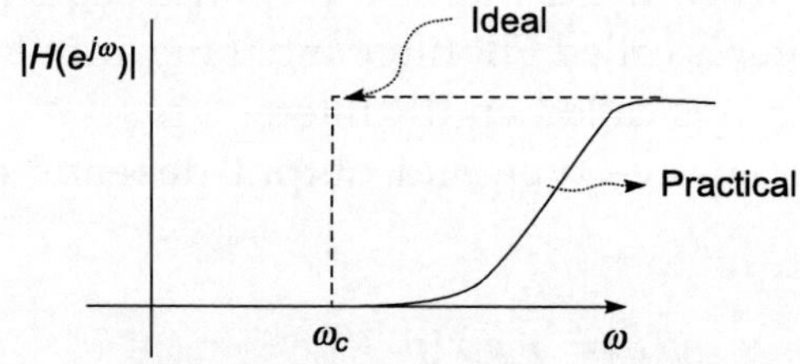

Figure 3.32(b) High-pass filter

Band-pass filter allows the signal between ω_{c1} and ω_{c2} to pass and rejects the remaining signal. Magnitude responses of ideal and practical band pass filter are shown in Figure 3.32(c).

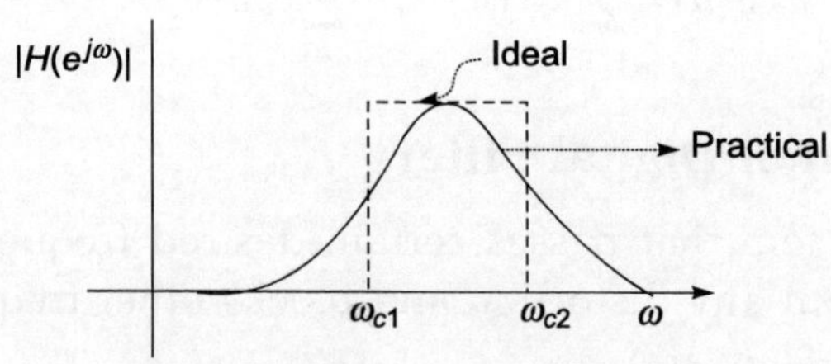

Figure 3.32(c) Band pass filter

Band-reject filter rejects the signal between ω_{c1} and ω_{c2} and passes the other signal. Magnitude responses of ideal and practical band reject filter are shown in Figure 3.32(d).

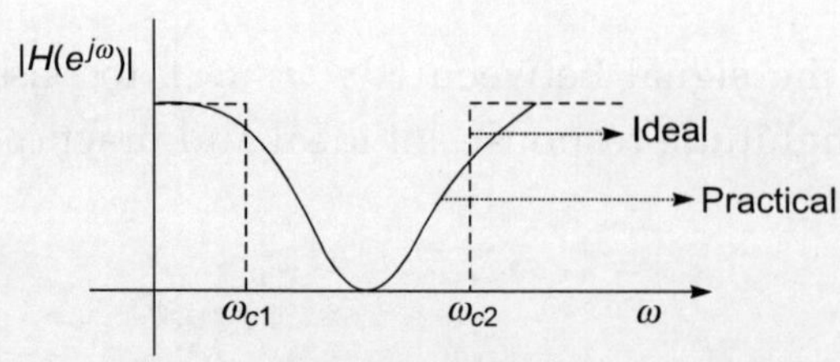

Figure 3.32(d) Band reject filter

3.14.2 Relationship between Analog and Digital Frequencies

Relationship between analog frequency (F in Hz and Ω in rad./sec) and digital frequency ω, period T, and sampling interval T_s are summarized in Table 3.2.

TABLE 3.2 Relationship between frequency, period and sampling interval

Analog frequency (F) in Hz	If T is period of the signal in sec, then $F = \frac{1}{T}$ Hz
Analog frequency (Ω) in rad./sec (angular frequency)	If F is frequency of the signal in Hz or T is period of the signal in sec, then $\Omega = 2\pi F$ rad/sec.
Digital frequency (ω) in radians	If T_s is the sampling interval in sec, then digital frequency, $\omega = 2\pi F T_s = \Omega T_s$ rad.

Example 3.39 What is the angular frequency and digital frequency of the sinusoid signal, if sinusoid signal frequency 1 kHz is sampled with $T_s = 0.1$ msec.

Solution:

Angular frequency, $\Omega = 2\pi F = 2\pi \times 1000 = 2000\pi$ rad/sec

Digital frequency, $\omega = 2\pi F T_s = 2\pi 1000 \times 0.1 \times 10^{-3} = 0.2\pi$ rad

3.14.3 Design of Linear Phase FIR Filters using Windows

We know that from the Fourier transform, impulse response, $h(n)$, and frequency response, $H(e^{j\omega})$, are related by

$$H(e^{j\omega}) = \sum_{n=-\infty}^{\infty} h(n) e^{-j\omega n} \tag{3.55}$$

$$h(n) = \frac{1}{2\pi} \int_{-\pi}^{\pi} H(e^{j\omega}) e^{j\omega n} d\omega \tag{3.56}$$

The impulse response, $h(n)$, obtained by Eq. (3.55) is infinite duration, since we are designing FIR filter, the impulse response, should be made finite. Let $h_d(n)$ represent the impulse response of a desired FIR filter. Then, $h_d(n)$, the impulse response of a desired FIR filter, is given by

$$h_d(n) = \begin{cases} h(n) & 0 \le n \le N-1 \\ 0 & \text{otherwise} \end{cases} \tag{3.57}$$

The frequency response of desired FIR filter is obtained by truncating Eq. (3.55) to length N.

$$H_d\left(e^{j\omega}\right)=\sum_{n=0}^{N-1} h_d(n)e^{-j\omega n} \tag{3.58}$$

Hence, frequency response of desired FIR filter is obtained by direct truncation of the infinite series which leads to Gibbs phenomenon. Truncation of the infinite series results in FIR filters with undesirable oscillations in the pass band and stop band as shown in Figure 3.33.

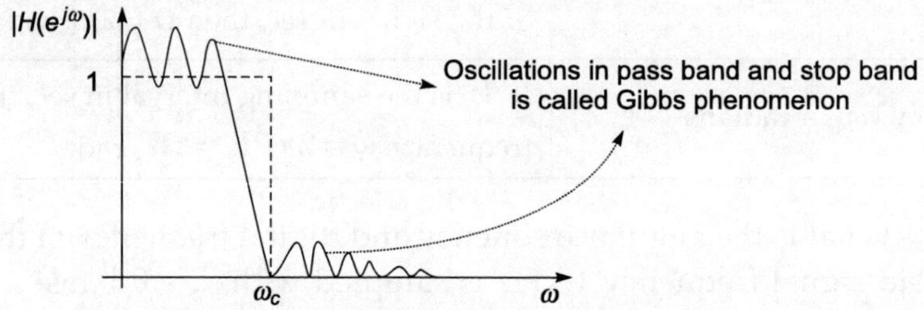

Figure 3.33 Oscillations in pass band and stop band

A practical approach to avoid oscillations in pass band and stop band is to multiply $h_d(n)$ by a window function $w(n)$, whose duration is finite

$$h(n)=h_d(n)w(n) \tag{3.59}$$

Most commonly used windows are rectangular, Hamming, Hanning and Kaiser windows. These windows are discussed below.

Rectangular window, denoted by $w_R(n)$, is defined by

$$w_R(n)=\begin{cases}1 & \text{for } 0,1,\ldots N-1\\ 0 & \text{otherwise}\end{cases} \tag{3.60}$$

Impulse response of the FIR filter is obtained by multiplying $h_d(n)$ with $w_R(n)$

$$h(n)=h_d(n)w_R(n)$$

The Fourier transform of rectangular window is given by

$$w_R(e^{j\omega})=\sum_{n=0}^{N-1} w_R(n)e^{-j\omega n}$$

$$w_R(e^{j\omega})=\frac{1-e^{-j\omega N}}{1-e^{-j\omega}}=\frac{e^{\frac{-j\omega N}{2}}}{e^{\frac{-j\omega}{2}}}\left[\frac{e^{\frac{j\omega N}{2}}-e^{\frac{-j\omega N}{2}}}{e^{\frac{j\omega}{2}}-e^{\frac{-j\omega}{2}}}\right]=e^{-\frac{j\omega(N-1)}{2}}\left[\frac{\sin\left(\frac{\omega N}{2}\right)}{\sin\left(\frac{\omega}{2}\right)}\right]$$

The window function has a magnitude response

$$|w_R(e^{j\omega})| = \left| \frac{\sin\left(\frac{\omega N}{2}\right)}{\sin\left(\frac{\omega}{2}\right)} \right| \tag{3.61}$$

The magnitude response of low-pass filter designed using rectangular window is shown in Figure 3.34. When rectangular window is used, oscillations are present in the stop band. For this reason the rectangular window is less of practical use.

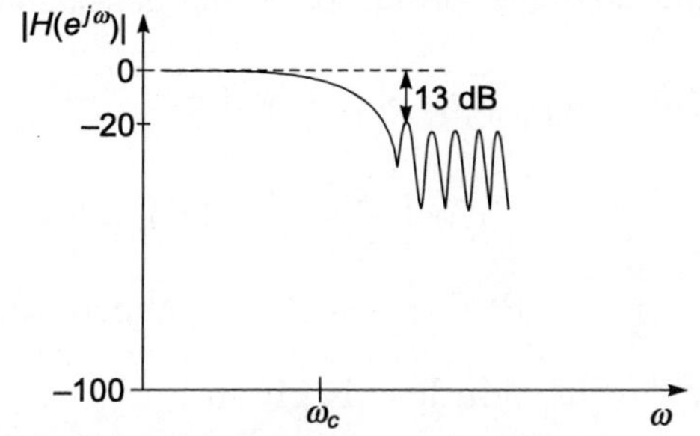

Figure 3.34 Magnitude response of the FIR low-pass filter designed using rectangular window

Hamming Window, denoted by $w_{\text{Ham}}(n)$, is defined by

$$w_{\text{Ham}}(n) = \begin{cases} 0.54 - 0.46\cos\left(\frac{2\pi n}{N-1}\right) & 0 \le n \le N-1 \\ 0 & \text{otherwise} \end{cases} \tag{3.62}$$

Fourier transform of Hamming window is given by

$$w_{\text{Ham}}(e^{j\omega}) = \sum_{n=0}^{N-1} w_{\text{Ham}}(n) e^{-j\omega n}$$

$$w_{\text{Ham}}(e^{j\omega}) = \sum_{n=0}^{N-1} \left[0.54 - 0.46\cos\left(\frac{2\pi n}{N-1}\right)\right] e^{-j\omega n}$$

$$w_{\text{Ham}}(e^{j\omega}) = 0.54 \frac{\sin\left(\frac{\omega N}{2}\right)}{\sin\left(\frac{\omega}{2}\right)} + 0.23 \frac{\sin\left(\frac{\omega N}{2} - \frac{N\pi}{N-1}\right)}{\sin\left(\frac{\omega}{2} - \frac{\pi}{N-1}\right)} + 0.23 \frac{\sin\left(\frac{\omega N}{2} + \frac{N\pi}{N-1}\right)}{\sin\left(\frac{\omega}{2} + \frac{\pi}{N-1}\right)} \tag{3.63}$$

The magnitude response of Hamming window is shown in Figure 3.35.

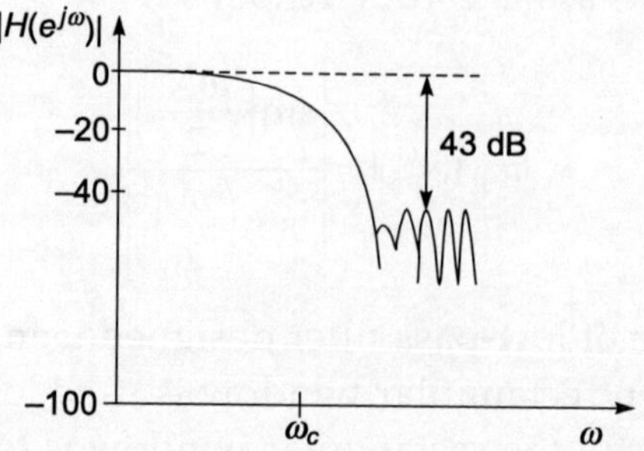

Figure 3.35 Magnitude response of the FIR low-pass filter designed using Hamming window

Hanning window, denoted by $w_{\text{Ham}}(n)$, is defined by

$$w_{\text{Han}}(n) = \begin{cases} 0.5 - 0.5\cos\left(\dfrac{2\pi n}{N-1}\right) & 0 \le n \le N-1 \\ 0 & \text{otherwise} \end{cases} \tag{3.64}$$

Fourier transform of Hamming window is given by

$$w_{\text{Han}}(e^{j\omega}) = \sum_{n=0}^{N-1} w_{\text{Han}}(n) e^{-j\omega n}$$

$$w_{\text{Han}}(e^{j\omega}) = \sum_{n=0}^{N-1} \left[0.5 - 0.5\cos\left(\frac{2\pi n}{N-1}\right)\right] e^{-j\omega n}$$

$$w_{\text{Han}}(e^{j\omega}) = 0.5\frac{\sin\left(\dfrac{\omega N}{2}\right)}{\sin\left(\dfrac{\omega}{2}\right)} + 0.25\frac{\sin\left(\dfrac{\omega N}{2} - \dfrac{N\pi}{N-1}\right)}{\sin\left(\dfrac{\omega}{2} - \dfrac{\pi}{N-1}\right)} + 0.25\frac{\sin\left(\dfrac{\omega N}{2} + \dfrac{N\pi}{N-1}\right)}{\sin\left(\dfrac{\omega}{2} + \dfrac{\pi}{N-1}\right)} \tag{3.65}$$

Magnitude response of Hanning window is shown in Figure 3.36.

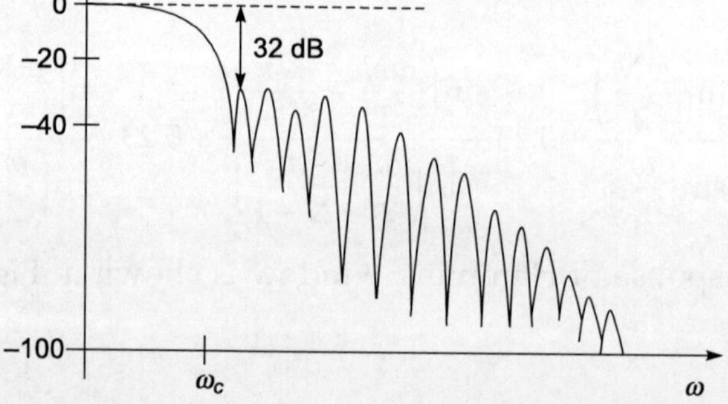

Figure 3.36 Magnitude response of the Fir low-pass filter designed using Hanning

Examples of FIR Filter Design using Windows

Low-pass filter passes low frequency component to the output while attenuating high frequency component. The frequency response of FIR filter for symmetric sequence is given by

$$H(e^{j\omega}) = |H(e^{j\omega})| e^{-j\omega\left(\frac{N-1}{2}\right)}$$

To design an ideal low-pass filter with cut-off frequency ω_c and if $\tau = \frac{N-1}{2}$, the desired frequency response is given by

$$H_d(e^{j\omega}) = \begin{cases} e^{-j\omega\tau} & -\omega_c \leq \omega \leq \omega_c \\ 0 & \text{otherwise} \end{cases} \tag{3.66}$$

The impulse response sequence is obtained by taking inverse Fourier transform of $H\left(e^{j\omega}\right)$

$$h(n) = \frac{1}{2\pi} \int_{-\omega_c}^{\omega_c} H(e^{j\omega}) e^{j\omega n} \, d\omega \tag{3.67}$$

The impulse response sequence for an ideal low-pass filter is given by

$$h_d(n) = \frac{1}{2\pi} \int_{-\omega_c}^{\omega_c} e^{-j\omega\tau} e^{j\omega n} d\omega$$

$$h_d(n) = \frac{1}{2\pi} \int_{-\omega_c}^{\omega_c} e^{-j\omega(\tau - n)} d\omega$$

$$h_d(n) = \frac{1}{2\pi j} \left[\frac{e^{-j\omega_c(\tau - n)}}{(n-\tau)} - \frac{e^{j\omega_c(\tau - n)}}{(n-\tau)} \right]$$

$$h_d(n) = \frac{1}{\pi(n-\tau)} \left[\frac{e^{j\omega_c(n-\tau)} - e^{-j\omega_c(n-\tau)}}{2j} \right]$$

Hence,

$$h_d(n) = \begin{cases} \frac{1}{\pi(n-\tau)} \sin \omega_c(n-\tau) & \text{for } n \neq r \\ \frac{\omega_c}{\pi} & \text{for } n = \tau \end{cases} \tag{3.68}$$

Example 3.40 Design a low-pass FIR filter for $N = 5$ and $\omega_c = 0.6$ radian.

Solution: For low-pass FIR filter, filter coefficients are given by Eq. 3.68. Substituting, $N = 5$, $\tau = 2$ and $\omega_c = 0.6$ radian in Eq. 3.68, we obtain

$$h_d(n) = \begin{cases} \dfrac{1}{\pi(n-\tau)} \sin \omega_c (n-\tau) & \text{for } n \neq \tau \\ \dfrac{\omega_c}{\pi} & \text{for } n = \tau \end{cases}$$

$$h_d(0) = h_d(4) = \frac{\sin(0.6)(2)}{2\pi} = 0.1483$$

$$h_d(1) = h_d(3) = \frac{\sin(0.6)(1)}{\pi} = 0.1797$$

$$h_d(2) = \frac{0.6}{\pi} = 0.1909$$

Using rectangular window:

$$h(n) = h_d(n) w_R(n) = h_d(n)$$

Using Hamming window:

$$h(n) = h_d(n) w_{\text{Ham}}(n)$$

For $N = 5$, $w_{\text{Ham}}(n)$ is given by

$$w_{\text{Ham}}(n) = \begin{cases} 0.54 - 0.46 \cos\left(\dfrac{2\pi n}{N-1}\right) & 0 \leq n \leq 4 \\ 0 & \text{otherwise} \end{cases}$$

$$w_{\text{Ham}}(0) = w_{\text{Ham}}(4) = 0.08$$

$$w_{\text{Ham}}(1) = w_{\text{Ham}}(3) = 0.54$$

$$w_{\text{Ham}}(2) = 1$$

$$h(n) = h_d(n) w_{\text{Ham}}(n)$$

$$h(0) = h(4) = 0.01186$$

$$h(1) = h(3) = 0.09703$$

$$h(2) = 0.1909$$

Using Hanning window:

$$h(n) = h_d(n) w_{\text{Han}}(n)$$

For $N = 5$, $w_{\text{Ham}}(n)$ is given by

$$w_{\text{Han}}(n) = \begin{cases} 0.5 - 0.5\cos\left(\dfrac{2\pi n}{N-1}\right) & 0 \le n \le 4 \\ 0 & \text{otherwise} \end{cases}$$

$$w_{\text{Han}}(0) = w_{\text{Han}}(4) = 0$$

$$w_{\text{Han}}(1) = w_{\text{Han}}(3) = 0.5$$

$$w_{\text{Han}}(2) = 1$$

$$h(n) = h_d(n) w_{\text{Han}}(n)$$

$$h(0) = h(4) = 0$$

$$h(1) = h(3) = 0.08985$$

$$h(2) = 0.1909$$

Example 3.41 Design a low-pass FIR filter for $N = 6$ and $\omega_c = 0.25$ radian.

Solution: For low-pass FIR filter, filter coefficients are given by Eq. 3.68. Substituting, $N = 6$, $\tau = 2.5$ and $\omega_c = 0.25$ radian in Eq. 3.68, we get

$$h_d(n) = \frac{1}{\pi(n-\tau)} \sin \omega_c \left(n - \frac{N-1}{2}\right)$$

$$h_d(0) = h_d(5) = \frac{\sin(0.25)(-2.5)}{-2.5\pi} = 0.0744$$

$$h_d(1) = h_d(4) = \frac{\sin(0.25)(-1.5)}{-1.5\pi} = 0.0777$$

$$h_d(2) = h_d(3) = \frac{\sin(0.25)(-0.5)}{-0.5\pi} = 0.07937$$

Using rectangular window:

$$h(n) = h_d(n) w_R(n)$$

Using Hamming window:

$$h(n) = h_d(n) w_{\text{Ham}}(n)$$

where, $w_{\text{Ham}}(n)$ for $N = 6$ is given by

$$w_{\text{Ham}}(n) = \begin{cases} 0.54 - 0.46\cos\left(\dfrac{2\pi n}{N-1}\right) & 0 \le n \le 5 \\ 0 & \text{otherwise} \end{cases}$$

$$w_{\text{Ham}}(0) = w_{\text{Ham}}(5) = 0.08$$
$$w_{\text{Ham}}(1) = w_{\text{Ham}}(4) = 0.397$$
$$w_{\text{Ham}}(2) = w_{\text{Ham}}(3) = 0.91215$$
$$h(n) = h_d(n)\, w_{\text{Ham}}(n)$$
$$h(0) = h(5) = 0.00595$$
$$h(1) = h(4) = 0.03085$$
$$h(2) = h(3) = 0.0724$$

Using Hanning window:

$$h(n) = h_d(n)\, w_{\text{Han}}(n)$$

where, $w_{\text{Han}}(n)$ for $N = 6$ is given by

$$w_{\text{Han}}(n) = \begin{cases} 0.5 - 0.5\cos\left(\dfrac{2\pi n}{N-1}\right) & 0 \le n \le 5 \\ 0 & \text{otherwise} \end{cases}$$

$$w_{\text{Han}}(0) = w_{\text{Han}}(5) = 0$$
$$w_{\text{Han}}(1) = w_{\text{Han}}(4) = 0.34549$$
$$w_{\text{Han}}(2) = w_{\text{Han}}(3) = 0.90451$$
$$h(n) = h_d(n)\, w_{\text{Han}}(n)$$
$$h(0) = h(5) = 0$$
$$h(1) = h(4) = 0.02684$$
$$h(2) = h(3) = 0.07179$$

High-pass filter passes high frequency component to the output while attenuating low frequency component. The frequency response of FIR filter for symmetric sequence is given by

$$H(e^{j\omega}) = \left|H(e^{j\omega})\right| e^{-j\omega\left(\frac{N-1}{2}\right)}$$

To design an ideal high-pass filter with cut-off frequency (ω_c) and if $\tau = \dfrac{N-1}{2}$, the desired frequency response is given by

$$H_d(e^{j\omega}) = \begin{cases} e^{-j\omega\tau} & \omega_c \le |\omega| < \pi \\ 0 & \text{otherwise} \end{cases} \tag{3.69}$$

The impulse response sequence is obtained by taking inverse Fourier transform of $H\left(e^{j\omega}\right)$

$$h(n) = \frac{1}{2\pi}\int_{-\omega_c}^{\omega_c} H(e^{j\omega})e^{j\omega n}d\omega$$

The impulse response sequence for an ideal high-pass filter is given by

$$h_d(n) = \frac{1}{2\pi}\left[\int_{-\pi}^{-\omega_c} e^{j\omega(n-\tau)}d\omega + \int_{\omega_c}^{\pi} e^{j\omega(n-\tau)}d\omega\right]$$

$$h_d(n) = \frac{1}{\pi(n-\tau)}\left[\frac{e^{-j\omega_c(n-\tau)} - e^{-j\pi(n-\tau)}}{2j} + \frac{e^{j\pi(n-\tau)} - e^{j\omega_c(n-\tau)}}{2j}\right]$$

$$h_d(n) = \frac{1}{\pi(n-\tau)}\left[\frac{e^{j\pi(n-\tau)} - e^{-j\pi(n-\tau)}}{2j} - \frac{e^{j\omega_c(n-\tau)} - e^{-j\omega_c(n-\tau)}}{2j}\right]$$

$$h_d(n) = \frac{1}{\pi(n-\tau)}[\sin\pi(n-\tau) - \sin\omega_c(n-\tau)] \quad n \ne \tau$$

$$h_d(n) = \frac{1}{2\pi}[-\omega_c + \pi + \pi - \omega_c] = 1 - \frac{\omega_c}{\pi} \qquad n = \tau$$

Hence,

$$h_d(n) = \begin{cases} \dfrac{1}{\pi(n-\tau)}[\sin\pi(n-\tau) - \sin\omega_c(n-\tau)] & \text{for } n \ne \tau \\ 1 - \dfrac{\omega_c}{\pi} & \text{for } n = \tau \end{cases} \tag{3.70}$$

Example 3.42 Design a high-pass FIR filter for $N = 5$ and $\omega_c = 0.6$ rad.

Solution: For a high-pass FIR filter, filter coefficients are given by Eq. 3.70

$$h_d(n) = \begin{cases} \dfrac{1}{\pi(n-\tau)}[\sin\pi(n-\tau) - \sin\omega_c(n-\tau)] & \text{for } n \ne \tau \\ 1 - \dfrac{\omega_c}{\pi} & \text{for } n = \tau \end{cases}$$

Substituting, $N = 5$, $\tau = 2$ and $\omega_c = 0.6$ radian in Eq. 3.70, we get

$$h_d(0) = h_d(4) = \frac{\sin(0-2)\pi - \sin(0-2)0.6}{\pi(0-2)} = -0.1483$$

$$h_d(1) = h_d(3) = \frac{\sin(1-2)\pi - \sin(1-2)0.6}{\pi(1-2)} = -0.1797$$

$$h_d(2) = 1 - \frac{0.6}{\pi} = 0.8090$$

Using rectangular window:

$$h(n) = h_d(n) w_R(n) = h_d(n)$$

$$h(0) = h(4) = -0.1483, \qquad h(1) = h(3) = -0.1797, \qquad h(2) = 0.8090$$

Using Hamming window:

$$h(n) = h_d(n) w_{\text{Ham}}(n)$$

Where $w_{\text{Ham}}(n)$ for $N = 5$ is given by

$$w_{\text{Ham}}(n) = \begin{cases} 0.54 - 0.46\cos\left(\dfrac{2\pi n}{N-1}\right) & 0 \le n \le 4 \\ 0 & \text{otherwise} \end{cases}$$

$$w_{\text{Ham}}(0) = w_{\text{Ham}}(4) = 0.08$$

$$w_{\text{Ham}}(1) = w_{\text{Ham}}(3) = 0.54$$

$$w_{\text{Ham}}(2) = 1$$

$$h(n) = h_d(n) w_{\text{Ham}}(n)$$

$$h(0) = h(4) = -0.01186$$

$$h(1) = h(3) = -0.09703$$

$$h(2) = 0.809$$

Using Hanning window:

$$h(n) = h_d(n) w_{\text{Han}}(n)$$

where, $w_{\text{Ham}}(n)$ for $N = 5$ is given by

$$w_{\text{Han}}(n) = \begin{cases} 0.5 - 0.5\cos\left(\dfrac{2\pi n}{N-1}\right) & 0 \le n \le 4 \\ 0 & \text{otherwise} \end{cases}$$

$$w_{\text{Han}}(0) = w_{\text{Han}}(4) = 0$$

$$w_{\text{Han}}(1) = w_{\text{Han}}(3) = 0.5$$

$$w_{\text{Han}}(2) = 1$$

$$h(n) = h_d(n) w_{\text{Han}}(n)$$

$$h(0) = h(4) = 0$$

$$h(1) = h(3) = -0.08985$$

$$h(2) = 0.809$$

Example 3.43 Design a high-pass FIR filter for N = 6 and $\omega_c = 0.65$ radian.

Solution: For a high-pass FIR filter, filter coefficients are given by Eq. 3.70.

$$h_d(n) = \frac{1}{\pi(n-\tau)}[\sin\pi(n-\tau) - \sin\omega_c(n-\tau)]$$

Substituting N = 6, τ = 2.5 and $\omega_c = 0.65$ radian in Eq. 11.45, we get

$$h_d(0) = h_d(5) = \frac{[\sin\pi(-2.5) - \sin 0.65(-2.5)]}{-2.5\pi} = 0.00019$$

$$h_d(1) = h_d(4) = \frac{[\sin\pi(-1.5) - \sin 0.65(-1.5)]}{-1.5\pi} = -0.3878$$

$$h_d(2) = h_d(3) = \frac{[\sin\pi(-0.5) - \sin 0.65(-0.5)]}{-0.5\pi} = 0.4333$$

Using rectangular window:

$$h(n) = h_d(n) w(n) = h_d(n)$$

$$h(0) = h(5) = 0.00019 \qquad h(1) = h(4) = -0.3878 \qquad h(2) = h(3) = 0.4333$$

Using Hamming window:

$$h(n) = h_d(n) w_{\text{Ham}}(n)$$

Where $w_{\text{Ham}}(n)$ for N = 6 is given by

$$w_{\text{Ham}}(n) = \begin{cases} 0.54 - 0.46\cos\left(\dfrac{2\pi n}{N-1}\right) & 0 \le n \le 5 \\ 0 & \text{otherwise} \end{cases}$$

$$w_{\text{Ham}}(0) = w_{\text{Ham}}(5) = 0.08$$
$$w_{\text{Ham}}(1) = w_{\text{Ham}}(4) = 0.3978$$
$$w_{\text{Ham}}(2) = w_{\text{Ham}}(3) = 0.9121$$
$$h(n) = h_d(n) w_{\text{Ham}}(n)$$
$$h(0) = h(5) = 0.0000152$$
$$h(1) = h(4) = -0.1542$$
$$h(2) = h(3) = 0.3952$$

Using Hanning window:

$$h(n) = h_d(n) w_{\text{Han}}(n)$$

where, $w_{\text{Han}}(n)$ for $N = 6$ is given by

$$w_{\text{Han}}(n) = \begin{cases} 0.5 - 0.5\cos\left(\dfrac{2\pi n}{N-1}\right) & 0 \le n \le 5 \\ 0 & \text{otherwise} \end{cases}$$

$$w_{\text{Han}}(0) = w_{\text{Han}}(5) = 0$$
$$w_{\text{Han}}(1) = w_{\text{Han}}(4) = 0.3454$$
$$w_{\text{Han}}(2) = w_{\text{Han}}(3) = 0.9045$$
$$h(n) = h_d(n) w_{\text{Han}}(n)$$
$$h(0) = h(5) = 0$$
$$h(1) = h(4) = -0.1339$$
$$h(2) = h(3) = 0.3919$$

Band-pass filter rejects both low frequency and high frequency components. The frequency response of an FIR filter for symmetric sequence is given by

$$H(e^{j\omega}) = \left|H(e^{j\omega})\right| e^{-j\omega\left(\frac{N-1}{2}\right)}$$

To design an ideal band-pass filter with cut-off frequencies ω_{c1} and ω_{c2} and if $\tau = \dfrac{N-1}{2}$, the desired frequency response is given by

$$H_d(e^{j\omega}) = \begin{cases} e^{-j\omega\tau} & \omega_{c1} \le |\omega| \le \omega_{c2} < \pi \\ 0 & \text{otherwise} \end{cases} \tag{3.71}$$

The impulse response sequence for an ideal band-pass filter is given by

$$h_d(n)=\frac{1}{2\pi}\left[\int_{-\omega_{c2}}^{-\omega_{c1}} e^{j\omega(n-\tau)}d\omega+\int_{\omega_{c1}}^{\omega_{c2}} e^{j\omega(n-\tau)}d\omega\right]$$

$$h_d(n)=\frac{1}{\pi(n-\tau)}\left[\frac{e^{-j\omega_{c1}(n-\tau)}-e^{-j\omega_{c2}(n-\tau)}}{2j}+\frac{e^{j\omega_{c2}(n-\tau)}-e^{-j\omega_{c1}(n-\tau)}}{2j}\right]$$

$$h_d(n)=\frac{1}{\pi(n-\tau)}\left[\frac{e^{j\omega_{c2}(n-\tau)}-e^{-j\omega_{c2}(n-\tau)}}{2j}-\frac{e^{j\omega_{c1}(n-\tau)}-e^{-j\omega_{c1}(n-\tau)}}{2j}\right]$$

$$h_d(n)=\frac{1}{\pi(n-\tau)}[\sin\omega_{c2}(n-\tau)-\sin\omega_{c1}(n-\tau)]\quad n\neq\tau$$

$$h_d(n)=\frac{1}{\pi}[\omega_{c2}-\omega_{c1}]\qquad n=\tau$$

Hence,

$$h_d(n)=\begin{cases}\frac{1}{\pi(n-\tau)}[\sin\omega_{c2}(n-\tau)-\sin\omega_{c1}(n-\tau)] & \text{for } n\neq\tau\\ \frac{1}{\pi}[\omega_{c2}-\omega_{c1}] & \text{for } n=\tau\end{cases}\tag{3.72}$$

Example 3.44 Design a band-pass FIR filter for N = 7, $\omega_{c1}=0.4$ radian and $\omega_{c2}=0.5$ radian.

Solution: For a band-pass FIR filter, filter coefficients are given by Eq. 3.72. By taking, N = 7, τ = 3 and $\omega_{c1}=0.4$ radian and $\omega_{c2}=0.5$ radian.

$$h_d(n)=\begin{cases}\frac{1}{\pi(n-\tau)}[\sin\omega_{c2}(n-\tau)-\sin\omega_{c1}(n-\tau)] & \text{for } n\neq\tau\\ \frac{1}{\pi}[\omega_{c2}-\omega_{c1}] & \text{for } n=\tau\end{cases}$$

Substituting N = 7, τ = 3 and $\omega_{c1}=0.4$ radian and $\omega_{c2}=0.5$ radian in Eq. 3.78, we get

$$h_d(0)=h_d(6)=\frac{\sin(0-3)0.5-\sin(0-3)0.4}{\pi(0-3)}=0.0070$$

$$h_d(1)=h_d(5)=\frac{\sin(1-3)0.5-\sin(1-3)0.4}{\pi(1-3)}=0.0198$$

$$h_d(2) = h_d(4) = \frac{\sin(2-3)0.5 - \sin(2-3)0.4}{\pi(2-3)} = 0.0287$$

$$h_d(3) = \frac{(0.5-0.4)}{\pi} = 0.0318$$

Using rectangular window:

$$h(n) = h_d(n) w_R(n) = h_d(n)$$

Using Hamming window:

$$h(n) = h_d(n) w_{\text{Ham}}(n)$$

where, $w_{\text{Ham}}(n)$ for $N = 7$ is given by

$$w_{\text{Ham}}(n) = \begin{cases} 0.54 - 0.46\cos\left(\dfrac{2\pi n}{N-1}\right) & 0 \le n \le 6 \\ 0 & \text{otherwise} \end{cases}$$

$$\begin{aligned} w_{\text{Ham}}(0) &= w_{\text{Ham}}(6) = 0.08 \\ w_{\text{Ham}}(1) &= w_{\text{Ham}}(5) = 0.31 \\ w_{\text{Ham}}(2) &= w_{\text{Ham}}(4) = 0.77 \\ w_{\text{Ham}}(3) &= 1 \end{aligned}$$

$$\begin{aligned} h(n) &= h_d(n) w_{\text{Ham}}(n) \\ h(0) &= h(6) = 0.00056 \\ h(1) &= h(5) = 0.0061 \\ h(2) &= h(4) = 0.0221 \\ h(3) &= 0.0318 \end{aligned}$$

Using Hanning window:

$$h(n) = h_d(n) w_{\text{Han}}(n)$$

where, $w_{\text{Han}}(n)$ for $N = 7$ is given by

$$w_{\text{Han}}(n) = \begin{cases} 0.5 - 0.5\cos\left(\dfrac{2\pi n}{N-1}\right) & 0 \le n \le 6 \\ 0 & \text{otherwise} \end{cases}$$

$$\begin{aligned} w_{\text{Han}}(0) &= w_{\text{Han}}(6) = 0 \\ w_{\text{Han}}(1) &= w_{\text{Han}}(5) = 0.25 \end{aligned}$$

$$w_{\text{Han}}(2) = w_{\text{Han}}(4) = 0.75$$
$$w_{\text{Han}}(3) = 1$$
$$h(n) = h_d(n)w_{\text{Han}}(n)$$
$$h(0) = h(6) = 0$$
$$h(1) = h(5) = 0.00495$$
$$h(2) = h(4) = 0.021525$$
$$h(3) = 0.031$$

Example 3.45 Design a band-pass FIR filter for $N = 8$, $\omega_{c1} = 0.25$ radian and $\omega_{c2} = 0.7$ radian.

Solution: For a band-pass FIR filter, filter coefficients are given by Eq. 3.72. By taking, $N = 8$, $\tau = 3.5$ and $\omega_{c1} = 0.25$ radian and $\omega_{c2} = 0.7$ radian

$$h_d(n) = \frac{1}{\pi(n-\tau)}[\sin \omega_{c2}(n-\tau) - \sin \omega_{c1}(n-\tau)]$$

Substituting $N = 8$, $\tau = 3.5$ and $\omega_{c1} = 0.25$ radian and $\omega_{c2} = 0.7$ radian in the above equation, we get

$$h_d(0) = h_d(7) = \frac{\sin(0-3.5)0.7 - \sin(0-3.5)0.25}{\pi(0-3.5)} = -0.0118$$

$$h_d(1) = h_d(6) = \frac{\sin(1-3.5)0.7 - \sin(1-3.5)0.25}{\pi(1-3.5)} = 0.0507$$

$$h_d(2) = h_d(5) = \frac{\sin(2-3.5)0.7 - \sin(2-3.5)0.25}{\pi(2-3.5)} = 0.1063$$

$$h_d(3) = h_d(4) = \frac{\sin(3-3.5)0.7 - \sin(3-3.5)0.25}{\pi(3-3.5)} = 0.1389$$

Using rectangular window:

$$h(n) = h_d(n)w_R(n) = h_d(n)$$

Using Hamming window:

$$h(n) = h_d(n)w_{\text{Ham}}(n)$$

where, $w_{\text{Ham}}(n)$ for $N = 8$ is given by

$$w_{\text{Ham}}(n) = \begin{cases} 0.54 - 0.46\cos\left(\dfrac{2\pi n}{N-1}\right) & 0 \le n \le 7 \\ 0 & \text{otherwise} \end{cases}$$

$$w_{\text{Ham}}(0) = w_{\text{Ham}}(7) = 0.08$$

$$w_{\text{Ham}}(1) = w_{\text{Ham}}(6) = 0.253$$

$$w_{\text{Ham}}(2) = w_{\text{Ham}}(5) = 0.642$$

$$w_{\text{Ham}}(3) = w_{\text{Ham}}(4) = 0.954$$

$$h(n) = h_d(n)\, w_{\text{Ham}}(n)$$

$$h(0) = h(7) = -9.44 \times 10^{-4}$$

$$h(1) = h(6) = 0.0128$$

$$h(2) = h(5) = 0.0682$$

$$h(3) = h(4) = 0.1325$$

Using Hanning window:

$$h(n) = h_d(n)\, w_{\text{Han}}(n)$$

where, $w_{\text{Han}}(n)$ for $N = 8$ is given by

$$w_{\text{Han}}(n) = \begin{cases} 0.5 - 0.5\cos\left(\dfrac{2\pi n}{N-1}\right) & 0 \le n \le 7 \\ 0 & \text{otherwise} \end{cases}$$

$$w_{\text{Han}}(0) = w_{\text{Han}}(7) = 0$$

$$w_{\text{Han}}(1) = w_{\text{Han}}(6) = 0.1882$$

$$w_{\text{Han}}(2) = w_{\text{Han}}(5) = 0.611$$

$$w_{\text{Han}}(3) = w_{\text{Han}}(4) = 0.950$$

$$h(n) = h_d(n)\, w_{\text{Han}}(n)$$

$$h(0) = h(7) = 0$$

$$h(1) = h(6) = 9.541 \times 10^{-3}$$

$$h(2) = h(5) = 0.0649$$

$$h(3) = h(4) = 0.1319$$

Band-reject filter rejects both low frequency and high frequency components. The frequency response of an FIR filter for symmetric sequence is given by

$$H(e^{j\omega}) = |H(e^{j\omega})|e^{-j\omega\left(\frac{N-1}{2}\right)}$$

To design an ideal band-reject filter with cut-off frequencies ω_{c1} and ω_{c2} and if $\tau = \frac{N-1}{2}$, the desired frequency response is given by

$$H_d(e^{j\omega}) = \begin{cases} e^{-j\omega\tau} & 0 \le |\omega| \le \omega_{c1}, \omega_{c2} \le |\omega| \le \pi \\ 0 & \text{otherwise} \end{cases} \tag{3.73}$$

The impulse response sequence for an ideal band-reject filter is given by

$$h_d(n) = \frac{1}{2\pi}\left[\int_{-\pi}^{-\omega_{c2}} e^{j\omega(n-\tau)}d\omega + \int_{-\omega_{c1}}^{\omega_{c1}} e^{j\omega(n-\tau)}d\omega + \int_{\omega_{c2}}^{\pi} e^{j\omega(n-\tau)}d\omega\right]$$

$$= \frac{1}{\pi(n-\tau)}\left[\frac{e^{-j\omega_{c2}(n-\tau)} - e^{-j\pi(n-\tau)}}{2j} + \frac{e^{j\omega_{c1}(n-\tau)} - e^{-j\omega_{c1}(n-\tau)}}{2j} + \frac{e^{j\pi(n-\tau)} - e^{-j\omega_{c2}(n-\tau)}}{2j}\right]$$

$$= \frac{1}{\pi(n-\tau)}\left[\frac{e^{j\omega_{c1}(n-\tau)} - e^{-j\omega_{c1}(n-\tau)}}{2j} - \frac{e^{j\omega_{c2}(n-\tau)} - e^{-j\omega_{c2}(n-\tau)}}{2j} + \frac{e^{j\pi(n-\tau)} - e^{-j\pi(n-\tau)}}{2j}\right]$$

$$h_d(n) = \frac{1}{\pi(n-\tau)}[\sin\omega_{c1}(n-\tau) - \sin\omega_{c2}(n-\tau) + \sin\pi(n-\tau)] \quad n \neq \tau$$

$$h_d(n) = \frac{1}{\pi}[\pi - \omega_{c2} + \omega_{c1}] \qquad n = \tau$$

$$\text{Hence, } h_d(n) = \begin{cases} \frac{1}{\pi(n-\tau)}[\sin\omega_{c1}(n-\tau) - \sin\omega_{c2}(n-\tau) + \sin\pi(n-\tau)] & \text{for } n \neq \tau \\ \frac{1}{\pi}[\pi - \omega_{c2} + \omega_{c1} & \text{for } n = \tau \end{cases} \tag{3.74}$$

Example 3.46 Design a band-reject FIR filter for N = 5, $\omega_{c1} = 0.35$ radian and $\omega_{c2} = 0.65$ radian.

Solution: For a band-pass FIR filter, filter coefficients are given by Eq. 3.74.

$$h_d(n) = \begin{cases} \frac{1}{\pi(n-\tau)}[\sin\omega_{c1}(n-\tau) - \sin\omega_{c2}(n-\tau) + \sin\pi(n-\tau)] & \text{for } n \neq \tau \\ \frac{1}{\pi}[\pi - \omega_{c2} + \omega_{c1} & \text{for } n = \tau \end{cases}$$

Substituting, $N = 5$, $\tau = 2$ and $\omega_{c1} = 0.35$ radian and $\omega_{c2} = 0.65$ radian in Eq. 3.81, we get

$$h_d(0) = h_d(4) = \frac{\sin(0-2)0.35 - \sin(0-2)0.65 + \sin(0-2)\pi}{\pi(0-2)} = -0.0508$$

$$h_d(1) = h_d(3) = \frac{\sin(1-2)0.35 - \sin(1-2)0.65 + \sin(1-2)\pi}{\pi(1-2)} = -0.0834$$

$$h_d(2) = \frac{1}{\pi}[\pi - \omega_{c2} + \omega_{c1}] = \frac{1}{\pi}[\pi - 0.65 + 0.35] = 0.9045$$

Using rectangular window:

$$h(n) = h_d(n) w_R(n) = h_d(n)$$

Using Hamming window:

$$h(n) = h_d(n) w_{\text{Ham}}(n)$$

where, $w_{\text{Ham}}(n)$ for $N = 5$ is given by

$$w_{\text{Ham}}(n) = \begin{cases} 0.54 - 0.46\cos\left(\dfrac{2\pi n}{N-1}\right) & 0 \le n \le 4 \\ 0 & \text{otherwise} \end{cases}$$

$$w_{\text{Ham}}(0) = w_{\text{Ham}}(4) = 0.08$$
$$w_{\text{Ham}}(1) = w_{\text{Ham}}(3) = 0.54$$
$$w_{\text{Ham}}(2) = 1$$
$$h(n) = h_d(n) w_{\text{Ham}}(n)$$
$$h(0) = h(4) = -0.004064$$
$$h(1) = h(3) = -0.045036$$
$$h(2) = 0.9045$$

Using Hanning window:

$$h(n) = h_d(n) w_{\text{Han}}(n)$$

where, $w_{\text{Han}}(n)$ for $N = 5$ is given by

$$w_{\text{Han}}(n) = \begin{cases} 0.5 - 0.5\cos\left(\dfrac{2\pi n}{N-1}\right) & 0 \le n \le 4 \\ 0 & \text{otherwise} \end{cases}$$

$$w_{\text{Han}}(0) = w_{\text{Han}}(4) = 0$$

$$w_{\text{Han}}(1) = w_{\text{Han}}(3) = 0.5$$

$$w_{\text{Han}}(2) = 1$$

$$h(n) = h_d(n) w_{\text{Han}}(n)$$

$$h(0) = h(4) = 0$$

$$h(1) = h(3) = -0.0417$$

$$h(2) = 0.9045$$

Example 3.47 Design a band-reject FIR filter for $N = 6$, $\omega_{c1} = 0.4$ radian and $\omega_{c2} = 0.9$ radian.

Solution: For a band-pass FIR filter, filter coefficients are given by Eq. 3.74.

$$h_d(n) = \frac{1}{\pi(n-\tau)}[\sin\omega_{c1}(n-\tau) - \sin\omega_{c2}(n-\tau) + \sin\pi(n-\tau)]$$

Substituting, $N = 6$, $\tau = 2.5$ and $\omega_{c1} = 0.4$ radian and $\omega_{c2} = 0.9$ radian in Eq. 13.47, we get

$$h_d(0) = h_d(5) = \frac{\sin(0-2.5)0.4 - \sin(0-2.5)0.9 + \sin(0-2.5)\pi}{\pi(0-2.5)} = 0.1353$$

$$h_d(1) = h_d(4) = \frac{\sin(1-2.5)0.4 - \sin(1-2.5)0.9 + \sin(1-2.5)\pi}{\pi(1-2.5)} = -0.2994$$

$$h_d(2) = h_d(3) = \frac{\sin(2-2.5)0.4 - \sin(2-2.5)0.9 + \sin(2-2.5)\pi}{\pi(2-2.5)} = 0.4861$$

Using rectangular window:

$$h(n) = h_d(n) w_R(n) = h_d(n)$$

Using Hamming window:

$$h(n) = h_d(n) w_{\text{Ham}}(n)$$

where, $w_{\text{Ham}}(n)$ for $N = 6$ is given by

$$w_{\text{Ham}}(n) = \begin{cases} 0.54 - 0.46\cos\left(\dfrac{2\pi n}{N-1}\right) & 0 \le n \le 5 \\ 0 & \text{otherwise} \end{cases}$$

$$w_{\text{Ham}}(0) = w_{\text{Ham}}(5) = 0.08$$

$$w_{\text{Ham}}(1) = w_{\text{Ham}}(4) = 0.397$$

$$w_{\text{Ham}}(2) = w_{\text{Ham}}(3) = 0.9121$$

$$h(n) = h_d(n)w_{\text{Ham}}(n)$$

$$h(0) = h(5) = 0.0108$$

$$h(1) = h(4) = -0.1188$$

$$h(2) = h(3) = 0.4433$$

Using Hanning window:

$$h(n) = h_d(n)w_{\text{Han}}(n)$$

where, $w_{\text{Han}}(n)$ for $N = 6$ is given by

$$w_{\text{Han}}(n) = \begin{cases} 0.5 - 0.5\cos\left(\dfrac{2\pi n}{N-1}\right) & 0 \le n \le 5 \\ 0 & \text{otherwise} \end{cases}$$

$$w_{\text{Han}}(0) = w_{\text{Han}}(5) = 0$$

$$w_{\text{Han}}(1) = w_{\text{Han}}(4) = 0.3454$$

$$w_{\text{Han}}(2) = w_{\text{Han}}(3) = 0.9045$$

$$h(n) = h_d(n)w_{\text{Han}}(n)$$

$$h(0) = h(5) = 0$$

$$h(1) = h(4) = -0.1034$$

$$h(2) = h(3) = 0.4396$$

3.14.4 Design of Butterworth IIR Filters

In this section, we will discuss the design of digital Butterworth filter using the bilinear transformation method. IIR digital filters are designed with the following steps.

- Digital filter specifications are converted into analog filter specifications.
- Using filter order N and band edge frequency $\Omega_p = 1$ rad/sec, find $H(S)$ for normalized Butterworth low-pass filter.
- Design of low-pass, high-pass, band-pass and band-reject Butterworth IIR filters from normalized low-pass IIR filter
- Analog transfer function is then converted into digital transfer function of the filter

Step 1: Procedure for converting digital cut-off frequency to analog cut-off frequency.

If cut-off frequency (F_c) is given in Hz and sampling frequency (F_s) is given in Hz, then convert F_c into ω_c in radians by

$$\omega_c = \frac{2\pi \times F_c}{F_s} \tag{3.75}$$

Using the bilinear transformation method cut-off frequency Ω_c in rad/sec is related to digital frequency ω_c in rad by

$$\Omega_c = \frac{2}{T}\tan\frac{\omega_c}{2} \tag{3.76}$$

Step 2: Procedure for finding $H(S)$ normalized Butterworth low-pass filter

Butterworth low-pass filters are called normalized when band edge frequency $\Omega_p = 1$ rad/sec. For $\Omega_p = 1$ rad/sec and for different filter order N, generalized system function $H(S)$ for N^{th} order butterworth filter is given by

$$H(S) = \frac{1}{S^N + b_{N-1}S^{N-1} + \ldots + b_1 S + 1} \tag{3.77}$$

TABLE 3.3 Butterworth polynomials for different values of filter order N.

N	b_1	b_2	b_3	b_4	b_5	b_6	b_7	b_8
1	1							
2	$\sqrt{2}$							
3	2	2						
4	2.613	3.414	2.613					
5	3.236	5.236	5.236	3.236				
6	3.864	7.464	9.141	7.464	3.864			
7	4.494	10.103	14.606	14.606	10.103	4.494		
8	5.126	13.138	21.848	25.691	21.848	12.138	5.126	1

Table 3.3 gives the values of Butterworth polynomials, $b_{N-1} \ldots b_1$, for different values of filter order and for $\Omega_c = 1$ rad/sec. Using Table 3.3 system function $H(S)$ for $N = 1$, 2 and 3 are given by the following equations

For $\quad N = 1,\ H(S) = \dfrac{1}{S+1}$

For $\quad N = 2,\ H(S) = \dfrac{1}{S^2 + \sqrt{2}S + 1}$

For $\quad N = 3,\ H(S) = \dfrac{1}{S^3 + 2S^2 + 2S + 1}$

Step 3: Design of low-pass, high-pass, band-pass and band-reject IIR filters from normalized low-pass IIR filter.

In normalized low-pass Butterworth filters pass band edge frequency $\Omega_p = 1$ rad/sec. In this step, design of low-pass or high-pass or band-pass or a band-stop filters from a low-pass filter with pass band edge frequency (Ω_p) is performed. To convert low-pass filter with pass band edge frequency (Ω_p), to low-pass filter with different cut-off frequency (Ω_l) or high-pass filter with cut-off frequency (Ω_H) or band pass filter with cut-off frequencies (Ω_l) and (Ω_l) or band-reject filter with cut-off frequencies Ω_l and Ω_u, then the transformation that converts normalized low-pass filter to low-pass filter or high-pass filter or band-pass filter or band-reject filter is given by Table 3.4.

TABLE 3.4 Transformations for analog filters from normalized low-pass filter

Type of Filter	Cut-off frequencies of new filter	Transformation
Low-pass	Ω_l	$s \rightarrow \frac{\Omega_P}{\Omega_l} s \rightarrow \frac{1}{\Omega_l} s$
High-pass	Ω_H	$s \rightarrow \frac{\Omega_P \Omega_H}{s} \rightarrow \frac{\Omega_H}{s}$
Band-pass	Ω_l, Ω_u	$s \rightarrow \Omega_P \frac{s^2 + \Omega_l \Omega_u}{s(\Omega_u - \Omega_l)} \rightarrow \frac{s^2 + \Omega_l \Omega_u}{s(\Omega_u - \Omega_l)}$
Band-stop	Ω_l, Ω_u	$s \rightarrow \Omega_P \frac{s(\Omega_u - \Omega_l)}{s^2 + \Omega_l \Omega_u} \rightarrow \frac{s(\Omega_u - \Omega_l)}{s^2 + \Omega_l \Omega_u}$

Step 4: Procedure for finding digital transfer function H(Z)

Digital transfer function is computed by using the bilinear transformation method. In this method S in analog transfer function is replaced by

$$S \rightarrow \frac{2}{T_s}\left[\frac{1 - Z^{-1}}{1 + Z^{-1}}\right] \tag{3.78}$$

Example 3.48 Design a digital low-pass Butterworth filter for cut-off frequency $F_c = 1000$ Hz, $N = 2$ and sampling frequency $F_s = 10{,}000$ Hz.

Solution:

Step 1: To obtain specifications of equivalent to digital filter

$$\omega_l = \frac{2\pi \times 1000}{10000} = 0.2\pi \, \text{rad}$$

The specifications of digital filter given in the problem are to be converted into equivalent analog filter specifications.

$$\Omega_l = \frac{2}{T_s}\tan\frac{\omega_l}{2} = 20000\tan\frac{0.2\pi}{2} = 6498.4 \text{ rad/sec}$$

Step 2: The system function of normalized low-pass filter for $N = 2$, by using Table 3.3 is given by

$$H_{nL}(S) = \frac{1}{S^2 + \sqrt{2}S + 1}$$

Step 3: Normalized low-pass to low-pass frequency transformation is given by

$$s \rightarrow \frac{\Omega_P}{\Omega_l} s$$

where, Ω_p is the pass band edge frequency of low-pass filter. In normalized low-pass filter Butterworth filter Ω_P = 1 rad/sec. Hence, frequency transformation from normalized low-pass to high-pass filter is given by

$$S \rightarrow \frac{S}{\Omega_l} = \frac{S}{6498.4}$$

Step 4: The system function of high-pass filter for $N = 2$, is given by

$$H_l(S) = \frac{1}{\left[\dfrac{S}{6498.4}\right]^2 + \sqrt{2}\left[\dfrac{S}{6498.4}\right] + 1}$$

$$H_l(S) = \frac{42.229 \times 10^6}{S^2 + 9190.12S + 42.229 \times 10^6}$$

Step 5: To obtain $H(Z)$ using bilinear transformation method

To obtain $H(Z)$, system function of the digital filter, using bilinear transformation method, S in analog transfer function is replaced by

$$S \to \frac{2}{T_s}\left[\frac{1-Z^{-1}}{1+Z^{-1}}\right]$$

f_s = 150 Hz, $T = \frac{1}{f_s} = \frac{1}{10000}$, hence is given by

$$S = 20000\left[\frac{1-Z^{-1}}{1+Z^{-1}}\right]$$

Substituting S value in $H_l(S)$, we get

$$H_l(Z) = \frac{42.229 \times 10^6}{\left[20000\left[\frac{1-Z^{-1}}{1+Z^{-1}}\right]\right]^2 + 9190.12 \times 20000\left[\frac{1-Z^{-1}}{1+Z^{-1}}\right] + 42.229 \times 10^6}$$

$$H_l(Z) = \frac{0.0674\left(1+2Z^{-1}+Z^{-2}\right)}{\left(1-1.142Z^{-1}+0.412Z^{-2}\right)} \therefore H_l\left(e^{j\omega}\right) = \frac{0.0674\left(1+2e^{-j\omega}+e^{-2j\omega}\right)}{\left(1-1.142e^{-j\omega}+0.412e^{-2j\omega}\right)}$$

Example 3.49 For cut-off frequency F_c = 50 Hz and sampling frequency F_c = 150 Hz, design a second order digital high-pass Butterworth filter.

Solution:

Step 1: To obtain specifications of equivalent to digital filter

$$\omega_H = \frac{2\pi \times 50}{150} = 0.66\pi \text{ rad}$$

The specifications of digital filter given in the problem are to be converted into equivalent analog filter specifications.

$$\Omega_H = \frac{2}{T_s}\tan\frac{\omega_H}{2} = 300\tan\frac{0.66\pi}{2} = 507.27 \text{ rad/sec}$$

Step 2: The system function of normalized low-pass filter for N = 2, by using Table 3.3 is given by

$$H_{nL}(S) = \frac{1}{S^2 + \sqrt{2}S + 1}$$

Step 3: Low-pass to high-pass frequency transformation is given by

$$S \to \frac{\Omega_P \Omega_H}{S}$$

where, Ω_P is the pass band edge frequency of low-pass filter. In normalized low-pass filter Butterworth filter Ω_P = 1 rad/sec. Hence, frequency transformation from normalized low-pass to high-pass filter is given by

$$S \to \frac{\Omega_H}{S} = \frac{507.27}{S}$$

Step 4: For $N = 2$, system function of high-pass filter is given by

$$H_H(S) = \frac{1}{\left[\frac{507.27}{S}\right]^2 + \sqrt{2}\left[\frac{507.27}{S}\right] + 1}$$

$$H_H(S) = \frac{S^2}{S^2 + 717.385S + 2.57 \times 10^3}$$

Step 5: Obtain $H(Z)$ using the bilinear transformation method

To obtain system function $H(Z)$ of the digital filter, using the bilinear transformation method, in analog transfer function $H_H(S)$, S is replaced by

$$S \to \frac{2}{T_s}\left[\frac{1-Z^{-1}}{1+Z^{-1}}\right]$$

$F_s = 150$ Hz, $T_s = \frac{1}{F_s} = \frac{1}{150}$, hence S is given by

$$S = 300\left[\frac{1-Z^{-1}}{1+Z^{-1}}\right]$$

Substituting S value in $H_H(S)$, we get

$$H_H(Z) = \frac{\left[300\left[\frac{1-Z^{-1}}{1+Z^{-1}}\right]\right]^2}{\left[300\left[\frac{1-Z^{-1}}{1+Z^{-1}}\right]\right]^2 + 717.38 \times 300\left[\frac{1-Z^{-1}}{1+Z^{-1}}\right] + 2.57 \times 10^5}$$

$$H_H(Z) = \frac{0.160\left(1-2Z^{-1}+Z^{-2}\right)}{1+0.594Z^{-1}+0.234Z^{-2}} \;\therefore H_H\left(e^{j\omega}\right) = \frac{0.160\left(1-2e^{-j\omega}+e^{-2j\omega}\right)}{1+0.594e^{-j\omega}+0.234e^{-2j\omega}}$$

Example 3.50 Design a digital band-pass Butterworth filter for cut-off frequency $F_l = 70$ Hz, $F_u = 200$ Hz, $N = 1$ and sampling frequency, $F_s = 500$ Hz.

Solution:

Step 1: To obtain specifications of equivalent to digital filter

$$\omega_l = \frac{2\pi \times 70}{500} = 0.879 \text{ rad}$$

$$\omega_u = \frac{2\pi \times 200}{500} = 2.513 \text{ rad}$$

The specifications of digital filter given in the problem are to be converted into equivalent analog filter specifications. These are also called pre-warping.

$$\Omega_l = \frac{2}{T}\tan\frac{\omega_l}{2} = 1000\tan\frac{0.879}{2} = 470.16 \text{ rad/sec}$$

$$\Omega_u = \frac{2}{T}\tan\frac{\omega_u}{2} = 1000\tan\frac{2.513}{2} = 3076.24 \text{ rad/sec}$$

Step 2: The system function of normalized low-pass filter for $N = 1$, by using Table 3.3 is given by

$$H_{nL}(S) = \frac{1}{S+1}$$

Step 3: Low-pass to band pass frequency transformation is given by

$$S \to \frac{\Omega_P\left[S^2 + \Omega_l\Omega_u\right]}{S[\Omega_u - \Omega_l]}$$

where, Ω_P is the pass band edge frequency of low-pass filter. In normalized low-pass filter Butterworth filter $\Omega_P = 1$ rad/sec. Hence, frequency transformation from normalized low-pass to band-pass filter is given by

$$S \to \frac{\left[S^2 + \Omega_l\Omega_u\right]}{S[\Omega_u - \Omega_l]} = \frac{\left[S^2 - (470.16)(3076.24)\right]}{S[3076.24 - 470.16]} = \frac{[S^2 + 1.446 \times 10^6]}{2606.08S}$$

Step 4: The system function of band-pass filter for $N = 1$, is given by

$$H_{BP}(S) = \frac{1}{\dfrac{[S^2 + 1.446 \times 10^6]}{2606.08S} + 1}$$

$$H_{BP}(S) = \frac{2606.08S}{S^2 + 2606.08S + 1.446 \times 10^6}$$

Step 5: To obtain *H*(*Z*) using bilinear transformation method

To obtain *H*(*Z*), system function of the digital filter, using bilinear transformation method, *S* in analog transfer function is replaced by

$$S \to \frac{2}{T}\left[\frac{1-Z^{-1}}{1+Z^{-1}}\right]$$

$F_s = 500$ Hz, $T = \frac{1}{F_s} = \frac{1}{500}$, hence *S* is given by

$$S = 1000\left[\frac{1-Z^{-1}}{1+Z^{-1}}\right]$$

Substituting, *S* value in $H_{BP}(S)$, we get

$$H_{BP}(Z) = \frac{2606.08\left[1000\left[\frac{1-Z^{-1}}{1+Z^{-1}}\right]\right]}{\left[1000\left[\frac{1-Z^{-1}}{1+Z^{-1}}\right]\right]^2 + 2606.08 \times 1000\left[\frac{1-Z^{-1}}{1+Z^{-1}}\right] + 1.446 \times 10^6}$$

$$H_{BP}(Z) = \frac{0.5158\left(1-Z^{-2}\right)}{1+0.176Z^{-1}-0.0316Z^{-2}} \therefore H_{BP}\left(e^{j\omega}\right) = \frac{0.5158\left(1-e^{-2j\omega}\right)}{1+0.176e^{-j\omega}-0.0316e^{-2j\omega}}$$

3.15 DECIMATION AND INTERPOLATION PROCESS

Multirate DSP system uses multiple sampling rates within the system. In many applications of digital signal processing, it is necessary to process data at more than one sampling rate, this has led to the development of a new sub-area in DSP known as multirate processing. Hence, multirate DSP system uses multiple sampling rates within the system. Multirate processing is basically a technique for changing the sampling frequency of a signal digitally. Multirate processing is useful when two systems working at different sampling rates have to communicate, sampling rates must be made compatible. Whenever a signal at one rate has to be used by a system that expects a different rate, the rate has to be increased or decreased, and some processing is required to do so. The process of converting a signal from a given sampling rate to a different sampling rate is called sampling rate conversion. Different sampling rate can be obtained using an upsampler and downsampler. The two basic operations in a multirate system are decreasing (decimation) and increasing (interpolation) the sampling-rate of a signal. Decimation and interpolation are the fundamental operations in multirate signal processing to change the sampling rate of a sequence.

3.15.1 Decimation Process

Decimation is used to decrease the sampling rate or number of samples per second. To decimate a digital signal $x(n)$ by a factor M means to reduce its sampling rate M times. The relation between the original and decimated signal is given by

$$y(n) = x(nM) \tag{3.79}$$

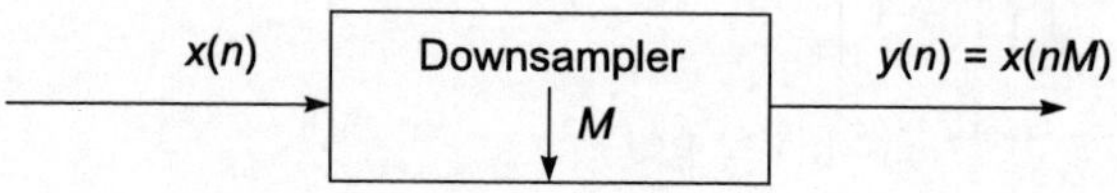

Figure 3.37 Downsampling

The decimation operation is shown in Figure 3.37. Sampling rate reduction is achieved by discarding $M - 1$ samples for every M samples. In the case of $M = 3$, two samples out of every three samples of $x(n)$ are discarded. In effect, decimation process is a data compression operation. For example, let $x(n) = \left\{ \underset{\uparrow}{1}, 2, 3, 4, 5, 6, 7, 8, 1, 2, 3, 4, 5, 6 \right\}$ be decimated by a factor 3. Then the decimated sequence, $y(n) = \left\{ \underset{\uparrow}{1}, 4, 7, 2, 5 \right\}$. Factor by which the signal is decimated is called the decimation factor.

Figure 3.38 shows the spectrum of $y(n)$ if $x(n)$, is decimated by a factor M, the spectrum $Y(e^{j\omega})$ is expanded by M, and repeated with period 2π. All the frequencies

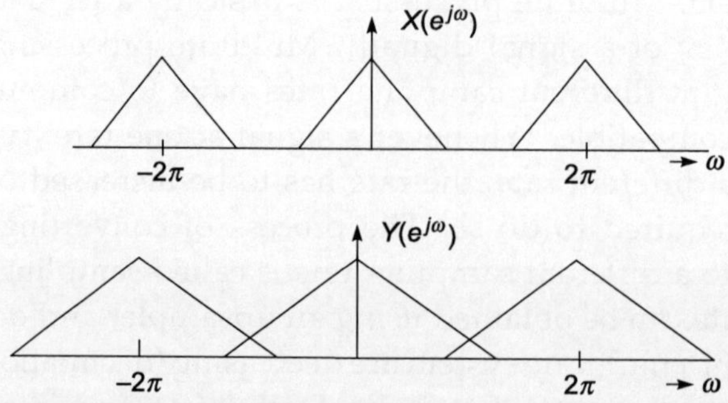

Figure 3.38 Spectrum of the input signal $x(n)$, (b) spectrum of the signal decimated by a factor of 2

above $\frac{\pi}{M}$ are going to be aliased as shown in Figure 3.35. Hence, the bandwidth of the signal $x(n)$ must be limited to the interval $\left[\frac{-\pi}{M}, \frac{\pi}{M}\right]$ to prevent sampling theorem violation or to avoid aliasing after decimation. Therefore, the decimation operation is generally preceded by a low-pass filter as shown in Figure 3.39.

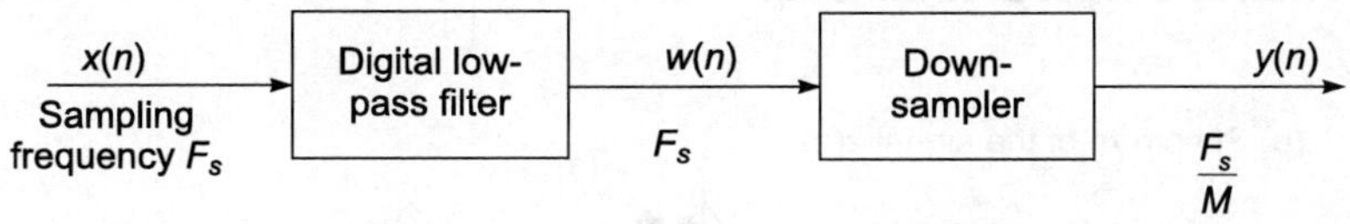

Figure 3.39 Decimation process using a decimation factor, M

$$w(n) = \sum_{k=0}^{N-1} h(k)x(n-k) \tag{3.80}$$

$$y(n) = w(nM) = \sum_{k=0}^{N-1} h(k)x(nM-k) \tag{3.81}$$

In the above equation, N is the order of the low-pass filter. Equation 3.86 specifies computation of only those samples that need to be kept, i.e., every M^{th} sample. Hence, decimated signal is obtained by taking the convolution of Mth sample of the input signal $x(n)$ and the filter impulse response, $h(n)$.

3.15.2 Interpolation Process

In interpolation process is used to increase the sampling rate or number of samples per second. Therefore, in interpolation or upsampling process, the sampling frequency is increased by an integer factor. Interpolation process using an interpolate factor (L) is as shown in Figure 3.40.

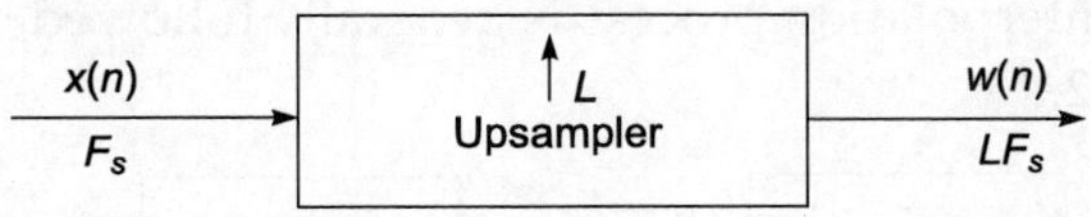

Figure 3.40 Upsampling

The relation between original and interpolated signal is given by

$$y(n) = x\left(\frac{n}{L}\right) \tag{3.82}$$

Therefore to interpolate or to upsample a digital signal $x(n)$ by a factor L is to include $L-1$ zeros between two samples. For example, let $x(n)=\left\{\begin{matrix}0,1,5,10\\ \uparrow\end{matrix}\right\}$ be interpolated using $L = 3$. The output sequence $y(n)$ is obtained by inserting three zeros between two samples. Then the interpolated sequence $y(n)=\left\{\begin{matrix}0,0,0,1,0,0,5,0,0,1,0,0,0\\ \uparrow\end{matrix}\right\}$.

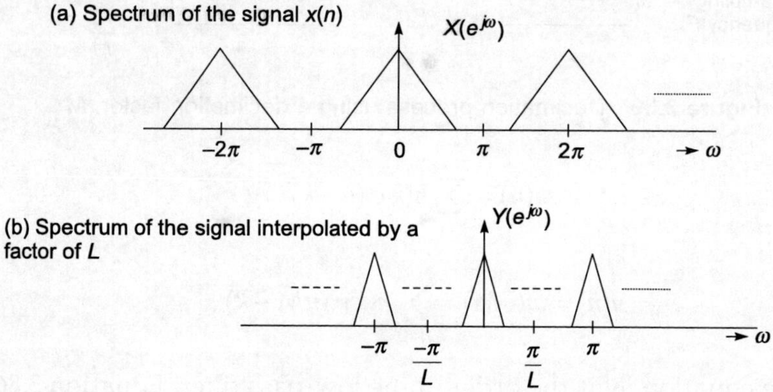

Figure 3.41

Interpolating by a factor L alone introduces artifacts in frequency domain as image frequencies as shown in Fig. 3.41. Spectrum of the original signal is periodic with period 2π and the spectrum of the interpolated signal has period $\frac{2\pi}{L}$. The spectrum of the interpolated signal must have the same shape as the spectrum of $x(n)$. All the image frequencies are outside the interval $\left[\frac{-\pi}{L},\frac{\pi}{L}\right]$ and hence they can be eliminated by filtering. Thus, the interpolation process is generally followed by a low-pass filter as shown in Figure 3.42.

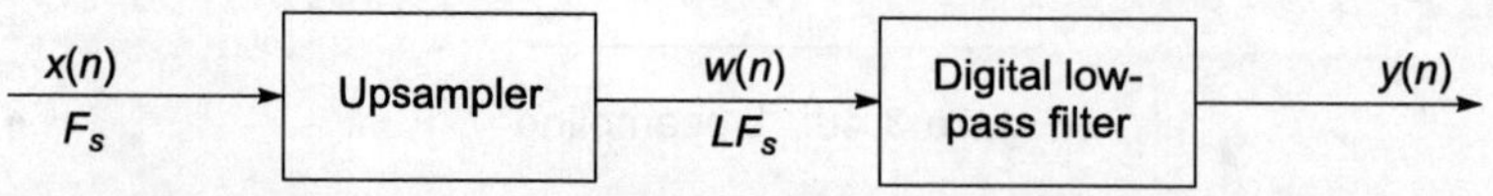

Figure 3.42 Interpolation process using an interpolation

$$w(n)=\begin{cases} x\left(\dfrac{n}{L}\right) & n=0,\pm L,\pm 2L \\ 0 & \text{for other values of } n \end{cases} \tag{3.83}$$

$$y(n) = \sum_{k=-\infty}^{\infty} h(k)w(n-k) \tag{3.84}$$

Example 3.51 Obtain the decimated signal, $w(n)$, by a factor 3 from the input signal, $x(n) = \left\{\begin{matrix} 0, 4, 5, 3, 2, 1, 5, 0, 5, 2, 1, 3, 4, 5\ldots \\ \uparrow \end{matrix}\right\}$. Obtain output $y(n)$ by interpolating the signal by a factor 3.

Solution: Given

$$x(n) = \left\{\begin{matrix} 0, 4, 5, 3, 2, 1, 5, 0, 5, 2, 1, 3, 4, 5\ldots \\ \uparrow \end{matrix}\right\}$$

If signal $x(n)$ is decimated by a factor 3, $w(n)$ is given by

$$w(n) = \left\{\begin{matrix} 0, 3, 5, 2, 4, \ldots \\ \uparrow \end{matrix}\right\}$$

If signal $w(n)$ is interpolated by a factor 3, $y(n)$ is given by

$$y(n) = \left\{\begin{matrix} 0, 0, 0, 3, 0, 0, 5, 0, 0, 2, 0, 0, 4, 0, 0\ldots \\ \uparrow \end{matrix}\right\}$$

The signal $x(n)$, $w(n)$ and $y(n)$ are shown in Figure 3.43.

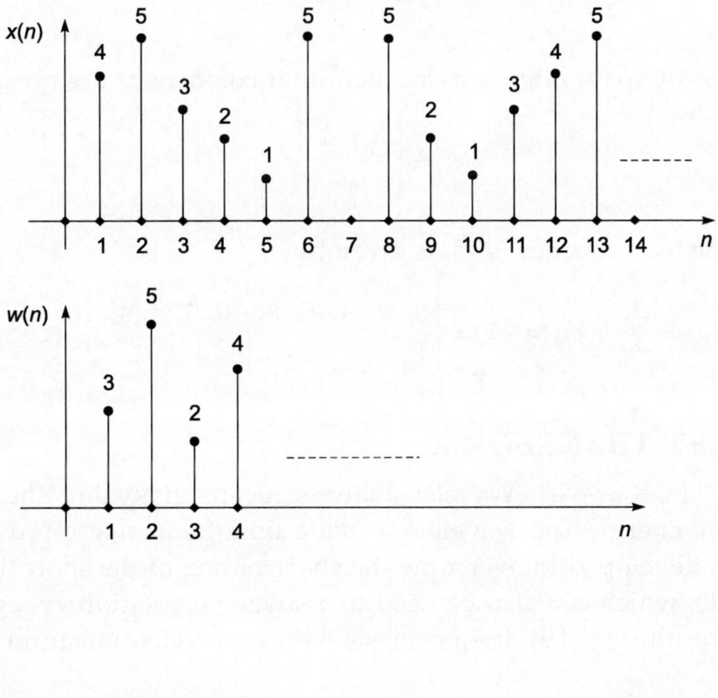

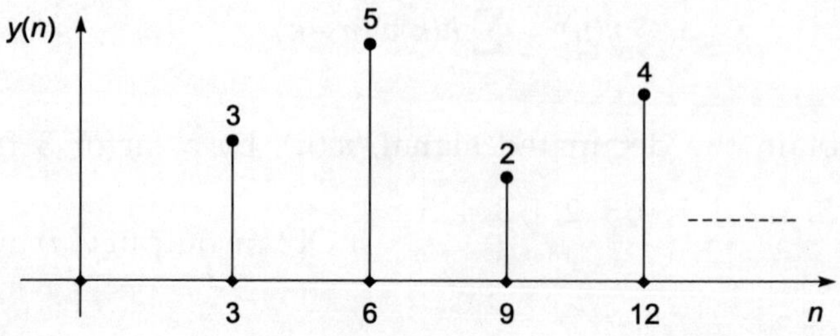

Figure 3.43

Linear interpolation method is used to estimate sample values between pairs of adjacent sample values of a discrete time sequencer. In this method first insert $(L-1)$ zeros between samples of the original sample sequence. Convolving sequence, $h(n)$, is derived based on linear interpolation process. If $h(n)$ is a second order linear filter, then it uses only two adjacent samples to interpolate a sample. If more adjacent samples are to be used to interpolate a sample, then higher order linear filter must be used.

Example 3.52 Consider a sample sequence $x(n) = \left\{\begin{matrix}0, & 3, & 6, & 9, & 12\\ \uparrow & & & & \end{matrix}\right\}$. Using the linear interpolation method, increase the sampling rate for $L = 2$.

Solution: After inserting zeros to increase the sampling rate, we get

$$x(n) = \left\{\begin{matrix}0, & 0, & 4, & 0, & 8, & 0, & 12, & 0, & 16\\ \uparrow & & & & & & & & \end{matrix}\right\}$$

If second order low-pass filter is used, then filter coefficients are given by

$$h(n) = \left\{\begin{matrix}\frac{1}{2}, & 1, & \frac{1}{2}\\ & \uparrow & \end{matrix}\right\}$$

Hence, interpolated sequence, $y(n)$, is given by

$$y(n) = \sum_{k=-1}^{1} h(k)x(n-k) = \left\{\begin{matrix}0, & 2, & 4, & 6, & 8, & 10, & 12, & 14, & 16\\ \uparrow & & & & & & & & \end{matrix}\right\}$$

3.16 WAVELET TRANSFORM

Wavelet means "small wave". Wavelet analysis means analyzing short duration signals with finite energy. The wavelets provide time-frequency representation of the signal. It was developed to overcome the shortcoming of the short time Fourier transforms (STFT), which can also be used to analyze non-stationary signals. STFT gives a constant resolution at all frequencies, whereas wavelet transform uses multi-resolution techniques by which different frequencies are analyzed with different resolutions.

3.16.1 Fourier Transform Interpretation

In 19th century, the French mathematician J. Fourier showed that any periodic function can be expressed as an infinite sum of periodic complex exponential functions of different frequencies. It is defined by the following equations.

$$X(\Omega) = \int_{-\infty}^{\infty} x(t)e^{-j\Omega t}\,dt \tag{3.85}$$

$$x(t) = \frac{1}{2\pi}\int_{-\infty}^{\infty} X(\Omega)e^{j\Omega t}\,d\Omega \tag{3.86}$$

In the above equations, t stands for time, $\Omega = 2\pi F$, Ω stands for angular frequency in rad./sec, F stands for frequency in Hz. x denotes the signal in time domain and X denotes the signal in frequency domain. Equation 3.90 is called the Fourier transform of $x(t)$ and Eq. 3.91 is called the inverse Fourier transform of $X(\Omega)$. Since the integration is from minus infinity to plus infinity over time, whether the frequency component (F) appears at time t_1 or t_2, it will have the same effect on integration. Hence, the Fourier transform is not suitable if the signal has time varying frequency, i.e., the signal is non-stationary. Hence, Fourier transform gives information whether a certain frequency component exists or not. This information is independent of where in time this component appears. Hence, Fourier transform is suitable if the signal is stationary. For example,

$$x(t) = \cos(2\pi 200t) + \cos(2\pi 300t) + \cos(2\pi 400t)$$

is a stationary signal. In the above equation, signal has frequencies of 200, 300 and 400 Hz at any given time instant. Hence, in stationary signals, all frequency components exist throughout the duration of the signal.

3.16.2 Short Time Fourier Transform or Windowed Fourier Transform

In defining a time dependent Fourier representation, the signal is divided into small segments using window function, where these portions (segments) of the signal can be assumed to be stationary. This approach of the Fourier transform is called short time Fourier transform (STFT). The window function (W) is chosen, such that width of the window function must be equal to the segment of the signal where its stationarity is valid. The following is the definition of the time dependent Fourier transform or short time Fourier transform.

$$X(t', \Omega) = \int [x(t) \cdot W^*(t - t')] e^{-j\Omega t} dt \tag{3.87}$$

In the above equation $x(t)$ is the signal, $W(t)$ is the window function and * is the complex conjugate. Hence, STFT of the signal is nothing but the FT of the signal multiplied by a window function. If we use a window of infinite length, we get the Fourier transform, which gives perfect frequency resolution, but no time information. In order to obtain the stationarity, we have to have short window function, in which the signal is stationary. The narrower we make the window, the better the time resolution, and better the assumption of stationarity, but poorer the frequency resolution.

Narrow window	Good time resolution and poor frequency resolution
Wide window	Good frequency resolution and poor time resolution

Hence, narrow window gives good time resolution but poor frequency resolution. Wide window gives good frequency resolution but poor time resolution and wide window may violate the condition of stationarity. Hence, in STFT, the problem of resolution is a result of choosing a window function.

3.16.3 Wavelet Transform Implementation

Wavelets are fundamental building block functions similar to trigonometric sine and cosine functions. Sine wave or cosine wave is an oscillating function of time or space and is periodic as shown in Figure 3.44(a). Wavelet functions also oscillate about zero, as in a sine or cosine wave. In contrast wavelets are localized waves as shown in Figure 3.44(b). A **wavelet** is a wave-like oscillation with an amplitude that starts out at zero, increases, and then decreases back to zero. Wavelets can be combined, using a "shift, multiply and sum" technique called convolution with portions of an unknown signal to extract information from the unknown signal.

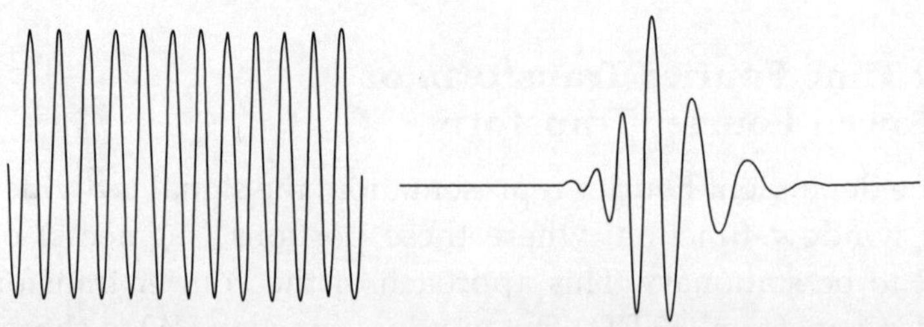

Figure 3.44 Demonstrations of (a) a wave, and (b) a wavelet.

3.16.3.1 *Continuous wavelet transform*

The continuous wavelet transform was developed as an alternative approach to short time Fourier transforms to overcome the resolution problem. The width of the window is changed as the transform is computed for every single spectral component, which is probably the most significant characteristic of the wavelet transform. The wavelet analysis is done in a similar way to the STFT analysis, i.e., signal to be analyzed is multiplied with wavelet function instead of window function in the STFT and then the transform is computed for each segment generated. However, unlike STFT, in wavelet transform, the width of the wavelet function changes with each spectral component. At high frequencies wavelet transform gives good time resolution and poor frequency resolution. Wavelet transform at low frequencies gives poor time resolution and good frequency resolution. The continuous-time wavelet transform of $x(t)$ with respect to wavelet, $\psi(t)$, is defined as follows:

$$W(a,b) = \int_{-\infty}^{\infty} x(t)\psi_{a,b}^{*}(t)dt \tag{3.88}$$

$$\psi_{a,b}^{*}(t) = \frac{1}{\sqrt{a}}\psi\left(\frac{t-b}{a}\right) \tag{3.89}$$

where, a and b are real and * denotes complex conjugation. Where the normalizing factor, $\frac{1}{\sqrt{a}}$, ensures that the energy stays the same for all a and b. The transformed signal is a function of two variables a and b, where, b represents the time shift or translation and a represents the time scaling or dilation. The term translation is used in the same sense as it was used in the STFT, it is related to location of window and hence this term corresponds to time information in the transform domain. The scale parameter (a) is defined as $|1/\text{frequency}|$ and corresponds to frequency information. Transforming function, $\psi(t)$, is called the mother wavelet. Similarly, high scale ($a = 1$) corresponds to detailed information of hidden part of the signal. Scaling either dilates (expands) or compresses a signal. Larger scale (low frequency) corresponds to dilated (or stretched out) signals and provides detailed information hidden in the signal. Smaller scales (high frequencies) correspond to compressed signals and provide global information about the signal.

3.16.3.2 *Inverse continuous wavelet transform*

The continuous wavelet transform is a reversible transform, provided the wavelet satisfies the condition

$$C = \int_{-\infty}^{\infty} \frac{|\psi(\omega)|^2}{|\omega|} d\omega \tag{3.90}$$

and is such that $-\infty < C < \infty$ then

$$x(t) = \frac{1}{C} \int_{a=-\infty}^{\infty} \int_{b=-\infty}^{\infty} \frac{1}{|a|^2} W(a,b)\psi_{a,b}(t) da db \tag{3.91}$$

In the above equation C is a constant and it depends on the wavelet used.

3.16.3.3 Discrete wavelet series

The discrete wavelet series is obtained by discretizing CWT, i.e., sampled version of CWT. Computation of wavelet series consumes significant amount of time and resources, depending on the resolution required. Discrete wavelet is not continuously scalable and translatable, but can only be scaled and translated to discrete steps. This is achieved by modifying the wavelet representation

$$W(a,b) = \int x(t) \frac{1}{\sqrt{a}} \psi\left(\frac{t-b}{a}\right) dt \tag{3.92}$$

The dilation (a) takes the value ($a = 2^{-j}$) and translation ($b = k2^{-j}$), where, j and k are integers representing the set of discrete dilation and translation. Substituting a and b values in Eq. 3.92 , we get

$$W(j,k) = \int x(t) 2^{\frac{j}{2}} \psi\left(2^j t - k\right) dt \tag{3.93}$$

In discrete wavelet function, we deal with basically two sets of functions, i.e., scaling function, $\phi_{j,k}(t)$, and wavelet function, $\psi_{j,k}(t)$.

The scaling function or father wavelet is given by the equation

$$\phi_{j,k}(t) = 2^{\frac{j}{2}} \phi\left(2^j t - k\right) \tag{3.94}$$

The wavelet function or mother wavelet is given by the equation

$$\psi_{j,k}(t) = 2^{\frac{j}{2}} \psi\left(2^j t - k\right) \tag{3.95}$$

The scale factor 2^j is also referred to as the dilation factor, while $2^j k$ is called the location.

A continuous time signal, $x(t)$, is approximated by the orthogonal wavelet series given by the equation

$$x(t) = \sum_k S_{J,k}\,\phi_{J,k}(t) + \sum_k d_{J,k}\,\psi_{J,k}(t) + \sum_k d_{J-1,k}\,\psi_{J-1,k}(t) \ldots + \sum_k d_{1,k}\,\psi_{1,k}(t)$$

The approximation or smooth coefficients ($S_{j,k}$) and detail coefficients ($d_{J,k}$, $d_{J-1,k}$, $d_{j-2,k} \ldots d_{1,k}$) are called wavelet transform coefficients and are given by the integral

$$S_{J,k} = \int \phi_{J,k}(t)x(t)dt \tag{3.96}$$

$$d_{j,k} = \int \psi_{j,k}(t)x(t)dt \quad \text{for} \quad j = 1, 2 \ldots J \tag{3.97}$$

3.16.3.4 *Discrete wavelet transform*

The discrete wavelet transform is very close to wavelet series, but it applies to discrete time signal, $x(n)$. Discrete Fourier transforms (DFT), which is a discrete version of the Fourier transform. Whereas discrete wavelet transform is not really a discrete version of the continuous wavelet transform. In discrete wavelet transform discrete wavelet coefficients are computed by using low-pass and high-pass filter banks.

Discrete wavelet transform (DWT), which is based on sub-band coding is found to yield a fast computation of wavelet transform. In 1983, Burt defined a technique very similar to sub-band coding and named it *pyramidal coding* which is also known as *multi-resolution analysis*. It is easy to implement and reduce the computation time and resources required. In continuous wavelet transform (CWT), the signals are analyzed using a set of basis function which relates to each other by simple scaling and translation. The continuous wavelet transform was computed by changing the scale of analysis window and shifting the window in time. In the case of DWT, a time scale representation of the digital signal is obtained using a digital filtering technique. The signal is passed through a series of high-pass and low-pass filters to analyze the high and low frequencies. As shown in Figure 3.45 (a) DWT is computed by successive low-pass and high-pass filters of the discrete signal. The low-pass filter coefficients are denoted by $g_0(n)$ and high-pass filter coefficients are denoted by $h_0(n)$. The low-pass filters coefficients, $g_0(n)$, are associated with the scaling function. The high-pass filters coefficients, $h_0(n)$, are associated with the wavelet function. The output of each low-pass filter is $a_k[n]$ for $k = 1, 2, \ldots, m$ or approximation components of the original signal for that level of the tree. The output of each high-pass filter is $d_k[n]$ for $k = 1, 2, \ldots, m$ or detailed components of the original signal.

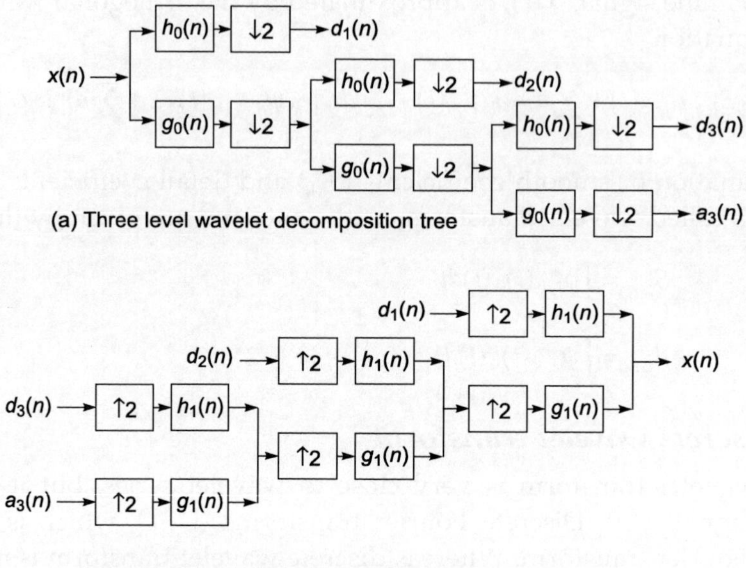

(a) Three level wavelet decomposition tree

(b) Three level wavelet reconstruction tree

Figure 3.45 Three-level wavelet trees

Figure 3.45(a) shows the three-level wavelet decomposition trees. This is called Mallat-tree decomposition or Mallat algorithm. The outputs of $g_0(n)$ and $h_0(n)$ are downsampled by 2. The operation $\downarrow 2$ denotes a decimation of a signal by a factor of 2. The low-pass output is further passed through low-pass and high-pass followed by a decimator. In discrete signals, frequency is expressed in terms of radians. If the original signal has the highest frequency of ω radians, then according to the sampling theorem, sampling frequency is twice the maximum frequency that exists in the signal. Hence, the sampling frequency of the signal is 2ω radians. After passing the signal through a half band low-pass filter, half of the samples are eliminated. Low-pass filter removes the high frequency information. Now the signal has the highest frequency $\left(\frac{\omega}{2}\right)$ instead of radians. Now the signals can be sampled at a frequency of radians, by discarding half the samples with no loss of information. Half the samples can be eliminated using decimation by two, which discards every other sample. The filtering and decimation process is continued until the desired level is reached. The outputs of high-pass and low-pass after decimation by two are given by

$$d_k[n] = \sum_k h_0(k)x(2n-k) \tag{3.98}$$

$$a_k[n] = \sum_k g_0(k)x(2n-k) \tag{3.99}$$

where, $x(n)$, the original signal to be decomposed, $h_0(n)$ and $g_0(n)$ are high-pass and low-pass filter coefficients, respectively. Suppose the original signal has 1024 sample points and frequency band of 0 to π radians. The output of high-pass filter followed by decimation by 2 contains 512 sample points and spans the frequencies from $\frac{\pi}{2}$ radians to π radians. These 512 samples are the first level DWT coefficients. The output of low-pass filter followed by decimation by 2 contains 512 sample points and spans the frequencies from 0 radians to $\frac{\pi}{2}$ radians. The signal is then passed through the low-pass and high-pass filters for further decomposition. The output of high-pass filter followed by decimation by 2 contains 256 samples spanning a frequency band $\frac{\pi}{4}$ radians to $\frac{\pi}{2}$ radians. These outputs constitute the second level of DWT coefficients. The low-pass filter output is filtered again for further decomposition. This process continues till two samples are left. The outputs of low-pass filters provides an approximation signal and output of high-pass filter provides detail information about the signal. Hence, output of low-pass and high-pass filters provides approximation and detail coefficients, respectively.

Reconstruction process is the reverse process of decomposition. Figure 3.45(b) shows the reconstruction of the original signal from the wavelet coefficients. At every level the approximation and detail coefficients are upsampled by 2, passed through low-pass and high-pass synthesis filters and then added. The operation $\uparrow 2$ denotes interpolation of a signal by a factor of 2. To obtain the original signal, this process is continued through the same number of levels as in the decomposition process.

3.16.3.5 Wavelet families

There are a number of basis functions that can be used as the mother wavelet for wavelet transformation. Since the mother wavelet produces all wavelet functions used in the transformation through translation and scaling, it determines the characteristics of the resulting wavelet transform. Therefore, the details of the particular application should be taken into account and the appropriate mother wavelet should be chosen in order to use the wavelet transform effectively. The Haar, Daubechies, Symlets and Coiflets are compactly supported orthogonal wavelets. These wavelets along with Meyer wavelets are capable of perfect reconstruction. The Meyer, Morlet and Mexican Hat wavelets are symmetric in shape. The wavelets are chosen based on their shape and their ability to analyze the signal in a particular application.

Haar wavelet was introduced by Harr in 1910. It is one of the oldest and simplest orthogonal wavelets with compact support. Therefore, any discussion of wavelets starts with it. It is a bipolar step function. The Haar wavelet is illustrated in Figure 3.46(a). The expression for Haar wavelet function, $\psi(t)$, is given by

$$\psi(t) = \begin{cases} 1 & 0 \le t \le \frac{1}{2} \\ -1 & \frac{1}{2} \le t \le 1 \\ 0 & \text{elsewhwere} \end{cases}$$

From Figure 3.43 (a) Haar wavelet is anti-symmetric with respect to $t = \frac{1}{2}$ and is a real function. The Haar wavelet is localised in the time domain, but it has poor localisation in the frequency domain.

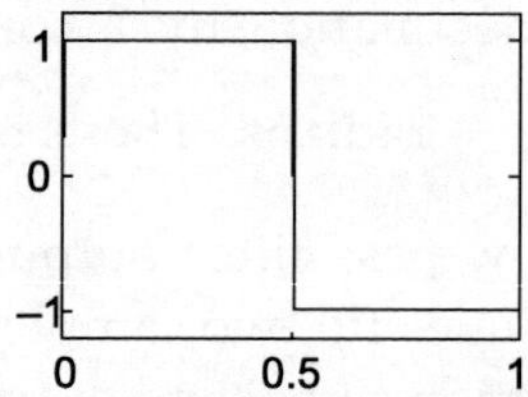

Figure 3.46(a) Haar wavelet

Morlet wavelet was developed by Morlet for analyzing the seismic signals. The Morlet wavelet is obtained by multiplying the Fourier basis with a Gaussian window. These wavelets are also called wavelets of constant shape. The expression for Morlet wavelet function, $\psi(t)$, is given by

$$\psi(t) = \exp\left(j\omega_0 t\right) \exp\left(-\frac{t^2}{2}\right)$$

By taking the Fourier transform, the spectrum of the Morlet wavelet which is given by

$$\psi(\omega) = \sqrt{\frac{\pi}{2}} \left\{ \exp\left[-\frac{(\omega-\omega_0)^2}{2}\right] + \exp\left[-\frac{(\omega-\omega_0)^2}{2}\right] \right\}$$

From the above expression, it is clear that the spectrum of the Morlet wavelet consists of two Gaussian functions shifted by ω_0 and $-\omega_0$. The Morlet wavelet is illustrated in Figure 3.46(b).

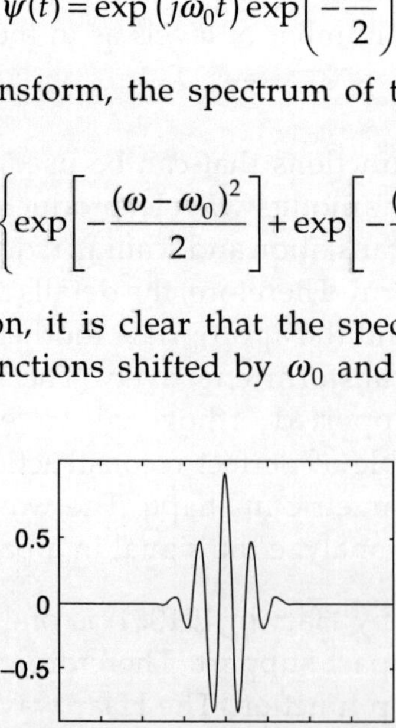

Figure 3.46(b) Morlet wavelet

Mexican-hat wavelet is obtained by taking the second order derivative of the negative Gaussian function. The expression of Mexican-hat wavelet is given by

$$\psi(t) = (1 - t^2)\exp\left(-\frac{t^2}{2}\right)$$

The Mexican-hat wavelet is illustrated in Fig. 3.46(c). By taking Fourier transform of the above equation, we will get the spectrum of the Mexican-hat wavelet and is given by

$$\psi(\omega) = -\omega^2 \exp\left(-\frac{\omega^2}{2}\right)$$

Squared magnitude of the Fourier transforms, $\psi(\omega)^2$, is shown in Figure 3.46 (c) it is seen that the wavelet is essentially a band pass function with the Fourier transform magnitude centred at approximately 2 rad./sec.

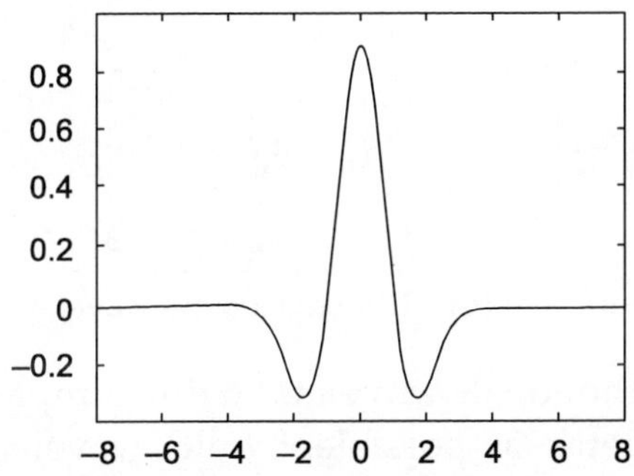

Figure 3.46(c) Mexican-hat wavelet

Shannon wavelet has poor time resolution, but its frequency localization is excellent expression of the Shannon wavelet is given below

$$\psi(t) = \frac{\sin(2\pi t) - \sin(\pi t)}{\pi t}$$

Because of the association of the sinc function with sampling theory and in honor of Claude Shannon, the scaling function is called Shannon scaling function and the wavelet is known as Shannon wavelet function.

Daubechies wavelets are the most popular wavelets. Daubechies wavelet bases are a family of orthogonal wavelets with compact support (Figure 3.46(d)). They are quiet asymmetric. She developed many wavelet systems with compactly supported scaling and wavelet functions. They represent the foundations of wavelet signal processing and are used in numerous applications. The Daubechies wavelets are a family of orthogonal wavelets defining a discrete wavelet transform and characterized by a maximal number of vanishing moments for some given support. With each wavelet type of this class, there is a scaling function (also called *father wavelet*) which generates an orthogonal multi-resolution analysis. The wavelet transform is also easy to put into practice using the fast wavelet transform. These are widely used in solving a broad range of problems. These are also called Maxflat wavelets as their frequency responses have maximum flatness at frequencies 0 and ϖ. This is a very desirable property in some applications. The peculiarity of this wavelet system is that, there is no explicit function, so we cannot draw it directly. The procedure followed by Daubechies to find out filter coefficients involves complex mathematics with use of spectral factorization and Riesz theorem. In this section, we will explore one wavelet system called Daub-4 type wavelet system.

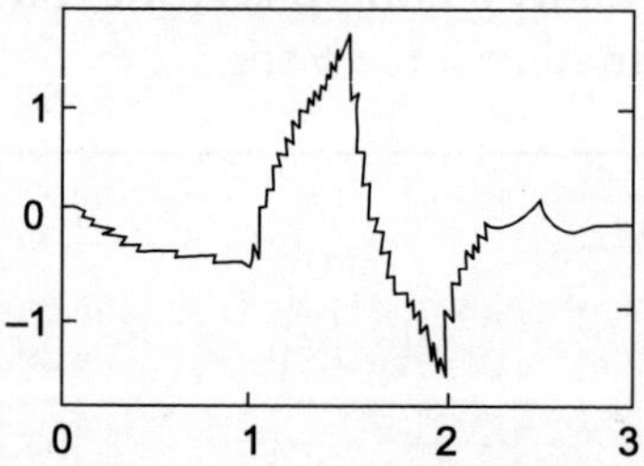

Figure 3.46(d) Daubechies-hat wavelet

Symlet Wavelets are orthogonal wavelets with compact support and were constructed to be as symmetric as possible. Symlet wavelets are shown in Figure 3.46(e). Daubechies proposes modification of her wavelet that increases their symmetry. These wavelets are nearly linear phase and filter length is $N = 2p$, where, p is the vanishing moment.

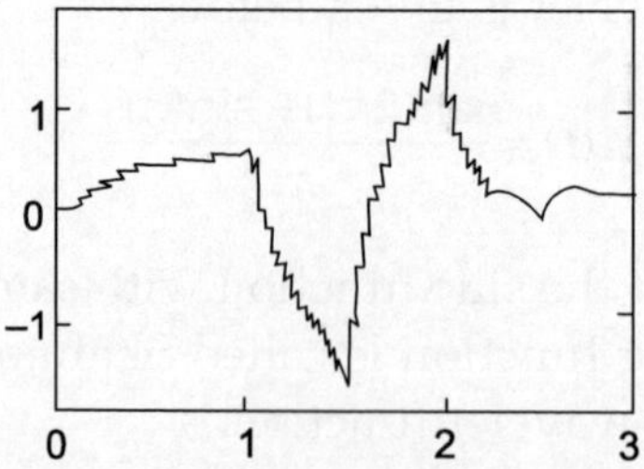

Figure 3.46(e) Symlet wavelet

Coiflet wavelets are discrete wavelets designed by Ingrid Daubechies and were constructed to be nearly symmetric. Coiflet wavelets are obtained by imposing vanishing moment conditions on both scaling and wavelet functions. The wavelet is near symmetric, their wavelet functions have $N/3$ vanishing moments and scaling functions $N/3 - 1$. Both the scaling function (low-pass filter) and the wavelet function (high-pass filter) must be normalized by a factor $\frac{1}{\sqrt{2}}$. The wavelet coefficients are derived by reversing the order of the scaling function coefficients and then reversing the sign of every second one,

$$B_k = (-1)^k C_{N-1-k}$$

where, k is the coefficient index, B is a wavelet coefficient and C a scaling function coefficient. N is the wavelet index, i.e., 6 for C6. Coiflet wavelets are shown in Figure 3.46(f).

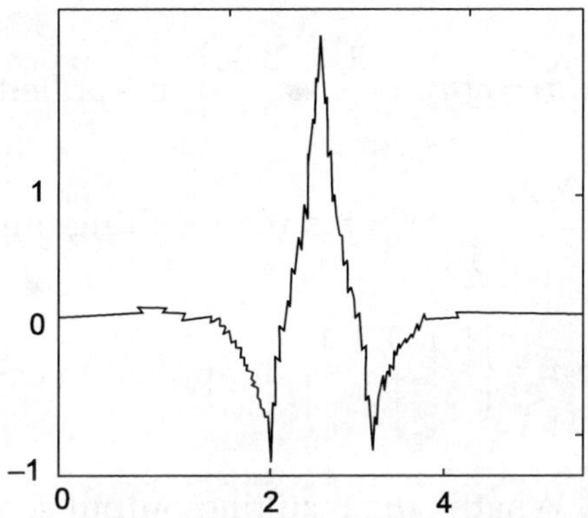

Figure 3.46(f) Coiflet wavelet

3.16.3.6 Applications of wavelets

There is a wide range of applications for wavelet transforms. They are applied in different fields ranging from signal processing to biometrics, and the list is still growing. The following are applications of wavelet transforms:

- data and image compression
- biometrics
- partial differential equation solving
- transient detection
- pattern recognition
- texture analysis
- noise reduction

Wavelets also find application in speech compression, which reduces transmission time in mobile applications. They are used in denoising, edge detection, feature extraction, speech recognition, echo cancellation and others. They are very promising for real-time audio and video compression applications.

3.17 SUMMARY

This chapter presents the mathematical derivations for convolution, correlation and transformation techniques. Design of FIR and IIR filters with the basics of digital signal processing.

EXERCISES

Review Questions

1. The input sequence, $x(n)=h(n)=\left\{\underset{\uparrow}{3},1,2,-1\right\}$, is applied to DSP, with response of the system, $h(n)=\left\{\underset{\uparrow}{3},2,1\right\}$. What is the resulting output sequence?

2. The input sequence, $x(n)=\left\{1,1,\underset{\uparrow}{0},1,1\right\}$, is applied to DSP, with response of the system, $h(n)=\left\{\underset{\uparrow}{-3},4\right\}$. What is the resulting output sequence?

3. For input sequence, $x(n)=\left\{\underset{\uparrow}{3},1,2,-1\right\}$, is applied to a discrete time processor with output, $y(n)=\left\{\underset{\uparrow}{9},9,11,2,0,-1\right\}$. What is the system response, $h(n)$?

4. For input sequence, $x(n)=\left\{1,1,\underset{\uparrow}{0},1,1\right\}$, is applied to a discrete time processor with output $y(n)=\left\{-3,1,4,\underset{\uparrow}{-3},1,4\right\}$. What is the system response $h(n)$?

5. Determine the correlation of sequence $x(n)=\left\{\underset{\uparrow}{2},1,2,4\right\}$ and $y(n)=\left\{\underset{\uparrow}{2},1,2,4\right\}$

6. Determine the correlation of sequence $x(n)=\left\{\underset{\uparrow}{1},2,3,4\right\}$ and $y(n)=\left\{\underset{\uparrow}{1},1,1,1\right\}$.

7. Compute correlation coefficient of the sequence $x(n)=\begin{Bmatrix}1,2,3\\ \uparrow\end{Bmatrix}$ and $y(n)=\begin{Bmatrix}1,2,3\\ \uparrow\end{Bmatrix}$

8. Compute correlation coefficient of the sequence $x(n)=\begin{Bmatrix}1,2,3\\ \uparrow\end{Bmatrix}$ and $y(n)=\begin{Bmatrix}-1,-2,-3\\ \uparrow\end{Bmatrix}$

9. Compute DFT of the sequence $x(n)=a^n$ for $0\le n\le N-1$

10. If $x(n)=\{\delta(n),\delta(n-3)\}$, find DFT of the periodic signal $x(n)$ with $N=4$.

11. If $X(k)=\begin{Bmatrix}2,1+j,0,1-j\\ \uparrow\end{Bmatrix}$, find DFT of the periodic signal $x(n)$ with $N=4$.

12. If $x(n)=\begin{Bmatrix}3,1,5,4,2,1,0,1\\ \uparrow\end{Bmatrix}$, find $X(k)$ by using decimation in time FFT algorithm.

13. If $x(n)=\begin{Bmatrix}3,1,5,4,2,1,0,1\\ \uparrow\end{Bmatrix}$, find $X(k)$ by using decimation in frequency FFT algorithm.

14. If $x(n)=\cos\left(\frac{n\pi}{2}\right)$, where, N = 4, find $X(k)$ using decimation in time and decimation in frequency FFT algorithm.

15. If $X(k)=\begin{Bmatrix}17,-1.12-j7.12,j3,3.121+j2.87,3,3.12-j2.87,-j3,-1.12+j7.12\\ \uparrow\end{Bmatrix}$.

    Find $x(n)$ by using decimation in frequency IFFT algorithm.

16. If $X(k)=\begin{Bmatrix}17,-1.12-j7.12,j3,3.121+j2.87,3,3.12-j2.87,-j3,-1.12+j7.12\\ \uparrow\end{Bmatrix}$.

    Find $x(n)$ by using decimation in time IFFT algorithm.

17. Compute low-pass FIR filter coefficients for N = 5 and $\omega_c=0.5$ radian using Hamming and Hanning windows.

18. Compute low-pass FIR filter coefficients for N = 4 and $\omega_c = 0.4$ radian using Hamming and Hanning windows.
19. Compute high-pass FIR filter coefficients for N = 5 and $\omega_c = 1.5$ radian using Hamming and Hanning windows.
20. Compute high-pass FIR filter coefficients for N = 4 and $\omega_c = 1.5$ radian using Hamming and Hanning windows.
21. Compute band-pass FIR filter coefficients for N = 7, $\omega_{c1} = 0.3$ radian and $\omega_{c2} = 0.65$ radian using Hamming and Hanning windows.
22. Compute band-pass FIR filter coefficients for N = 8, $\omega_{c1} = 0.3$ radian and $\omega_{c2} = 0.8$ radian using Hamming and Hanning windows.
23. Compute band-reject FIR filter coefficients for N = 5, $\omega_{c1} = 0.5$ radian and $\omega_{c2} = 0.75$ radian using Hamming and Hanning windows.
24. Compute band-reject FIR filter coefficients for N = 6, $\omega_{c1} = 0.25$ radian and $\omega_{c2} = 0.75$ radian using Hamming and Hanning windows.

ANSWERS

1. $y(n) = \left\{ \begin{matrix} 9, 9, 11, 2, 0, -1 \\ \uparrow \end{matrix} \right\}$

2. $y(n) = \left\{ \begin{matrix} -3, 1, 4, -3, 1, 4 \\ \qquad\qquad\quad \uparrow \end{matrix} \right\}$

3. $h(n) = \left\{ \begin{matrix} 3, 2, 1 \\ \uparrow \end{matrix} \right\}$

4. $h(n) = \left\{ \begin{matrix} -3, 4 \\ \uparrow \end{matrix} \right\}$

5. $r_{xy}(n) = \left\{ \begin{matrix} 8, 8, 12, 25, 12, 8, 8 \\ \uparrow \end{matrix} \right\}$

6. $r_{xy}(n) = \left\{ \begin{matrix} 1, 3, 6, 10, 9, 7, 4 \\ \uparrow \end{matrix} \right\}$

7. Correlation coefficient $= +1$

8. Correlation coefficient $= -1$

9. $X(k)=\dfrac{1-a^N e^{-j2\pi k}}{1-ae^{\frac{-j2\pi k}{N}}}=\dfrac{1-a^N}{1-ae^{\frac{-j2\pi k}{N}}}$ for $0\le k\le N-1$

10. $X(k)=\left\{\begin{matrix}2, 1+j, 0, 1-j\\ \uparrow\end{matrix}\right\}$

11. $x(n)=\left\{\begin{matrix}1, 0, 0, 1\\ \uparrow\end{matrix}\right\}$

12. $X(k)=\left\{\begin{matrix}17, -1.12-j7.12, j3, 3.121+j2.87, 3, 3.12-j2.87, -j3, -1.12+j7.12\\ \uparrow\end{matrix}\right\}$

13. $X(k)=\left\{\begin{matrix}17, -1.12-j7.12, j3, 3.121+j2.87, 3, 3.12-j2.87, -j3, -1.12+j7.12\\ \uparrow\end{matrix}\right\}$

14. $X(k)=\left\{\begin{matrix}0, 2, 0, 2\\ \uparrow\end{matrix}\right\}$

15. $x(n)=\left\{\begin{matrix}3, 1, 5, 4, 2, 1, 0, 1\\ \uparrow\end{matrix}\right\}$

16. $x(n)=\left\{\begin{matrix}3, 1, 5, 4, 2, 1, 0, 1\\ \uparrow\end{matrix}\right\}$

17. Using Hamming window:
$h(0)=h(4)=0.0107,\ h(1)=h(3)=0.082404,\ h(2)=0.15915$
Using Hanning window:
$h(0)=h(4)=0,\ h(1)=h(3)=0.0763,\ h(2)=0.15915$

18. Using Hamming window: $h(0)=h(3)=0.00958,\ h(1)=h(3)=0.0973$
Using Hanning window: $h(0)=h(4)=0,\ h(1)=h(3)=0.0948$

19. Using Hamming window:
$h(0)=h(4)=-0.001796,\ h(1)=h(3)=-0.17145,\ h(2)=0.5225$
Using Hanning window: $h(0)=h(4)=0,\ h(1)=h(3)=-0.15875,\ h(2)=0.5225$

20. Using Hamming window: $h(0)=h(3)=-0.030184,\ h(1)=h(2)=0.0973$
Using Hanning window: $h(0)=h(4)=0,\ h(1)=h(2)=0.1520$

21. Using Hamming window:
$h(0)=h(6)=0.001232,\ h(1)=h(5)=0.01965,\ h(2)=h(4)=0.0758,\ h(3)=0.1114.$
Using Hanning window:
$h(0)=h(6)=0,\ h(1)=h(5)=0.01585,\ h(2)=h(4)=0.07387,\ h(3)=0.1114$

22. Using Hamming window:
$h(0)=h(7)=-0.003872,\ h(1)=h(6)=0.007311,\ h(2)=h(5)=0.06766,$
$h(3)=h(4)=0.14567.$

Using Hanning window:
$h(0)=h(7)=0,\ h(1)=h(6)=0.005438,\ h(2)=h(5)=0.0644,\ h(3)=h(4)=0.1450.$

23. Using Hamming window:
$h(0)=h(4)=-0.0019864,\ h(1)=h(3)=-0.0347,\ h(2)=0.9204.$
Using Hanning window: $h(0)=h(4)=0,\ h(1)=h(3)=-0.03215,\ h(2)=0.9204.$

24. Using Hamming window:
$h(0)=h(5)=0.006424,\ h(1)=h(4)=-0.1293,\ h(2)=h(3)=0.44036.$
Using Hanning window:
$h(0)=h(5)=0,\ h(1)=h(4)=0.1125,\ h(2)=h(3)=0.4366.$

4 CHAPTER

Time Domain Analysis

This chapter provides insights on

- Short-time speech analysis and windowing concepts
- Short-time energy, average magnitude, and zero crossing rate computations
- Identification of speech vs silence discrimination using short-time energy and zero crossing rates
- Short-time autocorrelation and average magnitude difference function
- Pitch period estimation using the autocorrelation function
- DSP based pitch period estimation
- Pitch period estimation using parallel processing approach

4.1 INTRODUCTION

In chapter 2, we learnt about the fundamentals of speech and in chapter 3, we reviewed the basics of digital signal processing. This chapter will introduce time dependent processing techniques such as short-time energy, average magnitude, zero crossing rate, autocorrelation and average magnitude difference function. In this chapter, we will discuss the applications of energy and zero crossing rate in locating the beginning and end of a speech segment in a background of noise. Pitch period calculation using short-time autocorrelation and parallel processing approach are also discussed.

4.2 SHORT-TIME SPEECH ANALYSIS

The process of extracting properties or features of speech signal is called *speech analysis*. In the time domain methods, features such as energy, autocorrelation function, zero crossing rate, etc. are calculated. Most applications of speech processing (e.g., coding, synthesis, recognition) require these properties to accomplish their tasks.

In speech analysis, we directly extract relevant information by suppressing redundant aspects of the speech. By eliminating redundant and irrelevant aspects of the speech waveform simplifies data manipulation. Vocal tract shape and source characteristics change slowly with time and hence properties of speech signal change relatively slowly. The vocal tract shape and excitation signal may not alter for 10-20 ms. During slow speech, the vocal tract shape and type of excitation may not alter for durations up to 200 ms. Speech analysis assumes that the signal properties change relatively with time. Hence, in speech analysis, we divide the speech signal into segments by using a sliding window and each segment is analyzed separately. The selected segment of the window is referred to as speech frame.

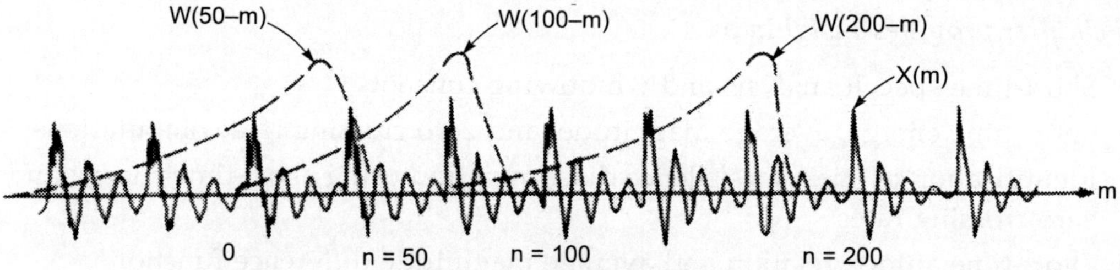

Figure 4.1 Speech signal with superimposed windows

For a given speech signal, $s(n)$, we define short-time signal, Q_n, of frame n as

$$Q_n = \sum_{m=-\infty}^{\infty} T\left[s(m)\right]w(n-m) \tag{4.1}$$

The speech signal is subjected to a transformation, $T[\ ]$. The resulting sequence is then multiplied by a window sequence positioned at a time corresponding to sample index, n. The product of $T[s(m)]$ and $w(n-m)$ is zero everywhere except in a small region, where Q_n corresponds to convolution of $T[s(n)]$ with $w(n)$, where window $w(n)$ represents a low-pass filter. General representation of short-time analysis is shown in Figure 4.2.

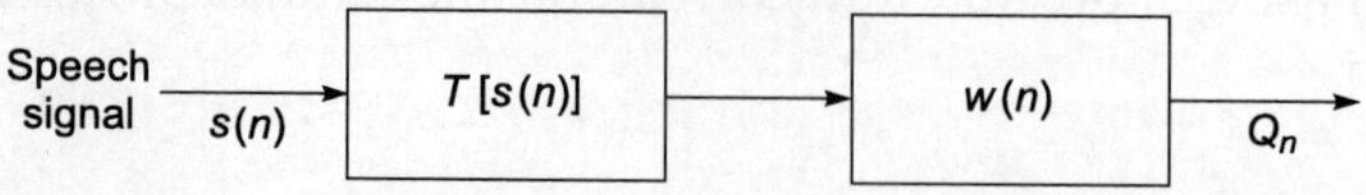

Figure 4.2 Representation of the short-time analysis principle

Windowing is the multiplication of a speech signal by a window, which yields a set of speech samples weighted by the shape of the window. By shifting $w(n)$, we can examine any part of $s(n)$ through the movable window. The size and shape of $w(n)$ depends on their effect in speech analysis. Three types of windows: rectangular, Hamming and Hanning are discussed in this section.

Rectangular Window

The simplest common window is rectangular window. It is denoted by $w_R(n)$ and it is defined by

$$W_R(n) = \begin{cases} 1 & \text{for} \quad 0, 1, \ldots N-1 \\ 0 & \text{otherwise} \end{cases} \tag{4.2}$$

The Fourier transform of a rectangular window is given by

$$w_R(e^{j\omega}) = \sum_{n=0}^{N-1} w_R(n) e^{-j\omega n} \tag{4.3}$$

$$w_R(e^{j\omega}) = \frac{1-e^{-j\omega n}}{1-e^{-j\omega}} = \frac{e^{\frac{-j\omega n}{2}}}{e^{\frac{-j\omega}{2}}}\left[\frac{e^{\frac{j\omega n}{2}} - e^{\frac{-j\omega n}{2}}}{e^{\frac{j\omega}{2}} - e^{\frac{-j\omega}{2}}}\right] = e^{-\frac{j\omega(N-1)}{2}}\left[\frac{\sin\left(\frac{\omega n}{2}\right)}{\sin\left(\frac{\omega}{2}\right)}\right] \tag{4.4}$$

Frequency response of the rectangular window is shown in Figure 4.3.

Hamming Windows

The Hamming window is represented by $w_{\text{Ham}}(n)$ and is defined by

$$w_{\text{Ham}}(n) = \begin{cases} 0.54 - 0.46\cos\left(\frac{2\pi n}{N-1}\right) & 0 \le n \le N-1 \\ 0 & \text{otherwise} \end{cases} \tag{4.5}$$

The Fourier transform of Hamming window is given by

$$w_{\text{Ham}}(e^{j\omega}) = \sum_{n=0}^{N-1} w_{\text{Ham}}(n) e^{-j\omega n} \tag{4.6}$$

$$w_{\text{Ham}}(e^{j\omega}) = \sum_{n=0}^{N-1}\left[0.54 - 0.46\cos\left(\frac{2\pi n}{N-1}\right)\right] e^{-j\omega n} \tag{4.7}$$

$$w_{\text{Ham}}(e^{j\omega}) = 0.54\frac{\sin\left(\frac{\omega n}{2}\right)}{\sin\left(\frac{\omega}{2}\right)} + 0.54\frac{\sin\left(\frac{\omega n}{2} - \frac{N\pi}{N-1}\right)}{\sin\left(\frac{\omega}{2} - \frac{\pi}{N-1}\right)} + 0.46\frac{\sin\left(\frac{\omega n}{2} + \frac{N\pi}{N-1}\right)}{\sin\left(\frac{\omega}{2} + \frac{\pi}{N-1}\right)} \tag{4.8}$$

Frequency response of the Hamming window is shown in Figure 4.3.

Hanning Window

The Hanning window is represented by $w_{\text{Han}}(n)$ and is defined by

$$w_{\text{Han}}(n) = \begin{cases} 0.5 - 0.5\cos\left(\dfrac{2\pi n}{N-1}\right) & 0 \le n \le N-1 \\ 0 & \text{otherwise} \end{cases} \tag{4.9}$$

The Fourier transform of Hanning window is given by

$$w_{\text{Han}}(e^{j\omega}) = \sum_{n=0}^{N-1} w_{\text{Han}}(n) e^{-j\omega n} \tag{4.10}$$

$$w_{\text{Han}}(e^{j\omega}) = \sum_{n=0}^{N-1} \left[0.5 - 0.5\cos\left(\frac{2\pi n}{N-1}\right)\right] e^{-j\omega n} \tag{4.11}$$

$$w_{\text{Han}}(e^{j\omega}) = 0.5\frac{\sin\left(\frac{\omega n}{2}\right)}{\sin\left(\frac{\omega}{2}\right)} + 0.5\frac{\sin\left(\frac{\omega n}{2} - \frac{N\pi}{N-1}\right)}{\sin\left(\frac{\omega}{2} - \frac{\pi}{N-1}\right)} + 0.5\frac{\sin\left(\frac{\omega n}{2} + \frac{N\pi}{N-1}\right)}{\sin\left(\frac{\omega}{2} + \frac{\pi}{N-1}\right)} \tag{4.12}$$

Frequency response of the Hanning window is shown in Figure 4.3.

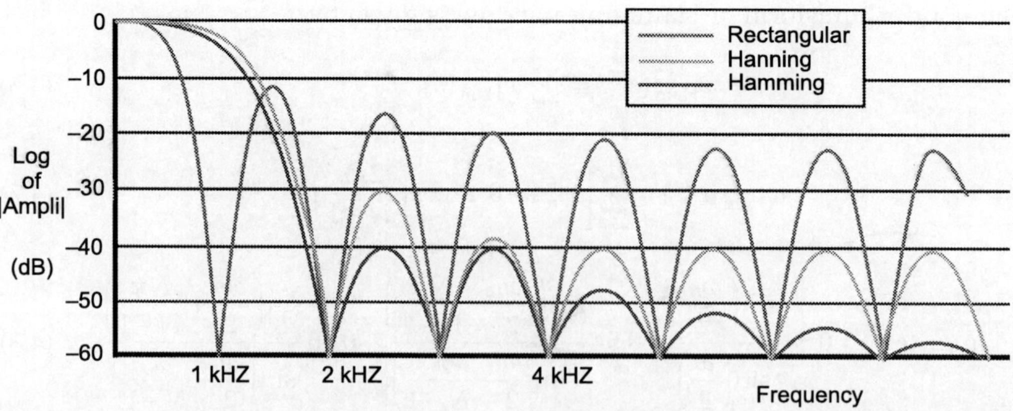

Figure 4.3 Frequency response of (a) rectangular, (b) Hanning, and (C) Hamming windows

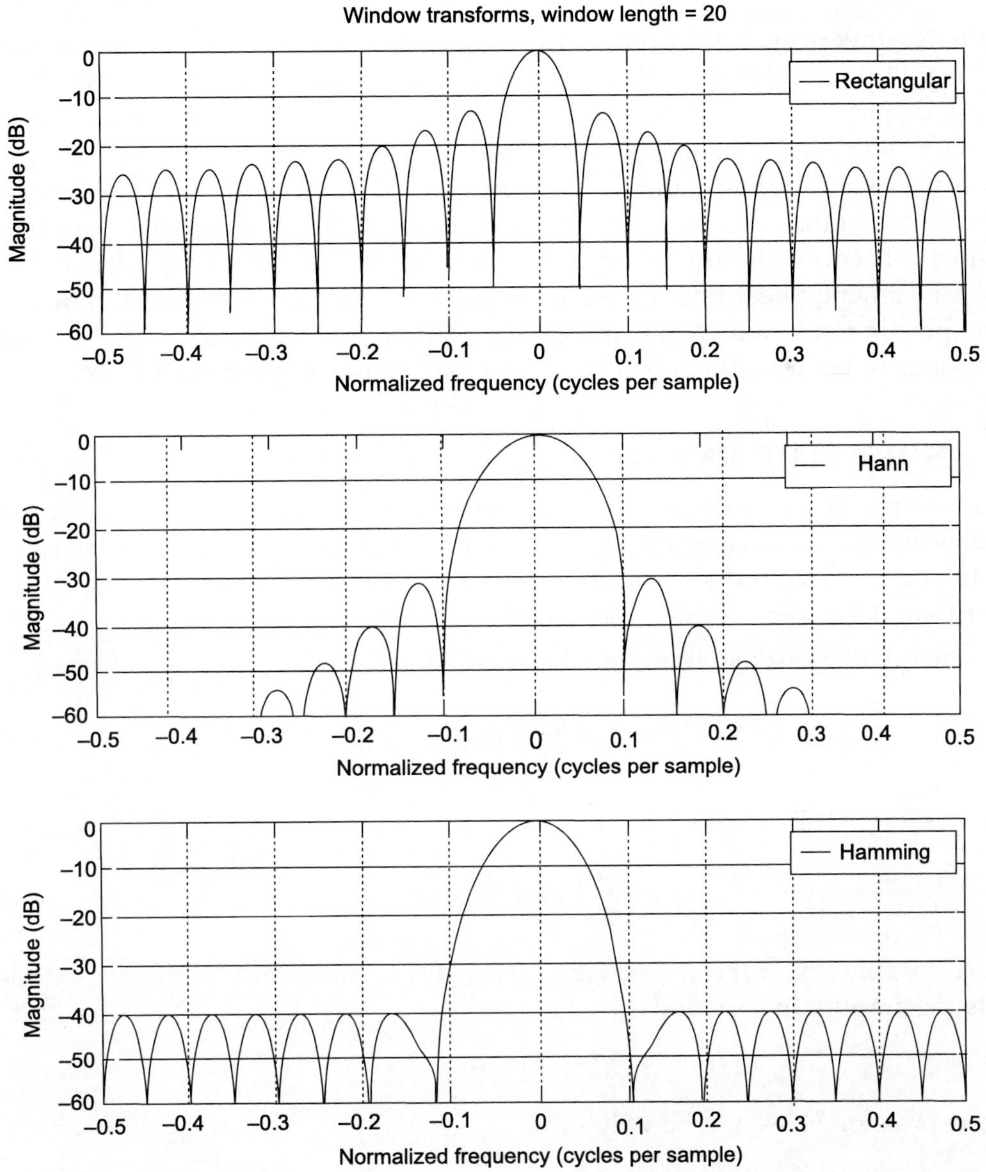

Figure 4.4 Fourier transform of (a) rectangular, (b) Hanning, (C) Hamming windows

As shown in Figure 4.4 the main lobe of the rectangular window is half as wide as the Hamming or Hanning window. Side lobes of Hamming or Hanning window are much lower than of the rectangular window. The first side lobe of Hanning is approximately 20 dB higher than the Hamming and side lobes rapidly decrease to very low level. Selecting the window shape and its resulting frequency response is a tradeoff between a narrow main lobe and low side lobes in the frequency domain. Narrow main lobe improves the frequency resolution (so that the resulting DFT magnitude, closely spaced narrow components are separated). Narrowband spectral displays are good for fundamental frequency (F_0) estimation. For F_0 estimation, however, windows must contain at least two pitch periods; so pitch analysis uses a long window and it is 30-50 ms. The side lobes add a noisy appearance to the DFT magnitude due to interference from adjacent harmonics and make it more difficult to discriminate low magnitude components. Wideband representation is better for viewing vocal tract parameters, which change rapidly and do not need fine frequency resolution. Wideband representation uses a window of 3 ms, which corresponds to band width of 300 Hz and smoothes the harmonic structure.

4.3 SHORT-TIME ENERGY

The short-time energy measurement of a speech signal can be used to determine voiced vs unvoiced speech. Short-time energy can also be used to detect the transition from unvoiced to voiced speech and vice versa. The energy of voiced speech is greater than the energy of unvoiced speech.

The energy of a discrete time signal is defined as

$$E = \sum_{m=-\infty}^{\infty} s^2(m) \tag{4.13}$$

Short-time energy is given by

$$E_n = \sum_{m=-\infty}^{\infty} [s(m)w(n-m)]^2 \tag{4.14}$$

Short-time energy of the speech signal provides a convenient representation that reflects the amplitude variations

Let $$h(n) = w^2(n)$$

The expression (4.14) can be written as

$$E_n = \sum_{m=-\infty}^{\infty} s^2(m)h(n-m) \tag{4.15}$$

Thus, the signal, $s^2(n)$, is filtered by a linear filter with impulse response $h(n)$ given by Eq. (4.15). $h(n)$ or window determines the nature of the short-time energy representation, Q_n (of Eq. (4.1)) corresponds to short-time energy and transformation $T[\ ]$ in the above equation is a squaring operation. The operations involved in Eq. (4.15) are represented in Figure 4.5.

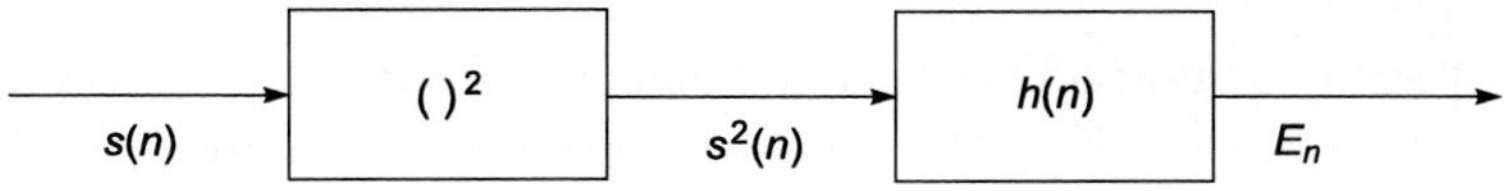

Figure 4.5 Representation of short-time energy

The window must be long enough to encompass several pitch periods to produce a smooth representation of the amplitude of the signal. At the same time the window must be short enough to reflect rapid changes in amplitude that occur at the voiced/unvoiced boundaries. The selection of the window size is a compromise, since a high pitched female or child's voice may have a pitch period as small as 16 samples at an 8 kHz. Sampling rate is up to 200 samples for a low pitched male voice. A window size of 160 samples or about 20 msec is a good compromise.

4.4 SHORT-TIME AVERAGE MAGNITUDE

The problem with the short-time energy function is that it is very sensitive to large signal levels since the sample values are squared. In a production application using fixed point math is a disadvantage. In addition, a multiply operation is required for each sample. So instead of using the short-time energy function, we use a related function, the short-time magnitude functions, shown in Eq. 4.16.

The average magnitude of a discrete time signal is defined as

$$M = \sum_{m=-\infty}^{\infty} |s(m)| \tag{4.16}$$

Short-time average magnitude is given by

$$M_n = \sum_{m=-\infty}^{\infty} |s(n)| w(n-m) \tag{4.17}$$

Q_n corresponds to short-time magnitude, if T in the above equation is absolute magnitude operation. The operations involved in Eq. (4.17) are represented in Figure 4.6. Hence, the signal, $|s(m)|$, is filtered with impulse response, $w(n)$.

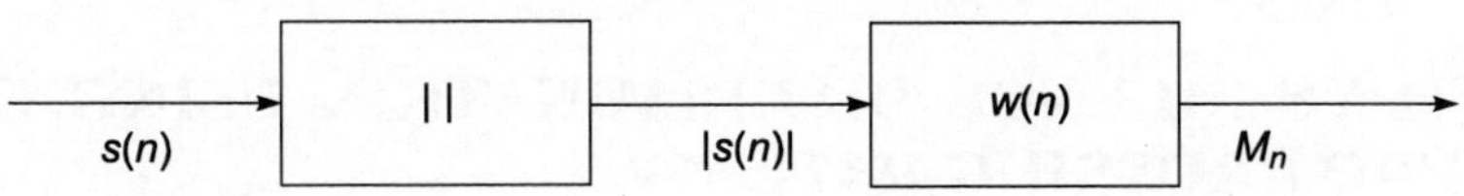

Figure 4.6 Representation of short-time average magnitude

4.5 SHORT-TIME ZERO CROSSING RATE

The short-time average zero crossing rate of a speech signal can be used in conjunction with the short-time average energy (or magnitude) to discriminate among voiced speech, unvoiced speech and silence. Short-time zero crossing rate is defined as the number of times the speech signal changes sign in a given time interval. In a signal $s(n)$ such as speech, a zero crossing occurs when $s(n) = 0$, i.e., the waveform crosses the time axis or changes algebraic sign. Hence, zero crossing is said to occur if successive samples have different algebraic signs. The short-time average crossing rate of a digital sample speech signal is defined as

$$Z_n = \sum_{m=-\infty}^{\infty} \frac{1}{2}\left\{\left|\operatorname{sgn}[s(m)] - \operatorname{sgn}[s(m-1)]\right| w(n-m)\right\} \tag{4.18}$$

where, $\operatorname{sgn}[s(n)] = 1 \quad s(n) \geq 0$

$\operatorname{sgn}[s(n)] = -1 \quad s(n) < 0$

Rectangular window,

$$w(n) = \frac{1}{N} \quad 0 \leq n \leq N-1$$

where, $w(n)$ is a rectangular window with N samples. Hence, if Z_n is scaled by $\frac{1}{N}$, we get zero crossing/sample. The operations involved in Eq. (4.18) are represented in Figure 4.7.

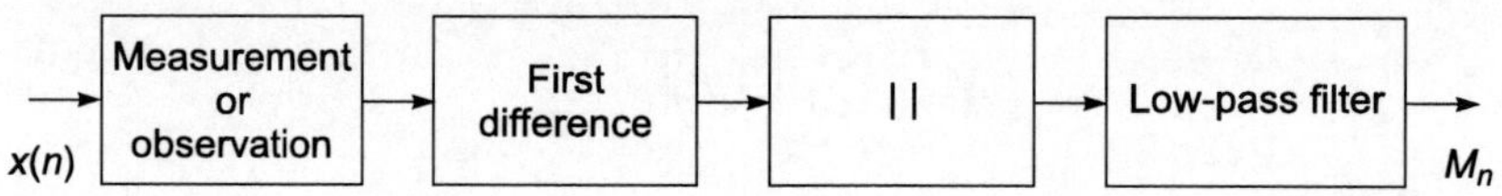

Figure 4.7 Representation of short-time average zero crossing

Zero crossing rate (ZCR) helps in voicing decision: zero crossing rate is high in unvoiced and it is low if speech signal is voiced. Voiced and unvoiced speeches have 1400 and 4900 zero crossing/sample, respectively. Voiced regions have higher energy and lower zero crossing rate, whereas unvoiced regions have lower energy and higher zero crossing rate.

4.6 SPEECH VS SILENCE DISCRIMINATION USING ENERGY AND ZERO CROSSING RATE

Speech vs silence discrimination is an automatic process to identify the boundaries of an isolated utterance, i.e. to separate speech from silence. It is useful to save the computation for processing. The problem of locating the beginning and end of a speech segment in a background of noise is of importance in many areas of speech processing. It is essential to locate the regions of speech signals that correspond to each word. In general, it is difficult to locate the beginning and end of a speech segment, if there are:

1. Weak fricatives ($|f|$, $|th|$, $|h|$) at the beginning or end
2. Weak plosives ($|p|$, $|t|$, $|k|$) at the beginning or end
3. Nasals at the end

Effective end point detection can be done with short-time energy and zero crossing rates. In the algorithm, zero crossing rate is computed with 10 ms window using equation

$$Z_n = \sum_{m=-\infty}^{\infty} \frac{1}{2}\left\{\left|\text{sgn}\left[s(m)\right] - \text{sgn}\left[s(m-1)\right]\right| w(n-m)\right\} \tag{4.19}$$

and energy or average magnitude is computed with 10 msec, window using equation

$$E_n = \sum_{m=-\infty}^{\infty} s^2(m)h(n-m) \tag{4.20}$$

$$M_n = \sum_{m=-\infty}^{\infty} |s(m)| w(n-m) \tag{4.21}$$

It is assumed that the first 100 msec of the interval does not contain any speech. The mean and standard deviation of the energy or average magnitude and zero crossing rates are computed for this interval. Using this statistical characterization zero crossing rate and energy thresholds are computed. Example for speech vs silence discrimination algorithm is depicted in Figure 4.8.

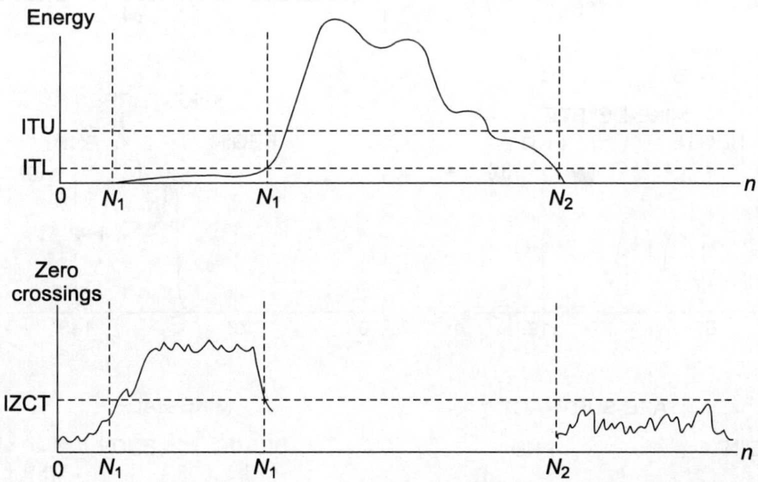

Figure 4.8 Typical example of energy and zero crossing measurement for a word

Speech vs silence discrimination is implemented using the following algorithm:

1. Check the first crossing of ITU and then for crossing of ITL.
2. ITU crossing indicates the beginning and end point as shown in Figure 4.8.
3. Then on ZC, move backward and forward from N_1 and N_2, respectively.

4. If the threshold IZCT is crossed more than three times, move backward to first time crossing which gives the boundary.
5. Similarly, if threshold is crossed more than three times, move forward to the first crossing which gives the boundary.

Figure 4.9 shows example of words with beginning and end of words.

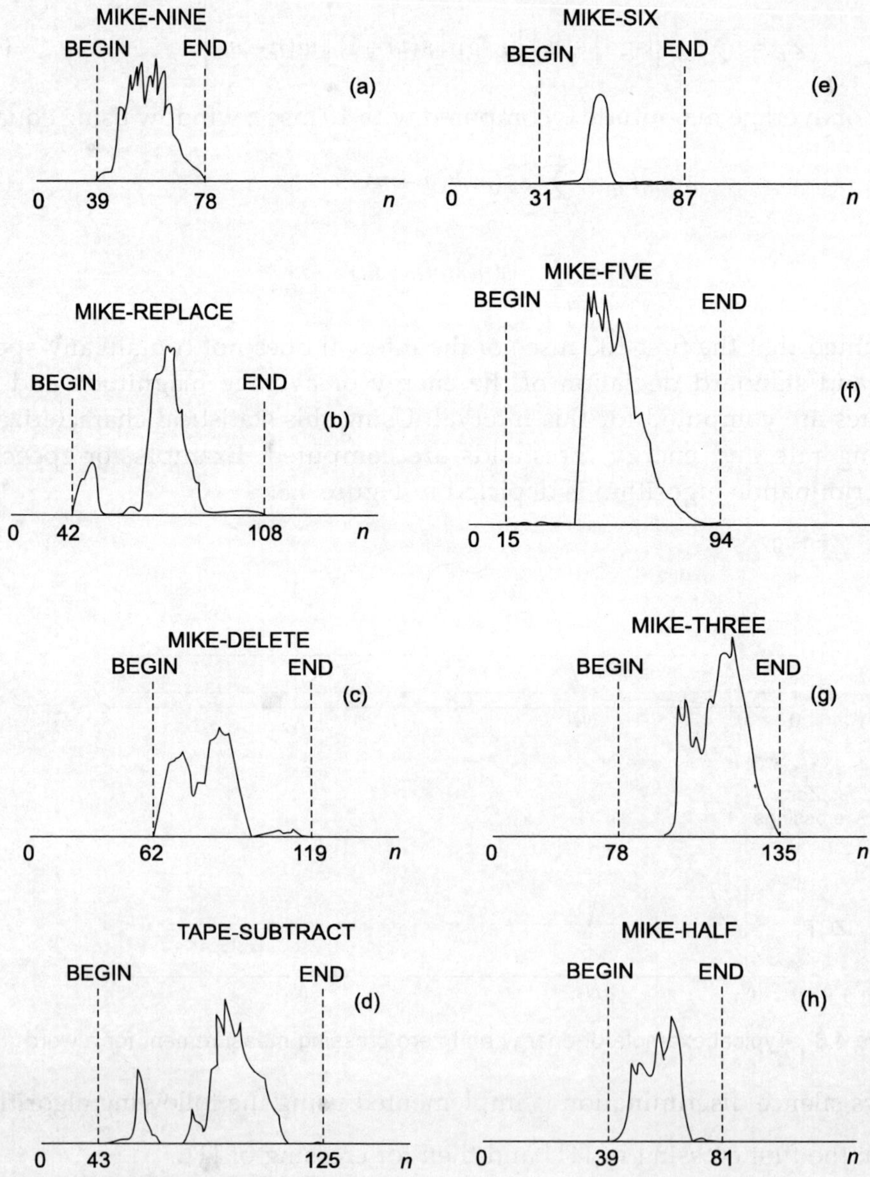

Figure 4.9 Sequence of energy plots showing how the end point algorithm performs over a variety of words (Rabiner and Sambur)

4.7 SHORT-TIME AUTOCORRELATION FUNCTION

Autocorrelation function is used to compare two samples of the same signal. In short-time autocorrelation function, with the help of sliding window we divide the speech signal into segments and then we make the analysis. The autocorrelation of a discrete time deterministic signal is given by

$$\phi(k) = \sum_{m=-\infty}^{\infty} s(m)s(m+k) \tag{4.22}$$

Short-time autocorrelation function is given by

$$R_n(k) = \sum_{m=-\infty}^{\infty} s(m)w(n-m)s(m+k)w(n-k-m) \tag{4.23}$$

where, index k is time shift parameter and $R_n(-k)$ is folded version of $R_n(k)$.

$$R_n(-k) = \sum_{m=-\infty}^{\infty} s(m)w(n-m)s(m-k)w(n+k-m) \tag{4.24}$$

Let $h_k(k) = w(n)w(n+k)$ and from properties $R_n(k) = R_n(-k)$

$$R_n(k) = \sum_{m=-\infty}^{\infty} s(m)s(m-k)h_k(n-m) \tag{4.25}$$

Thus, the value at time n of the kth autocorrelation is obtained by filtering the sequence $s(n)s(n-k)$ with a filter with impulse response $h_k(n)$. The equation can thus be interpreted as depicted in Figure 4.10.

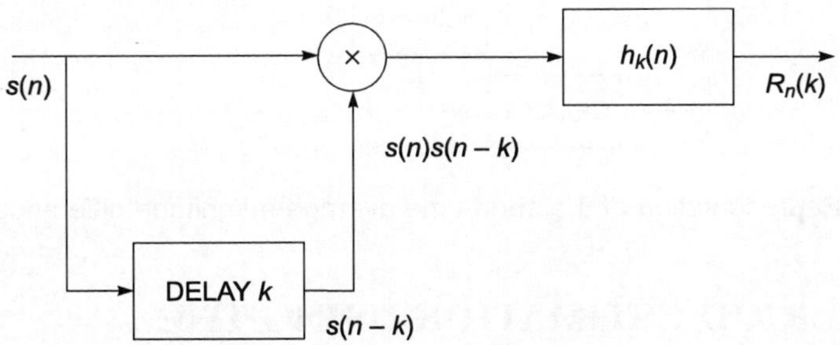

Figure 4.10 Representation of the short-time autocorrelation

The fact that variations in voiced signal are so evident suggests that time domain techniques should be capable of detecting pitch period of voiced signals. Most of the time domain pitch period estimation techniques use autocorrelation function (ACF). Autocorrelation function has many peaks, peaks at the pitch period has the greatest amplitude.

Example 4.1 Compute the pitch period if the sample rate is 8000 Hz and the autocorrelation function of the prominent peak is at 81.

$$\text{The pitch period} = \frac{81}{8000\ \text{Hz}} = 10.125\ \text{ms}$$

Pitch frequency is 98.8 Hz

4.8 AVERAGE MAGNITUDE DIFFERENCE FUNCTION

Average magnitude difference function is same as autocorrelation but faster to implement since integer arithmetic has no multiplies. Magnitude difference function is given by

$$d(n) = s(n) - s(n-k) \tag{4.26}$$

Short-time average magnitude (STAM) difference function is given by

$$= \sum_{m=-\infty}^{\infty} \left| s(n+m)w_1(m) - s(n+m-k)w_2(m-k) \right| \tag{4.27}$$

The equation can thus be interpreted as depicted in Figure 4.11.

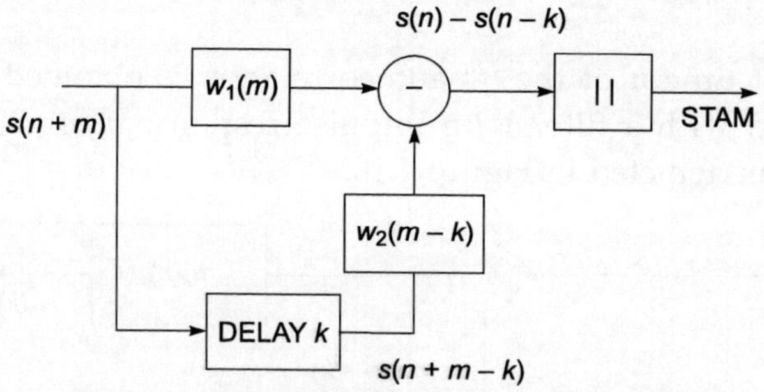

Figure 4.11 Representation of the short-time average magnitude difference function

4.9 PITCH PERIOD ESTIMATION USING THE AUTOCORRELATION FUNCTION

In this section, we will discuss three methods of implementation of autocorrelation pitch detectors. They are:

- No clipping
- Centre clipping
- Three level o/p clipper

1. No Clipping

In this method all the samples of speech are considered. The autocorrelation function may contain too much information, most of which is not related to the fundamental frequency. For pitch detection, speech signal is usually pre-processed to make the periodicity more prominent and to suppress other distracting features. Such techniques are often called "**spectrum flattening**". Then autocorrelation function is calculated. Speech waveform and correlation function are shown in Figure 4.12 (a).

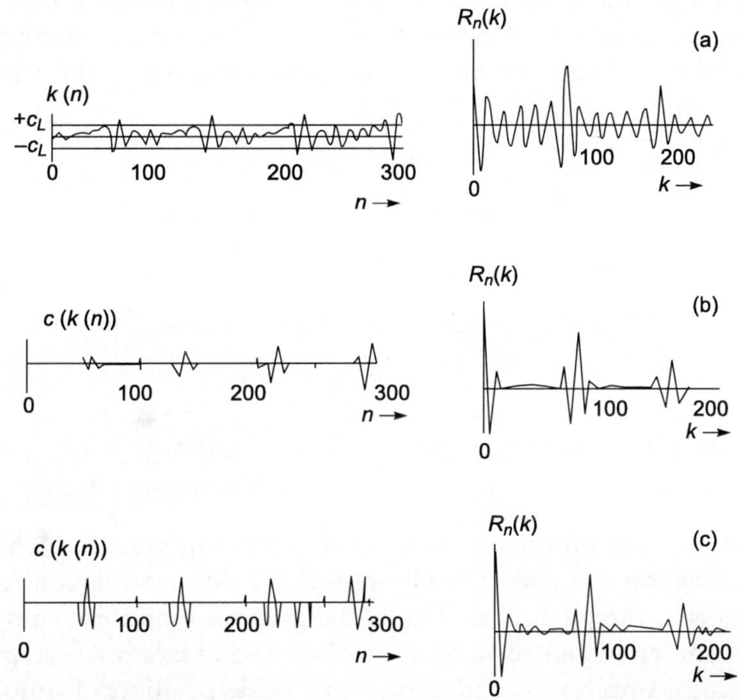

Figure 4.12 Waveforms and correlation function (a), no clipping (b), centre clipped, and (c) 3-level centre clipped.

2. Centre Clipping

Centre clipping is the most popular spectrum flattening technique. In this method Sondhi suggested, the clipping level is 30% of the maximum amplitude. The operation of centre clipper is depicted in Figure 4.12(b). In other algorithms the clipping level is 68% of the peak amplitude in both the first and the last third of the segment and set the clipping level at a fixed percentage of the minimum of these two maximum levels. The autocorrelation function is calculated. The resulting autocorrelation function has considerably fewer extraneous peaks compared to no clipping methods. Speech waveform and correlation function is shown in Figure 4.12 (b). Autocorrelation function has many peaks, peaks at the pitch period has the greatest amplitude and it is in the same position as shown in Figure 4.12 (a).

3. Three-Level o/p Clipper

In the previous methods, autocorrelation function requires large amount of computation. A simple modification of the centre clipping function leads to a great simplification in computation of the autocorrelation function. The modification is depicted in Figure 4.13. This function is called three-level centre clippers.

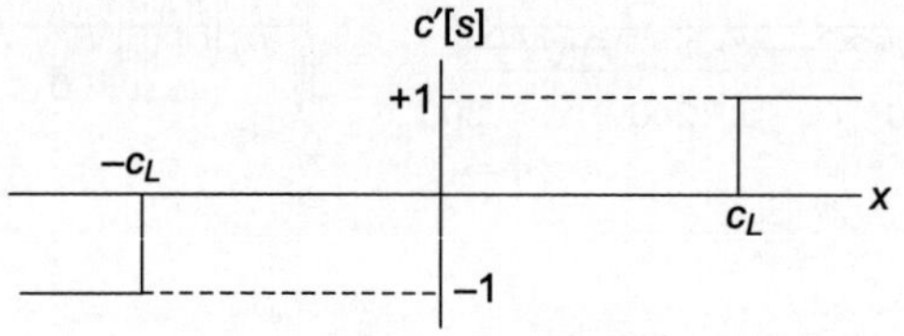

Figure 4.13 Level centre clipping function

$$\text{The output of the clipper} = \begin{cases} +1 & \text{if} \quad s(n) > c_L \\ -1 & \text{if} \quad s(n) < c_L \\ 0 & \text{otherwise} \end{cases} \tag{4.28}$$

Figure 4.13 shows the output of three-level centre clippers. Speech waveform and correlation function are shown in Figure 4.12 (c). Autocorrelation function has many peaks, peaks at the pitch period have the greatest amplitude and it is in the same position as shown in Figure 4.12 (a). As discussed in Example 4.1, pitch period is computed using sample rate and prominent peak position of autocorrelation function.

4.10 DSP BASED PITCH PERIOD ESTIMATION FOR SPEECH SIGNALS

Pitch period estimation is one of the important problems in speech processing. The algorithm for estimating the pitch period is based on short-time autocorrelation function. This proposed algorithm can be implemented on TMS 320C6713 processor is discussed in this section. The summary of the algorithm is given below.

- Speech signal is filtered with a 900 Hz low-pass analog filter.
- Speech signal is sampled at a rate of 10 kHz (10,000 samples/sec) and converted to digital form by using A/D converter.

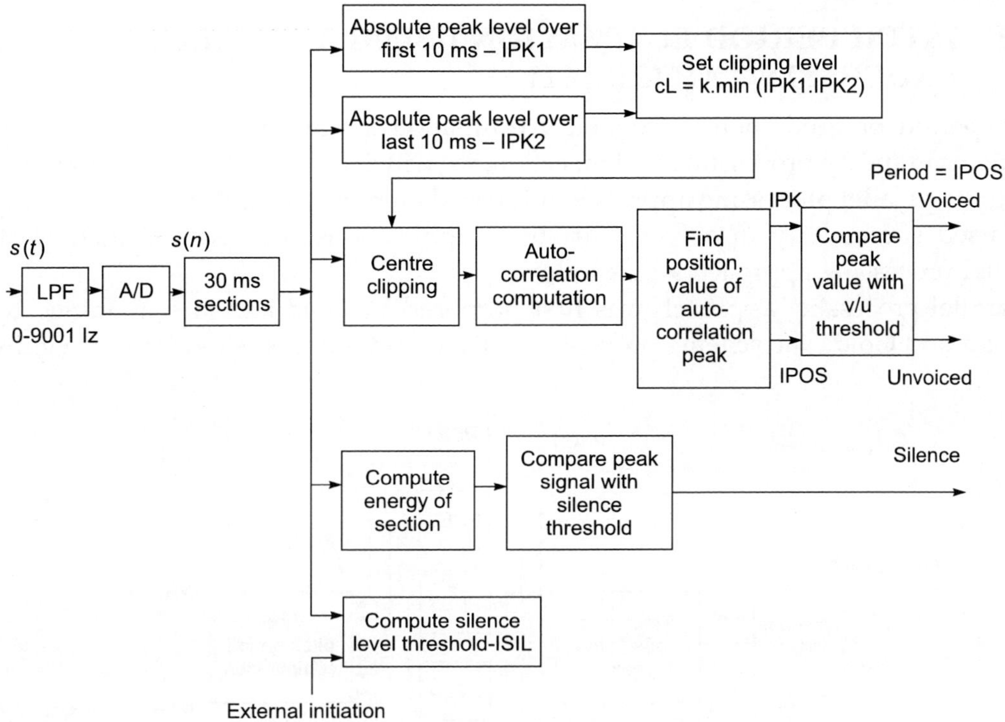

Figure 4.14 Clipping autocorrelation pitch detector.

- Segments of length 30 msec (300 samples) are selected at 10 msec interval.
- Absolute peak level over first 10 msec (IPK1) and over last 10 msec (IPK2) are calculated.
- Clipping level is determined as a fixed percentage (68%) of minimum absolute values of first and last 100 samples of peak signal level of speech segment.
- Using this clipping level, speech signal is processed. For speech signal above the clipping level autocorrelation function is computed.
- Largest peak of the autocorrelation is located. Pitch period is defined as location of the largest peak (I_{POS}). If peak (I_{PK}) falls below the threshold, the segment is classified as unvoiced sound and if it is above, it is classified as voiced sound.
- Energy of section is calculated, if energy lies below the silence level threshold, segment is classified as silence.

The complete implementation of the above algorithm is shown in Figure 4.14.

4.11 PITCH PERIOD ESTIMATION USING A PARALLEL PROCESSING APPROACH

Pitch period or fundamental frequency estimation is one of the most important problems in speech processing. In this section, we will discuss pitch period estimation using the parallel processing approach. It is based on purely time domain processing and used successfully in a wide variety of applications. The basic principles of parallel processing approach are as follows:

Parallel processing approach was first proposed by Gold and later modified by Rabiner and Gold. The scheme proposed by Gold and Rabiner is depicted in Figure 4.15.

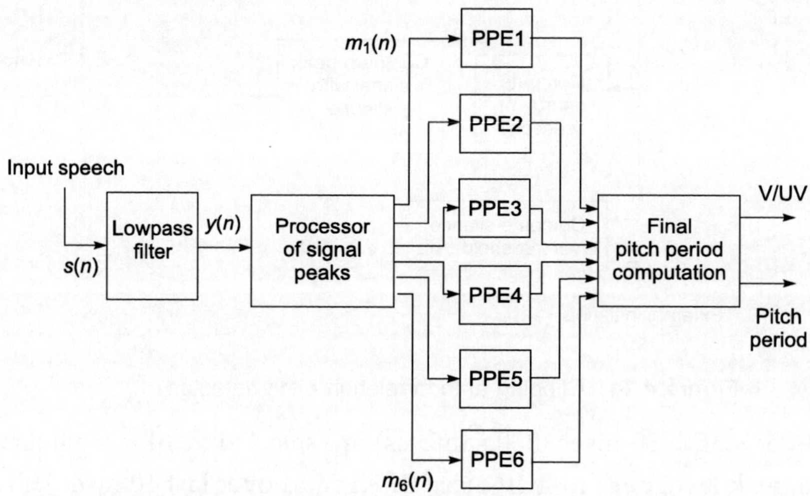

Figure 4.15 Pitch period estimation using a parallel processing approach

The speech waveform is sampled with a sampling frequency of 10 kHz ($T_S = 10^{-4}$ sec). The speech is then low pass filtered with a cutoff of about 900 Hz to produce a relatively smooth waveform. From their amplitude, peaks and valleys are located and several impulse trains are derived from the filter signal. Six different impulse trains $m_1(n)$, $m_2(n)$, $m_3(n)$, $m_4(n)$, $m_5(n)$ and $m_6(n)$ are used by Gold and Rabiner. Figure 4.16 shows six different impulse trains for a pure sine wave. The algorithm to compute impulse trains are as follows:

$m_1(n)$: An impulse equal to the peak amplitude occurs at the location of each peak.

$m_2(n)$: An impulse equal to the difference between the peak amplitude and the preceding valley amplitude occurs at each peak.

$m_3(n)$: An impulse equal to the difference between the peak amplitude and the preceding peak amplitude occurs at each peak. (if this result is negative the impulse is set to zero).

$m_4(n)$: An impulse equal to the negative of the amplitude at a valley occurs at each valley.

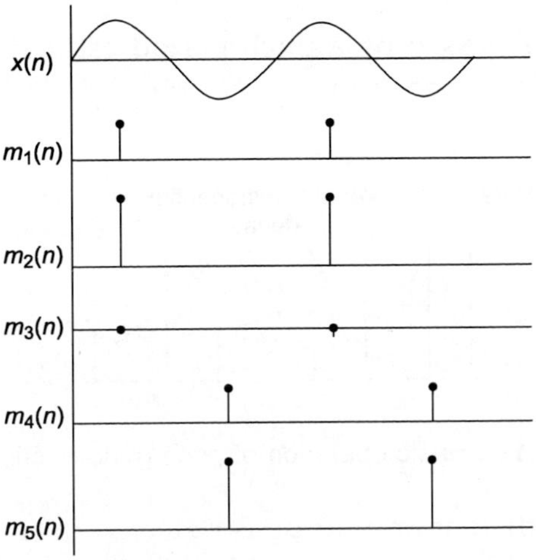

Figure 4.16 For sinusoid input impulse generated from peak and valleys

$m_5(n)$: An impulse equal to the negative of the amplitude at a valley plus the amplitude at the preceding peak occurs at each valley.

$m_6(n)$: An impulse equal to the negative of the amplitude at a valley plus the amplitude at the preceding local minimum occurs at each valley (if this result is negative the impulse is set to zero).

Figure 4.17 shows six different impulse trains for a speech wave. The algorithm to compute impulse trains are as discussed in sinusoid input.

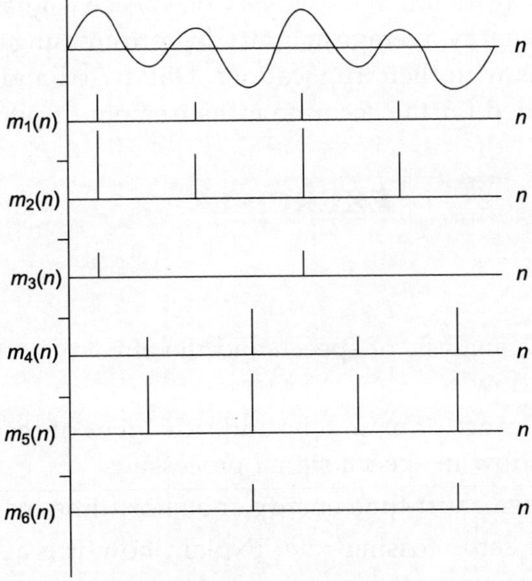

Figure 4.17 For speech input impulse generated from peak and valleys

Each impulse train is processed by a pitch period estimator (time varying linear system). The basic operation of each individual pitch period estimator is shown in Figure 4.18.

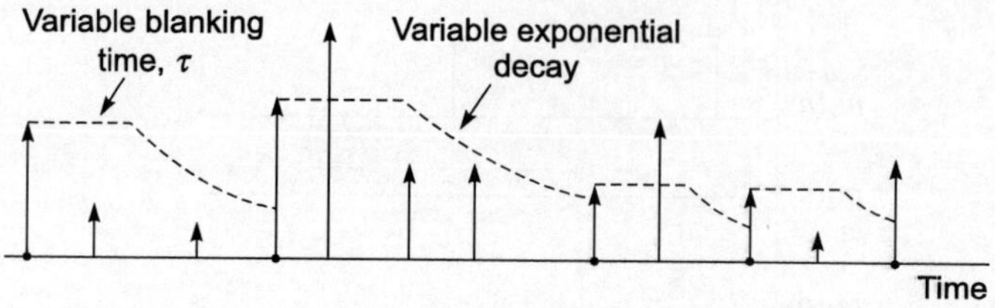

Figure 4.18 Basic operation of pitch period estimator

When an impulse of sufficient amplitude is detected in the input, output is reset to that impulse. A threshold is set for the blanking interval; during this interval no pulse can be detected. At the end of this interval the output begins to decay exponentially. When an impulse exceeds the level of the exponentially decaying output, the process is repeated. The rate of decay and the blanking interval depend upon the most recent estimates of the pitch period. This technique is applied to each of the six impulse train thereby obtaining six estimates of pitch period. These estimates are then combined and the value with the most occurrences is declared as the pitch period at that time.

4.12 SUMMARY

This chapter discusses in detail the analysis of speech signals in time domain, considering short-time energy, average magnitude, zero crossing rate, autocorrelation and difference functions with their applications. DSP based and parallel processing approach is also presented for the accurate measurement of pitch period.

EXERCISES

Review Questions

1. Explain short-time analysis of speech and the effect of rectangular, Hamming and Hanning window.
2. Explain the significance of main lobe and side lobes of rectangular, Hamming and Hanning window in speech signal processing.
3. Distinguish between short-time energy and short-time magnitude function.
4. Define short-time zero crossing rate. Explain how it is useful to distinguish between voiced and unvoiced speech sounds.

5. Distinguish between short-time magnitude function and average magnitude difference function.
6. Define short-time autocorrelation. Explain pitch period estimation using the autocorrelation method.
7. Explain pitch period estimation using no clipping, centre clipping and three-level output clipper.
8. Define short-time autocorrelation. Explain pitch period estimation using Sondhi's method
9. With necessary diagram, explain speech vs silence discrimination using energy and zero crossing rate.
10. With necessary diagram, explain pitch period estimation using parallel processing approach.

5 CHAPTER

Spectral and Homomorphic Speech Processing

This chapter provides insights on

- Pitch period estimation using FFT and autocorrelation function
- Formant estimation using log spectrum and power spectral density
- Short-time analysis and synthesis of speech
- Narrow and wideband spectrograms
- Fundamentals of homomorphic speech processing
- Cepstral domain methods to estimate pitch period and formants
- Mel and Bark scale
- Computation of Mel-frequency cepstrum coefficients (MFC), perceptual linear prediction (PLP) and log frequency power coefficients (LFPCs)

5.1 INTRODUCTION

In chapter 4, we learnt about the time domain analysis of speech and pitch period estimation using short-time autocorrelation function. This chapter will introduce pitch period estimation in spectral domain using FFT and autocorrelation function, formant estimation using log spectrum and power spectral density. We will be discussing short-time spectral analysis and synthesis of speech. In this chapter, we will learn how homomorphic processing converts convolved excitation signal and vocal tract parameters into an additive signal. After this we will discuss 3D wideband and narrowband spectrograms and their applications. Algorithms to compute cepstral coefficients, Mel-frequency cepstrum coefficients (MFC), perceptual linear prediction (PLP) coefficients and log frequency power coefficients (LFPCs) are also discussed which are important parameters for speech recognition and speaker identification.

5.2 PITCH PERIOD ESTIMATION IN SPECTRAL DOMAIN

As discussed in Chapter 2 the time span between successive vocal fold closures is referred to as pitch period or fundamental period, T_0. Pitch frequency is reciprocal of pitch period, $F_0 = \frac{1}{T_0}$. Pitch frequency is also referred to as fundamental frequency.

In chapter 4, pitch period estimation were discussed in time domain. In this section, we will discuss pitch frequency or fundamental frequency estimation using speech spectrum. Following methods are used for finding pitch frequency using speech spectrum

- Using FFT
- Using spectral autocorrelation

5.2.1 Pitch Period Estimation Using FFT

As discussed in chapter 3, fast Fourier transform (FFT) is one of the most important tools in digital signal processing. FFT algorithm converts discrete time signal, $s(n)$, to discrete spectrum $S(k)$ and there are basically two types of algorithms to achieve this and they are:

- Decimation in time FFT algorithm
- Decimation in frequency FFT algorithm

Both the algorithms require $\frac{N}{2}\log_2 N$ number of multiplications to convert discrete time signal, $s(n)$, to discrete spectrum $S(k)$, hence the main drawback of the frequency domain method is computational complexity. Algorithm for finding pitch period using FFT is as follows:

- Voiced speech signal is converted to digital signal, $s(n)$, with sampling frequency 22,100 Hz.
- Then speech signal, $s(n)$, is divided into a number of segments, such that each segment contains 512 samples.
- FFT of the segment is calculated. Since the sampling frequency is 22,100 Hz, frequency resolution is given by 43.16 Hz (each point corresponding to a frequency of 22,100 Hz/512).
- Detect the peaks in the spectrum. Fundamental frequency is the position of the first peak multiplied by frequency resolution. If the first peak occurs at 10th position, then the fundamental frequency is 43.16 Hz × 10 = 431.6 Hz.

5.2.2 Pitch Period Estimation using Spectral Autocorrelation

As discussed in Chapter 4, autocorrelation is one of the important tools to detect pitch period. Algorithm for finding pitch period using spectral autocorrelation method is as follows:

- Voiced speech signal is converted to digital signal, $s(n)$, with sampling frequency 22,100 Hz.
- Then speech signal, $s(n)$, is divided into a number of segments, such that each segment contains 512 samples.
- FFT of the segment is calculated.
- Then squared magnitude of the FFT is calculated to find the autocorrelation of this spectrum.
- Detect the peaks, spectral autocorrelation shows the spectral peaks at regular intervals. Distance between two peaks gives fundamental frequency.

5.3 FORMANTS ESTIMATION IN SPECTRAL DOMAIN

As discussed in chapter 2, vocal tract acts as a resonator. The resonant frequencies are called formant frequency or formants. Formants are denoted by $F_1, F_2, \ldots\ldots F_n$. Hence, the vocal tract filter is characterized by formant frequencies (or formants). The following methods are used for finding formant frequency using speech spectrum.

- Using log spectrum
- Using power spectral density estimate

5.3.1 Estimation of Formants Using Log Spectrum

If $e(n)$ is excitation signal and $h(n)$ is vocal tract impulse response, then short-time speech signal, $s(n)$, is obtained by taking the convolution of excitation signal, $e(n)$, which is a rapidly varying signal and impulse response of vocal tract, $h(n)$, which is a slowly varying function. Then windowed signal is obtained. For formants estimation use Hamming or Hanning window as discussed in section 4.1. Log magnitude operates on the spectrum and generates the amplitude in dB. Formant information lies in the envelope of the log magnitude of the spectrum. Algorithm for finding formats using log spectrum is as follows:

- Voiced speech signal is converted to digital signal, $s(n)$, with sampling frequency say, 22,100 Hz.
- Then speech signal, $s(n)$, is divided into a number of segments, such that each segment contains 512 samples.
- Windowed signal is obtained, i.e., pass each segment of speech signal through Hamming or Hanning window.

- FFT of the segment is calculated.
- Log magnitude of the FFT is calculated.

5.3.2 Estimation of Formants Using Power Spectral Density Estimate

Power spectral density estimate gives us the estimated average power in every spectral line and is defined as

$$P(k) = \frac{1}{N}\left|\sum_{n=0}^{N-1} s(n)e^{j\omega n}\right|^2 = \frac{1}{N}|S(k)|^2 \tag{5.1}$$

Estimated power spectral density is also known as periodogram. The periodogram is related to the estimate of the autocorrelation and is given by

$$r_n(\tau) = \frac{1}{N}\sum_{m=0}^{N-1-\tau} s_n(m)s_n(m+\tau) = \frac{1}{N}[s_n(\tau) * s_n(-\tau)] \tag{5.2}$$

$P(k)$ gives a good estimate for band limited and power signals sampled above the Nyquist rate and it yields poor estimate for noisy signals. Following methods are used for estimating formants using estimated power spectral density.

- Periodogram method
- PSD estimation by Welch method
- PSD estimation by Blackman-Tukey method

5.3.2.1 Periodogram method

Algorithm for finding formants using periodogram method is as follows:

- Voiced speech signal is converted to digital signal, $s(n)$.
- Then speech signal, $s(n)$, is divided into a number of segments.
- Find the autocorrelation of the signal of each segment.
- Find the Fourier transform of the signal. This provides the estimate of the power spectral density.

5.3.2.2 PSD estimation by Welch method

Welch method is based on the concept of averaging the results of overlapping periodogram estimates and finds widespread use in spectrum estimation. 50% to 75% overlap is common practice. Advantage of the Welch method over the periodogram method is that the variance of the power spectral density is further reduced. Algorithm for finding formants using Welch method is as follows:

- Voiced speech signal is converted to digital signal, $s(n)$.

- Then speech signal, $s(n)$, is divided into segments of size say 128. Let there be a 60% overlap between successive segments.
- Windowed signal is obtained, i.e., pass each segment of speech signal through Hamming window.
- Find the Fourier transform of the signal of each segment and then average, the power of each FFT coefficient of all the segments.
- Normalize (divide) this result by the power of the windowed signal given by $\Sigma |W(k)|^2$.

5.3.2.3 PSD estimation by Blackman-Tukey method

Blackman-Tukey method is based on the concept of finding the PSD from the windowed correlation. Algorithm for finding formants using Blackman-Tukey method is as follows:

- N sampled speech signal, $s(n)$, is zero padded to get $2N$-sample $y(n)$
- Find the autocorrelation of the signal, $y(n)$
- Windowed signal is obtained, i.e., pass autocorrelation of the signal, $y(n)$, through Hamming window.
- Compute the FFT of the resulting data which yields smooth periodogram.

By comparing results with periodogram method, we see that smoothing is achieved by the averaging effect of the autocorrelation process rather than by the averaging of several periodograms.

5.4 SHORT-TIME FOURIER ANALYSIS OF SPEECH

Discrete Fourier transform reflects the spectral contents of the entire signal over time, but does not indicate when the spectral content occurs. STFT is sliding windowed Fourier transform. STFT is one of the most straightforward approaches for performing time-frequency analysis and helps to understand the concept of time frequency analysis. Speech signals are quasi-stationary signals because vocal tract and source characteristic change slowly with time. In order to analyze speech signals which are not in general stationary, the signal has to be processed in segments. In order to do this, the speech signal is segmented into blocks and each block is analyzed separately. This is done by using a sliding window on the speech signal as shown in Figure 5.1. The purpose of the sliding window is to restrict analysis of the data within that window. The selected segment of the window is referred to as the speech frame. Window length affects time resolution and the frequency resolution of the STFT. Narrow window results in a fine time resolution, because narrow windows have short-time duration (but a wide bandwidth). Wide window results in fine frequency resolution because wide windows have long duration (but a short bandwidth). If spectral contents changes quickly which requires fine time resolution, i.e., window length should be small. In speech signal processing window length of 10-25 ms is common.

For a given speech signal, $s(n)$, we define short-time signal, Q_n, of frame, n, as

$$Q_n = \sum_{m=-\infty}^{\infty} s(m)w(n-m) \tag{5.3}$$

The product of $x(m)$ and $w(n-m)$ is zero everywhere except in a small region, where, Q_n corresponds to convolution of $s(n)$ with $w(n)$, where, $w(n)$ represents a low-pass filter.

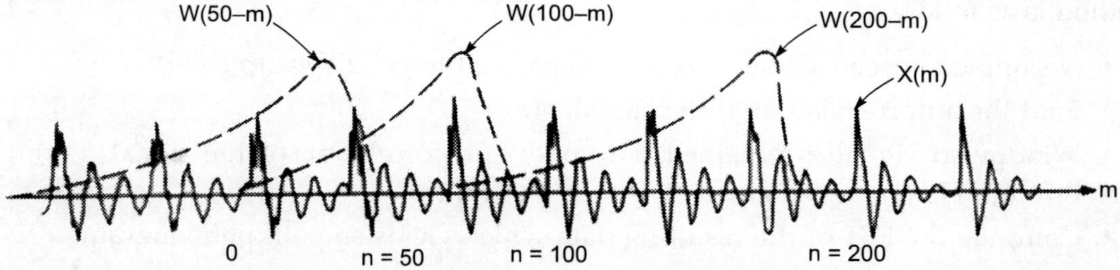

Figure 5.1 Speech signal with superimposed windows

The definition of the short-time Fourier transform (STFT) or sliding windowed Fourier transform or time dependent Fourier transform is given by

$$S_n(e^{j\omega}) = \sum_{m=-\infty}^{\infty} s(m)w(n-m)e^{-j\omega m} \tag{5.4}$$

where, n and ω are variables and $w(n - m)$ is a real window which determines the portion of $s(n)$ that is used in the computation of $S_n(e^{j\omega})$. The time dependent Fourier transform is a function of two variables: the time index, n, which is discrete and frequency variable (ω) which is continuous. $S_n(e^{j\omega})$ is the convolution of the sequence $s(n)e^{-j\omega n}$ and sequence $w(n)$. Hence, $s(n)e^{-j\omega n}$ is input, $w(n)$ plays the role of the impulse response of a linear shift invariant system and $S_n(e^{j\omega})$ is output. Figure 5.2(a) shows the linear filtering operation of short-time spectrum analysis for complex operations. Expressing complex $S_n(e^{j\omega})$ as

$$S_n(e^{j\omega}) = a_n(\omega) - jb_n(\omega) \tag{5.5}$$

Figure 5.2(b) shows the linear filtering operation of short-time spectrum analysis for real operations only.

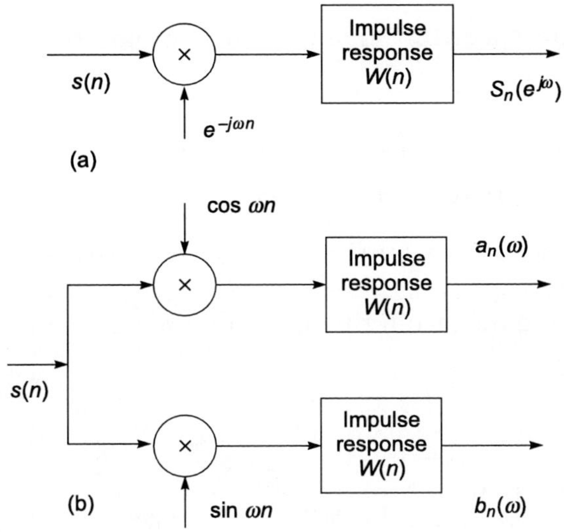

Figure 5.2 Linear filtering of short-time spectrum analysis: (a) for complex operations and, (b) for real operations

Another interpretation STFT, $S_n(e^{j\omega})$, is given by

Let $m = n - m$ in Eq. (5.4), Then Eq. (5.4) is given by

$$S_n(e^{j\omega}) = \sum_{m=-\infty}^{\infty} s(n-m)w(m)e^{-j\omega(n-m)} \tag{5.6}$$

$$S_n(e^{j\omega}) = \sum_{m=-\infty}^{\infty} w(m)s(n-m)e^{j\omega m}e^{-j\omega n} \tag{5.7}$$

$$S_n(e^{j\omega}) = e^{-j\omega n} \sum_{m=-\infty}^{\infty} w(m)s(n-m)e^{j\omega m} \tag{5.8}$$

In the above equation, let

$$\hat{S}_n(e^{j\omega}) = \sum_{m=-\infty}^{\infty} w(m)s(n-m)e^{j\omega m} \tag{5.9}$$

Hence, $\hat{S}_n(e^{j\omega})$ is obtained by taking the convolution of $w(n)e^{j\omega n}$ and $s(n)$ or since it is commutative, by taking convolution of $x(n)e^{j\omega n}$ and $w(n)$

Then

$$S_n(e^{j\omega}) = e^{-j\omega n} S_n(e^{j\omega}) \tag{5.10}$$

Hence, $s(n)$ is input, $w(n)e^{-j\omega n}$ plays the role of the impulse response of a linear shift invariant (LTI) system and $\hat{S}_n(e^{j\omega})$ is output of LTI system. After multiplying $\hat{S}_n(e^{j\omega})$ with $e^{-j\omega n}$, we obtain output $S_n\left(e^{j\omega}\right)$. Figure 5.3(a) shows the linear filtering operation of short-time spectrum analysis for complex operations for Eq. 5.10. Expressing complex $S_n\left(e^{j\omega}\right)$ as

$$S_n(e^{j\omega}) = e^{-j\omega n}\hat{S}_n(e^{j\omega})$$

$$S_n(e^{j\omega}) = (\cos\omega n - j\sin\omega n)(\hat{a}_n(\omega) - j\hat{b}_n(\omega))$$

$$S_n(e^{j\omega}) = [\hat{a}_n(\omega)\cos\omega n + \hat{b}_n(\omega)\sin\omega n] + j[\hat{b}_n(\omega)\cos\omega n - \hat{a}_n(\omega)\sin\omega n] \quad (5.11)$$

Figure 5.3(b) shows the linear filtering operation of short-time spectrum analysis for real operations only.

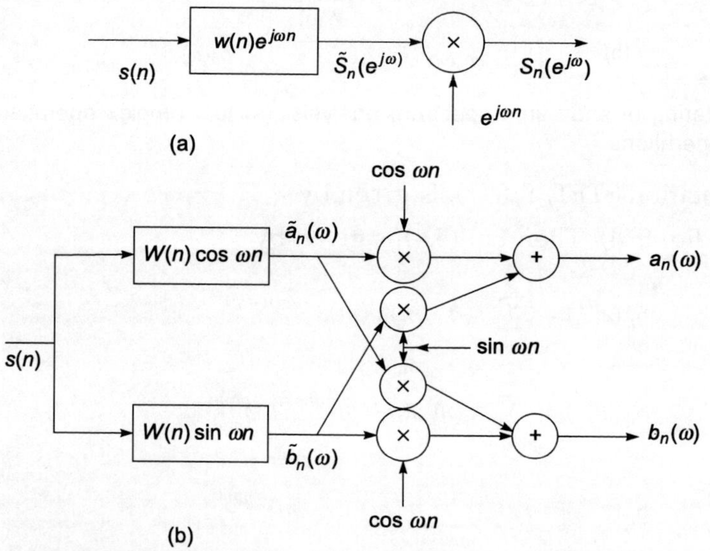

Figure 5.3 Second type interpretation of short-time spectral analysis (a) complex operations (b) real operations

In the first interpretation n is fixed and $S_n(e^{j\omega})$ is normal Fourier transform of the sequence $s(m)w(n-m)$. In the second interpretation $S_n(e^{j\omega})$ is a function of time index n with w is fixed. In the first interpretation LTI system is a low-pass filter and in the second interpretation LTI system is a band-pass filter.

5.5 MAGNITUDE OF SHORT-TIME FOURIER TRANSFORMS

In the first interpretation of short-time Fourier transform the magnitude is given by

$$S_n\left(e^{j\omega}\right) = a_n(\omega) - jb_n(\omega) \quad (5.12)$$

Hence, magnitude is given by

$$\left|S_n(e^{j\omega})\right| = \left[a_n^2(\omega) + b_n^2(\omega)\right]^{\frac{1}{2}} \tag{5.13}$$

In the second interpretation of short-time Fourier transform the magnitude is given by

$$S_n(e^{j\omega}) = e^{-j\omega n} S_n(e^{j\omega}) \tag{5.14}$$

$$S_n(e^{j\omega}) = (\cos\omega n - j\sin\omega n)(a_n(\omega) - jb_n(\omega)) \tag{5.15}$$

Hence, magnitude is given by

$$\left|S_n(e^{j\omega})\right| = \left[a_n^2(\omega) + b_n^2(\omega)\right]^{\frac{1}{2}} \tag{5.16}$$

Figure 5.4 (a) and (b) shows the implementation for obtaining the magnitude of short-time Fourier transform for both interpretations.

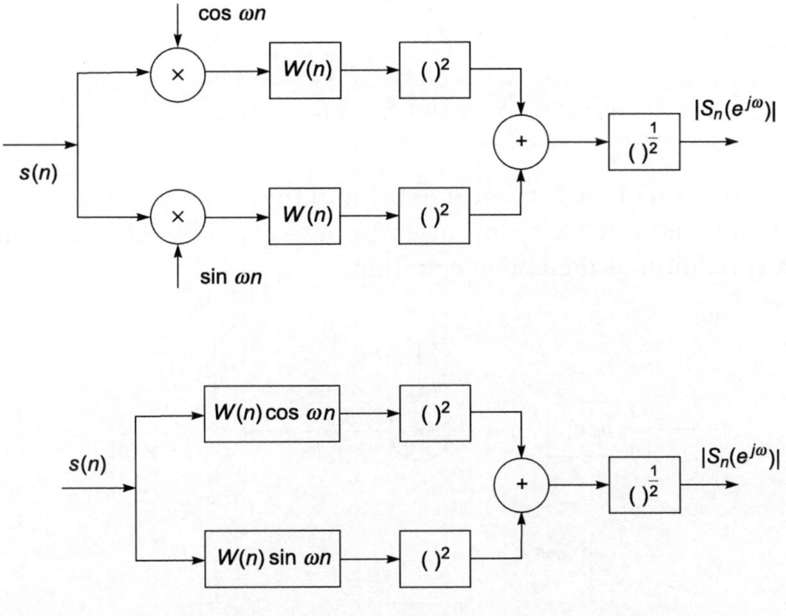

Figure 5.4 Implementation for obtaining the magnitude of short-time Fourier transform (a) using low-pass filter (b) using band-pass filter

5.6 FILTER BANK SUMMATION METHOD OF SHORT-TIME ANALYSIS AND SYNTHESIS

Short-time spectrum analysis and synthesis are implemented by filter bank. For any frequency, ω_k, short-time Fourier transforms is given by

$$S_n(e^{j\omega_k}) = \sum_{m=-\infty}^{\infty} s(m)w_k(n-m)e^{-j\omega_k m} \tag{5.17}$$

$$S_n(e^{j\omega_k}) = e^{-j\omega_k n} \sum_{m=-\infty}^{\infty} w_k(m)s(n-m)e^{j\omega_k m} \tag{5.18}$$

where, $w_k(n)$ is the window used at frequency ω_k.

Let $h_k(n) = w_k(n)e^{j\omega_k n}$

Equation (5.18) is expressed as

$$S_n(e^{j\omega_k}) = e^{-j\omega_k n} \sum_{m=-\infty}^{\infty} h_k(m)s(n-m) \tag{5.19}$$

where, $w_k(n)$ has the properties of low-pass filter and $h_k(n)$ has the properties of band-pass filter.

Let us define

$$y_k(n) = \sum_{m=-\infty}^{\infty} h_k(m)s(n-m) \tag{5.20}$$

$y_k(n)$ is the convolution of the sequence $h_k(n)$ and the sequence $s(n)$. Hence, $y_k(n)$ is output of a band-pass filter with impulse response given by Eq. (5.20). Figure (5.5) shows the interpretation of the above equation.

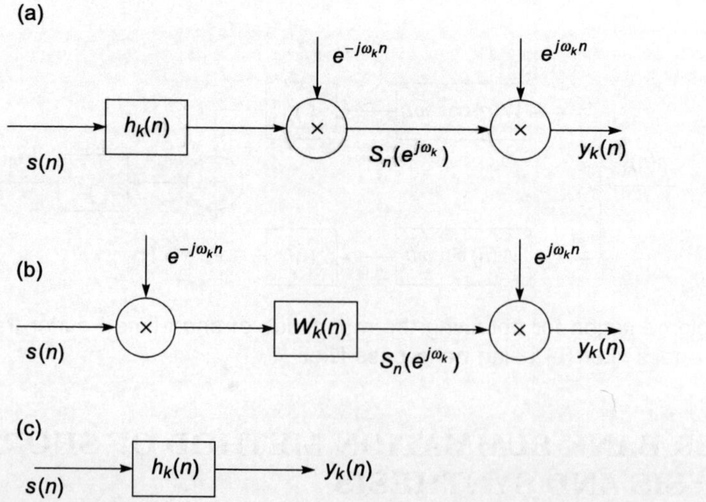

Figure 5.5 Different methods for implementing analysis and synthesis

Substituting $y_k(n)$ in Eq. (5.19), Eq. (5.19) becomes

$$S_n\left(e^{j\omega_k}\right) = e^{-j\omega_k n} y_k(n) \tag{5.21}$$

or

$$y_k(n) = e^{j\omega_k n} S_n\left(e^{j\omega_k}\right) \tag{5.22}$$

Substituting $S_n\left(e^{j\omega_k}\right)$ from (5.19) in the above equation, Eq. (5.22) becomes

$$y_k(n) = e^{-j\omega_k n} \sum_{m=-\infty}^{\infty} h_k(m)s(n-m)e^{j\omega_k n} \tag{5.23}$$

In analysis operation, we extract the features of speech signal and in synthesis operation we reproduce the speech signal using feature parameters. Using Eqs. (5.19) and (5.22) analysis and synthesis operations are shown by Figure 5.5 (a). Similarly, using Eqs. (5.17) and (5.22) analysis and synthesis operations are shown by Figure 5.5 (b).

If we consider N band-pass filter, where, ω_k is the centre frequency, then

$$\omega_k = \frac{2\pi k}{N} \quad \text{for} \quad k = 0, 1 \ldots\ldots\ldots N-1$$

Then linear system relating to $y_k(n)$ and $y(n)$ to $s(n)$ is shown in Figure (5.5). If the operations of analysis and synthesis are implied by Eq. (5.23), then analysis and synthesis operation are depicted in Figure 5.6.

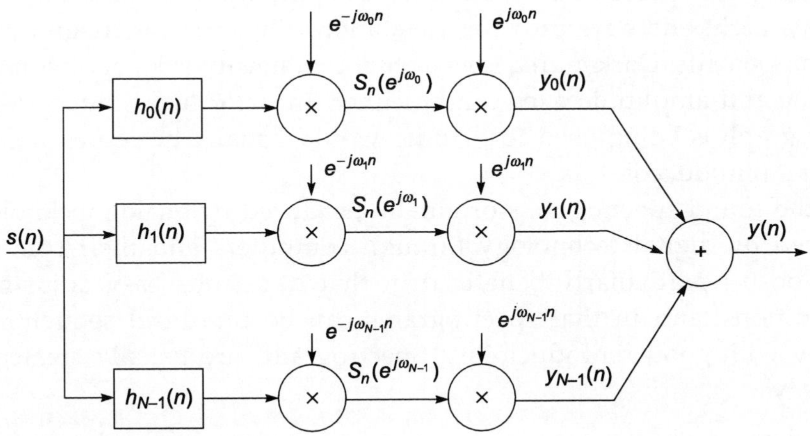

Figure 5.6 Analysis and synthesis operations for short-time Fourier transform analysis

Decimation reduces the number of samples and interpolator increases the number of samples. By including the decimator at the output of analysis and an interpolator at the input to the synthesis, complete filter bank implementation of analysis and synthesis operations for a single channel are depicted in Figure 5.7.

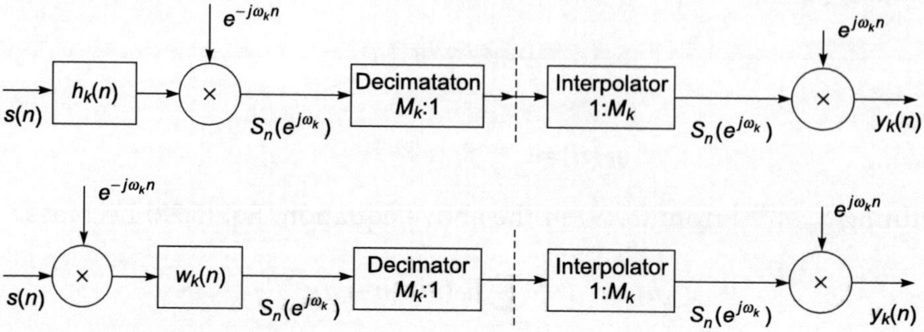

Figure 5.7 Analysis and synthesis operation for short-time spectral analysis with decimator and interpolator

5.7 SPECTROGRAM

Basic tool of spectral analysis is spectrogram. It converts a two-dimensional speech waveform (amplitude/time) into three-dimensional patterns (amplitude/frequency/time). The instrument that generates a spectrogram is called a **spectrograph**. Spectrograms are used to identify phonetic sounds, it is also used in music, sonar/radar, speech processing, seismology, etc. The most common format is a graph with two geometric dimensions: the horizontal axis represents time, the vertical axis is frequency; a third dimension indicating the amplitude of a particular frequency at a particular time is represented by the intensity. Amplitude is noted by darkness of the display. Light and dark area representations illustrate the magnitude of the frequency components. Darker area reveals higher intensity frequency components. The frequency and amplitude axes can be either linear or logarithmic, depending on what the graph is being used for. Audio would usually be represented with a logarithmic amplitude axis.

Many of the sound spectrograms originally produced by analog technology and now produced by digital technology through computers and their printers. The digital methods is particularly beneficial in that it permits easy adjustments of various conditions, and in that spectrograms can be produced sequentially and automatically with good reproducibility. Spectrograms are usually created in one of the two ways:

- Approximated as filter bank that results from a series of band-pass filters
- Calculated from the time signal using short-time Fourier transform (STFT)

In the 1940s, the sound spectrograph was created as shown in Figure 5.8. The band-pass filter method usually uses analog processing to divide the input signal into frequency bands; magnitude of each filter output controls a transducer that records the spectrogram as an image on paper.

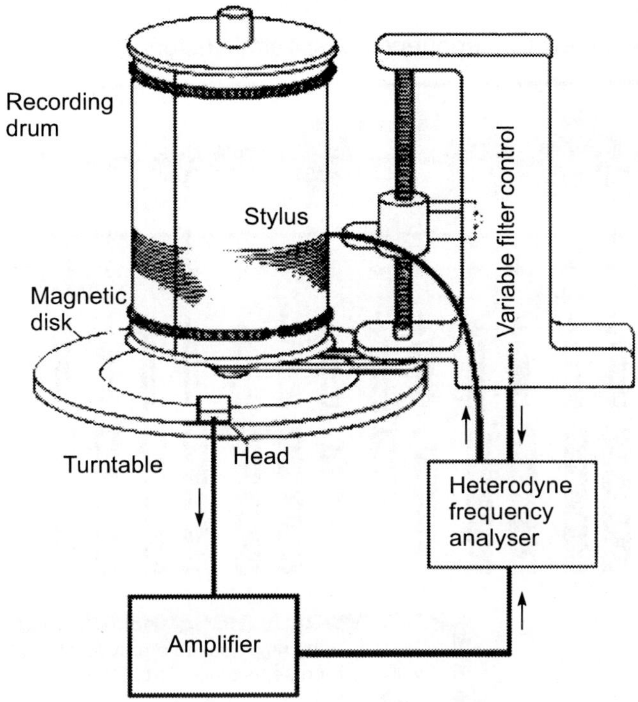

Figure 5.8 Sound spectrograph

Nowadays digital methods are used, in digital process spectrograms are created using STFT. The spectrogram of a signal, $s(t)$, can be estimated by computing the squared magnitude of the STFT of the signal $s(t)$ and it is given by

$$\text{Spectrogram}, (t, \omega) = |\text{STFT}\ (t, \omega)|^2 \tag{5.24}$$

In STFT digitally sampled data is broken up into blocks, and then Fourier transformed to calculate the squared magnitude of the frequency spectrum for each block. Spectrum represents measurement of magnitude versus frequency for a specific moment in time. Then it is rotated by 90 degrees, so that vertical axis represents frequency. Spectrogram generation is depicted in Figure 5.9. The spectrums are then "laid side by side" to form the image or three-dimensional surface. Horizontal axis represents time; the vertical axis is frequency, a third dimension indicating the amplitude of a particular frequency at a particular time. Amplitude is noted by darkness of the display. Light and dark area representation illustrates the magnitude of the frequency components.

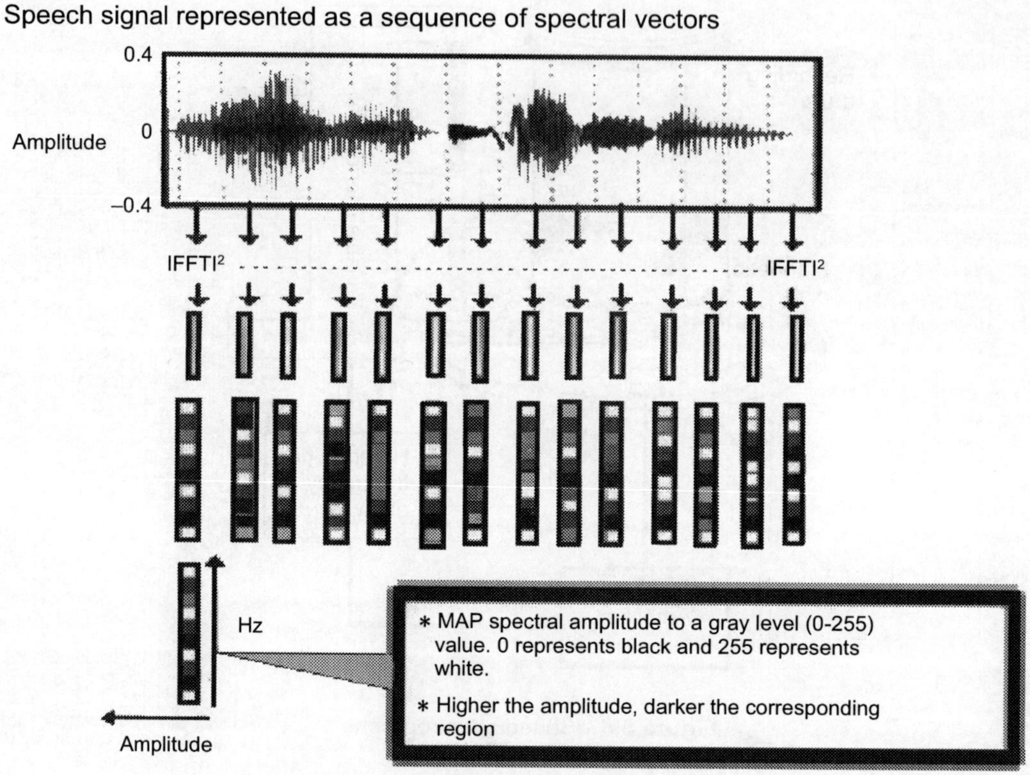

Figure 5.9 Spectrogram generation using STFT

From the above formula it is clear that spectrograms contain no information about the phase of the signal that it represents. Because of this reason it is not possible to reverse the process and to generate the original signal from a spectrogram. Another drawback of STFT is that it has a fixed resolution. Width of the window determines whether there is a good frequency resolution or good time resolution. A wide window gives better frequency resolution but poor time resolution. A narrower window gives good time resolution but poor frequency resolution. These are called narrowband and wideband transforms respectively.

Hence, depending on the window, there are two kinds of spectrograms, they are:

- Narrowband spectrograms
- Wideband spectrograms

Narrowband spectrogram: For a narrow band spectrogram, we use a "long window" with duration of typically at least two pitch periods as shown in Figure 5.10. Generally, it uses a window with 45 Hz bandwidth and thus duration of about 22 ms. These spectrograms have good frequency resolution and poor time resolution.

Wideband spectrogram: For a wideband spectrogram, we use a "short window" with a duration less than a single pitch period as shown in Figure 5.11. Generally, it uses a window with 300 Hz bandwidth and thus duration of about 3.3 ms. These spectrograms have good time resolution and poor frequency resolution.

The wideband and the narrowband spectrograms display a great deal of information about the properties of speech utterance. The temporal features, such as pitch period, can be seen as vertical striations in the wideband spectrogram. The spectral features such as pitch harmonics can be seen as horizontal striations in the narrowband spectrogram. Hence, wideband spectrograms are good for pitch period or fundamental frequency estimation and narrowband representations are better for viewing vocal tract parameters.

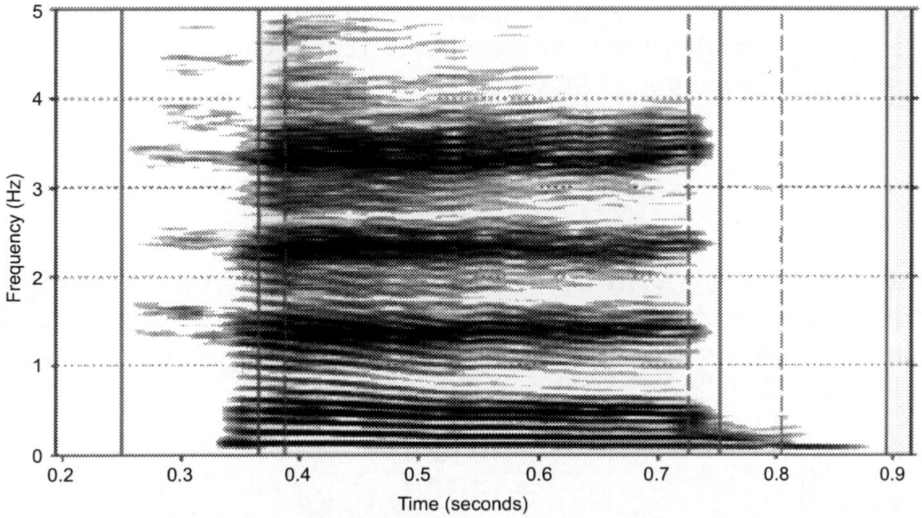

Figure 5.10 Narrowband spectrogram for the word: " heard "

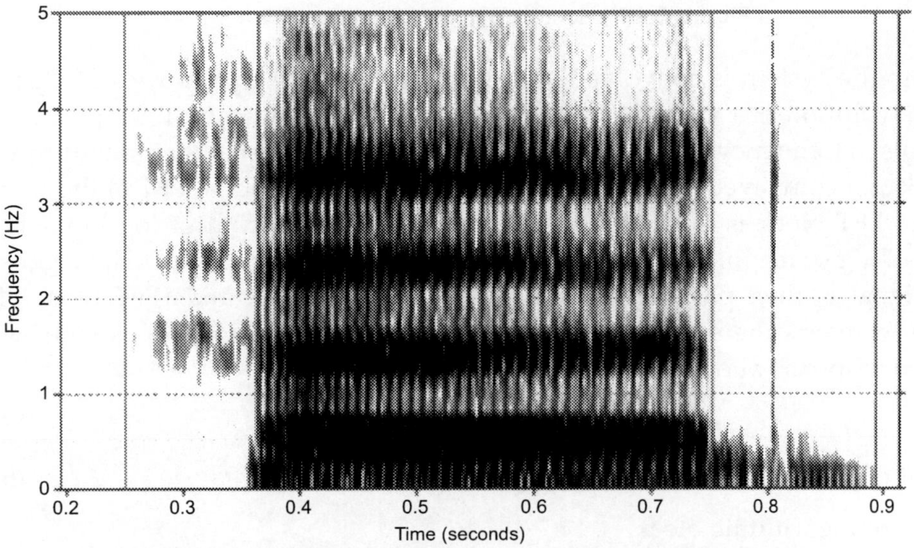

Figure 5.11 Wideband spectrogram for the word: " heard "

5.8 HOMOMORPHIC SPEECH PROCESSING

The term homomorphic processing is applied to a system that obeys the generalized principle of superposition. If $e_1(n)$ and $e_2(n)$ are inputs to a system and $s_1(n)$ and $s_2(n)$ are respective outputs and a is a scaling factor, then for a conventional linear system, principle of superposition is depicted in Figure 5.12 and can be stated as follows:

$$T[e_1(n)+e_2(n)] = T[e_1(n)]+T[e_2(n)] \tag{5.25}$$

$$s(n) = s_1(n)+s_2(n)$$

As discussed in section 2.5, speech signals vary slowly with time, hence short segment of speech signal is used throughout this book. Output of a linear time invariant system is given by

$$s(n) = e(n) * h(n) = \sum_{k=-\infty}^{\infty} h(n-k)e(k) \tag{5.26}$$

Symbol * indicates the operation of convolution.

A system that converts convolved components into an additive is called a homomorphic processing system. Likewise the inverse system, converts additive components into convolved components. Block schematic of canonical system is shown in Figure 5.12.

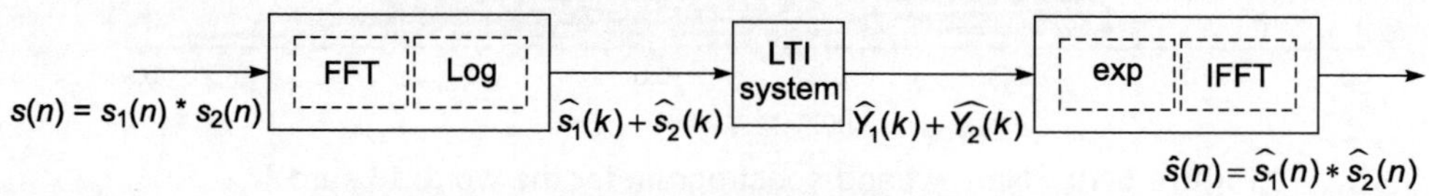

Figure 5.12 Homomorphic processing system

Input to the system is convolved speech signal. From the properties of DFT, we know, convolution of two signals in time domain is replaced by multiplication of two signals in frequency domain. Hence, convolved input signal is passed through a FFT block and convolved speech signal is converted to product of FFT of the signals. Output of FFT block is given to logarithmic block. The output of log block is the addition of log of magnitude part of FFT. Then signals are processed. The processed speech signal is then passed through exponentiation block and IFFT block. This operation converts the signals to time domain. Output of log block is sum of the vocal cord response and vocal tract response.

$$s(n) = s_1(n) * s_2(n) \tag{5.27}$$

Output of FFT block $\quad S(k) = S_1(k)S_2(k)$ (5.28)

Output of logarithmic block

$$\hat{S}[k] = \log[S(k)] = \log[S_1(k)] + \log[S_2(k)] = \hat{S}_1[k] + \hat{S}_2[k] \tag{5.29}$$

After processing

$$\hat{Y}[k] = \hat{Y}_1[k] + \hat{Y}_2[k] \tag{5.30}$$

Output of exponential block

$$Y(k) = e^{[\log[Y_1(k)] + \log[Y_2(k)]]} = Y_1(k) Y_2(k) \tag{5.31}$$

Output of IFFT block $\hat{s}(n) = \hat{s}_1(n) * \hat{s}_2(n)$ (5.32)

In processing, if we pass the signal through a low-pass filter we can eliminate vocal cord response and we get vocal tract response. Instead, if we pass the signal through high-pass filter we can eliminate vocal tract response and we get vocal cord response.

5.9 CEPSTRUM AND ITS APPLICATIONS

The cepstrum is a homomorphic transformation that converts a convolution

$$s(n) = e(n) * h(n) \tag{5.33}$$

into a sum in the cepstrum domain

$$\hat{s}[n] = \hat{e}[n] + \hat{h}[n] \tag{5.34}$$

The cepstrum is obtained by taking the inverse Fourier transform of the log spectrum. There are two types of cepstrums. They are:

- Complex cepstrum
- Real cepstrum

Let $s(n)$ denote the original signal. The complex cepstrum is defined as

$$\hat{s}[n] = \text{IDFT}\{\log(\text{DFT}\{s[n]\})\} \tag{5.35}$$

The real cepstrum is defined as

$$c[n] = \text{IDFT}\left\{\log\left|DFT\left(\{s[n]\}\right)\right|\right\} \tag{5.36}$$

where, DFT denotes the discrete Fourier transform.

The cepstrum analysis provides a methodology for separating excitation and vocal tract response.

where, $e(n)$ is excitation signal and $h(n)$ is vocal tract impulse response. Discrete Fourier transform representation is given by

$$S(k) = E(k) H(k) \tag{5.37}$$

Taking log on both the sides for absolute value, we obtain

$$\log|S(k)| = \log|E(k)| + \log|H(k)| \tag{5.38}$$

Log operation in the above equation separates the excitation and vocal tract spectrum. The cepstrum, which is the inverse DFT of $\log|S(k)|$ is given by

$$c[n] = \text{IDFT}\{\log|S(k)|\} = \text{IDFT}\{\log|S(k)|\} + \text{IDFT}\{\log|H(k)|\} \tag{5.39}$$

The term cepstrum is essentially a coined word which includes the meaning of the inverse transform of the spectrum. The unit of cepstrum is called quefrency and the quefrency becomes the time domain parameter. The first function on the right of Eq. (5.39) indicates the formation of a peak in the high-quefrency region and the second function represents a concentration in the low-quefrency region (0-4 ms). The process of separating the cepstral elements into two factors high-quefrency region and low-quefrency region is called liftering, which is derived from filtering and it is shown in Figure 5.13.

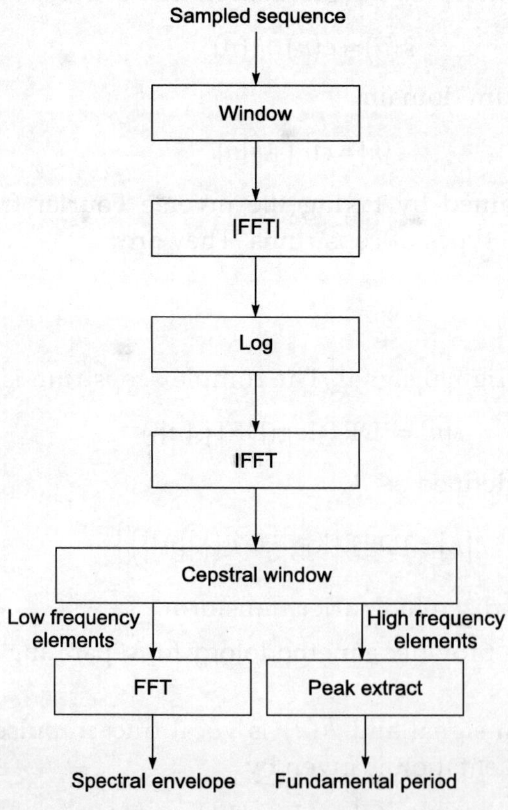

Figure 5.13 Practical implementation of cepstrum

The fundamental period or pitch period of the source, $e(n)$, can then be extracted from the peak at high-quefrency region. For pitch detection of voiced speech, one measures the interval between the origin and the first peak in the cepstrum. No pulses appear in the cepstrum for unvoiced speech. Thus, we can use cepstrum to detect voiced/unvoiced segments and estimate the pitch period. On the other hand, the discrete Fourier transform of the low-quefrency elements produces the logarithmic spectral envelope. These smoothed log spectra, display the resonant structure of particular input speech segment; i.e., the peaks in the spectrum corresponds essentially to the formant frequencies and formants can be estimated by locating the peaks. Spectrum and cepstrum for speech waveform is shown in Figure 5.14.

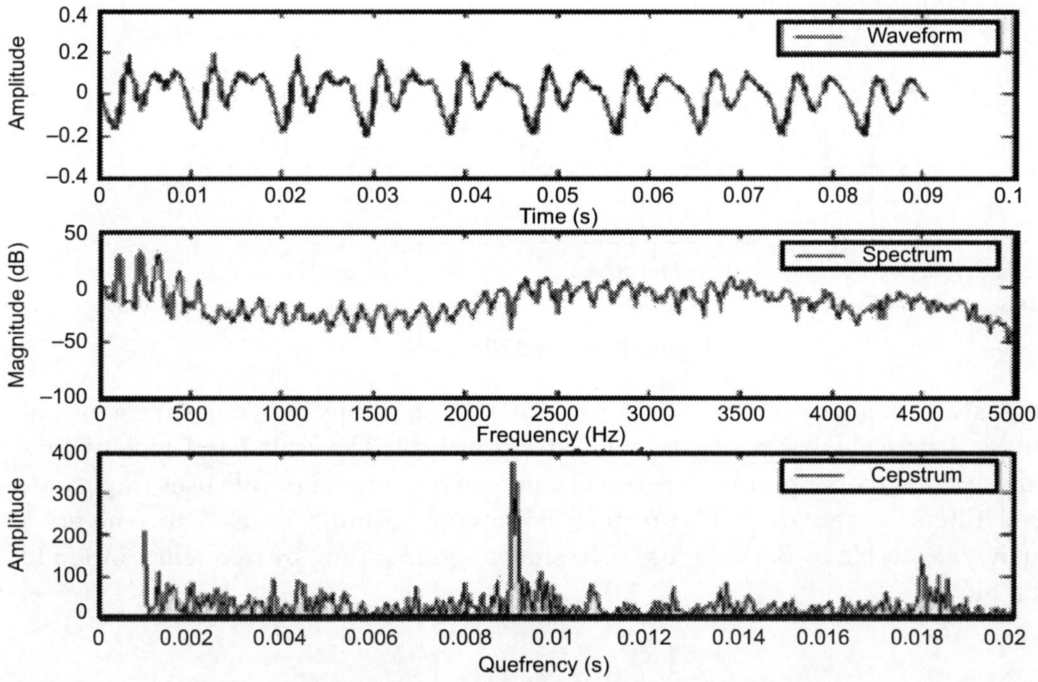

Figure 5.14 Spectrum and cepstrum for speech waveform

5.10 AUDITORY SYSTEM AS A FILTER BANK

American physicist Harvey Fletcher experimented extensively on human hearing. Experiments indicate the existence of an auditory filter in the vicinity of the tone, which effectively blocks the noise and detects the tone. The result showed that critical band filters must be designed, which represents different perceptual effects at different frequency bands. Commonly used frequency scales for design of critical band filters are **mel** and **Bark** scale. Mel scale uses triangular filters as shown in Fig. 5.15 and is based on pitch perception. Mel scale is the nearest match to the way that human ear perceives the sound. Mel scale is linear below 1000 Hz and nonlinear (logarithmic) after 1000 Hz. The centres of the triangle filter banks are spaced according to the mel scale:

$$F_{\text{MEL}} = 2595 \log_{10}\left(1 + \frac{F_{\text{LIN}}}{700}\right) \tag{5.40}$$

Figure 5.15 shows the mapping of frequency in Hz to mel scale

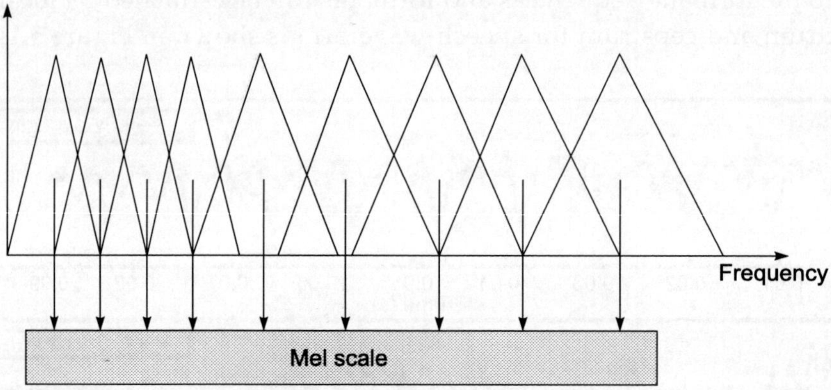

Figure 5.15 Mel filter bank

Bark scale is a phychoacoustical scale based on subjective measurements of loudness. This scale was proposed by Zwicker in 1961. The scale has 1 to 24 values and these values correspond to 24 critical bands of hearing. The scale uses trapezoid-shaped filters as shown in Figure 5.15 Following formula is used to convert a frequency (F) in Hz to Bark. Figure 5.16 shows the mapping of frequency in Hz to Bark scale

$$F_{\text{Bark}} = 6 \ln\left\{\frac{F}{600} + \left[\left\{\frac{F}{600}\right\}^2 + 1\right]^{0.5}\right\} \tag{5.41}$$

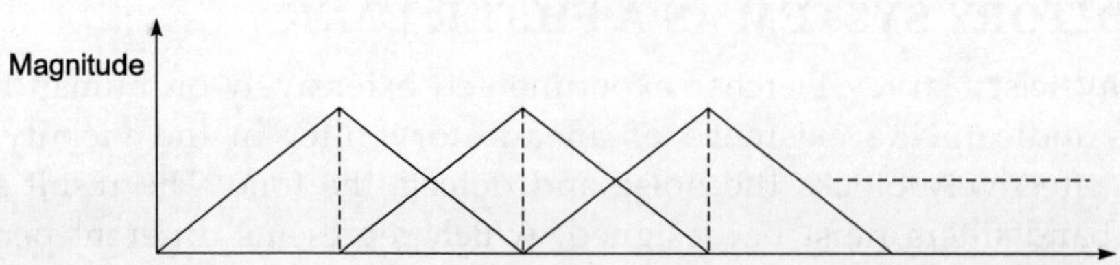

Figure 5.16 Bark filter bank

5.11 MEL-FREQUENCY CEPSTRUM (MFC)

Mel frequency cepstral coefficients (MFCC) are the most widely used features in speech recognition since mel scale is the nearest match to the way that human ear perceives the sound. Mel scale has linear frequency spacing below 1 kHz and logarithmic spacing above 1 kHz as shown in Figure 5.16. Hence, the difference between the cepstrum and the mel-frequency cepstrum is that in the MFC, the frequency bands are equally spaced on the mel scale, which approximates the human auditory system's response more closely than the linearly spaced frequency bands used in the normal cepstrum. This frequency warping can allow for better representation of sound, for example, in audio compression.

MFCC computation consists of the following steps:

- The features of the speech signal are framed into overlapping frames of 128 samples in length.
- FFT is calculated for each frame, and computes its squared magnitude. This gives the power spectrum.
- Map the powers of the spectrum obtained onto the mel scale using triangular overlapping windows, i.e., integrate the power spectrum within the overlapping critical band filter response. The integration is done using mel filters. It is made to pass through mel filter bank.

$$F_{\text{MEL}} = 2595 \log_{10}\left(1 + \frac{F_{\text{LIN}}}{700}\right) \tag{5.42}$$

- Take the logs of the powers at each of the mel frequencies, it compresses the spectral amplitude. The accumulated output from the kth filter bank is denoted as S_k. As human hearing exhibits logarithmic compression in the dynamic range of the log of the filter bank output, log (S_k), is usually taken to reflect this.
- Take the IFFT, it gives the MFCCs
- Perform smoothing and this is achieved by truncating the cepstrum. Typically, lower $L = 12$ MFCC coefficients are extracted out of 20 or more coefficients per frame.

MFCC features are derived as follows and the block diagram is shown in Figure 5.17.

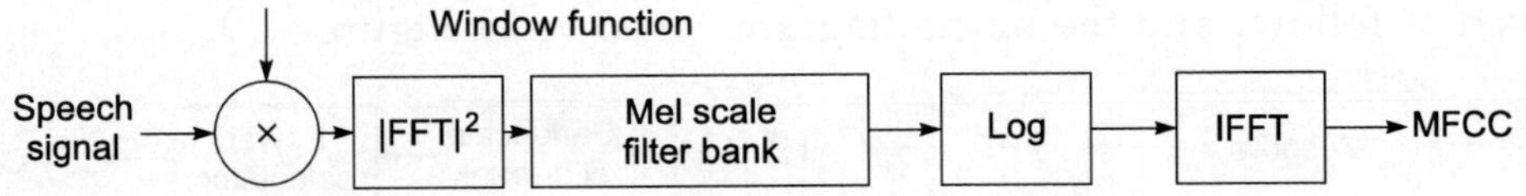

Figure 5.17 Block diagram for calculating MFCC

5.12 PERCEPTUAL LINEAR PREDICTION (PLP)

When we hear a sound, it is perceived by human ear. In this method perceptual property of the human ear is captured. PLP and MFCC algorithms are almost similar. In PLP, power spectrum of the speech signal is converted to Bark scale. PLP features are derived as follows and the block diagram is shown in Figure 5.18.

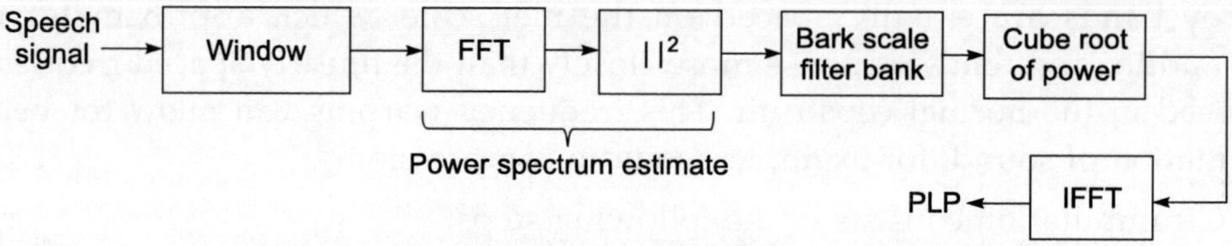

Figure 5.18 Block diagram for calculating PLP

PLP computation consists of the following steps:

- The features of the speech signal are framed into overlapping frames of 128 samples in length.
- FFT is calculated for each frame, compute its squared magnitude. This gives the power spectrum.
- Convert each frequency bin point to the corresponding Bark scale value.
- Calculate the cube root of the power spectrum for each Bark scale value. Multiply each with the weight corresponding to the height of the triangle at that position. Then integrate the cube root of the power spectrum bins over 24 filters, i.e., accumulate the cube root power point values.
- Take the IFFT, this gives the PLPs.
- Perform smoothing and this is achieved by autoregressive model. Typically, lower L = 12 PLP coefficients are extracted out of 20 or more coefficients per frame.

5.13 LOG FREQUENCY POWER COEFFICIENTS (LFPCs)

Log frequency power coefficients (LFPCs) describe the spectral shape of the speech signal. In this method, filter bank is designed to divide the speech signal into 12 frequency bands that match the critical perceptual bands of the human ear as per the Bark scale and then 12 LPFGCs are obtained for each frequency band. LFPCs features are derived as follows and the block diagram is shown in Figure 5.19.

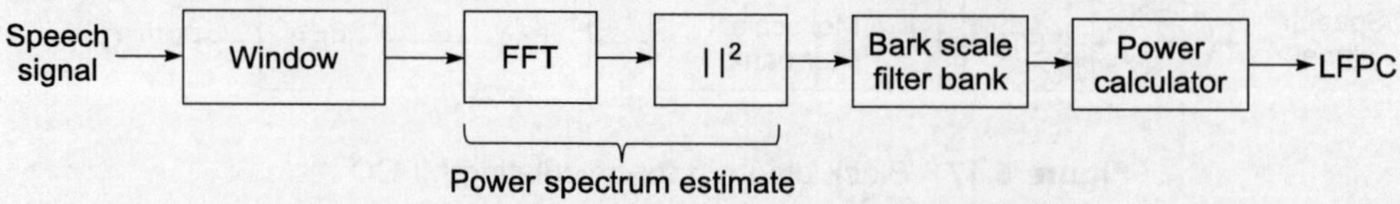

Figure 5.19 Block diagram for calculation of LFPC

LFPC computation consists of the following steps:

- The features of the speech signal are framed into overlapping frames of 128 samples in length.
- FFT is calculated for each frame, compute its squared magnitude. This gives the power spectrum.
- Convert each frequency bin point to the corresponding Bark scale value and design 14 filters of equal width on the Bark scale by using triangular filters.
- Then integrate the power spectrum falling in the width of the window for all the 14 filters, i.e., accumulated power of all the 14 windows will give 14 LFPCs.

5.14 SUMMARY

This chapter presents the frequency domain parameter estimation, spectrograms and homomorphic methods of speech processing. It involves feature extraction methods using mel and Bark scales. Detailed procedures are presented with the block diagrams.

EXERCISES

Review Questions

1. Explain pitch period estimation using FFT and the spectral autocorrelation method.
2. Define formant frequency, explain formant frequency estimation using log spectrum
3. Define formant frequency, explain formant frequency estimation using power spectral density estimate.
4. Explain different methods for formants estimation using estimated power spectral density.
5. Explain different types of interpretation of short-time spectral analysis.
6. With necessary block schematic, explain different types of interpretation of short-time spectral analysis for complex operations.
7. With necessary block schematic, explain the different types of interpretation of short-time spectral analysis for real operations.
8. With necessary block schematic, explain implementations, for obtaining magnitude of short-time Fourier transform using low-pass and band-pass filter.

9. Draw the block schematic and explain analysis and synthesis operations for short-time spectral analysis.
10. With necessary block schematic, explain, analysis and synthesis operation for short-time spectral analysis with decimator and interpolator.
11. What is spectrogram? How to evaluate the spectrogram of a speech segment?
12. What is spectrogram? Explain different types of spectrogram and its applications.
13. Explain difference between narrowband and wideband spectrograms and its applications.
14. With a block schematic, explain homomorphic processing system and justify the use of an LTI filter for a convolved signal.
15. With necessary block schematic, explain the method to separate the impulse response of vocal tract from speech signal.
16. What is cepstrum? Explain the measurement fundamental and formant frequency using a cepstrum.
17. Distinguish between mel and Bark scale and their effects in speech processing.
18. With necessary block schematic, explain different steps for MFCCs computation.
19. What are MFCCs? Explain the procedure for calculating MFCC.
20. What are PLPs? Explain the procedure for calculating PLP.
21. With necessary block schematic, explain the procedure to compute log frequency power coefficients.
22. What are LFPC and PLP? Compare the procedure for calculating LFPCs with PLP.

6

CHAPTER

LPC Analysis

This chapter provides insights on

- Pros and cons of linear predictive analysis
- Calculation of LPC coefficients using covariance, autocorrelation and lattice method
- Analysis of LPC using Burg's algorithm
- Line spectrum pair (LSP) analysis
- Pitch period and formant estimation using LPC coefficients
- LPC vocoder

6.1 INTRODUCTION

One of the most powerful speech analysis methods is linear predictive analysis method or auto-regressive (AR) modelling. The basic idea behind the linear predictive analysis is that speech samples can be approximated as a linear combination of past speech samples. This method is widely used because it is fast and simple, yet an effective way of estimating the main parameters of speech signals. This chapter covers detailed discussion on computation of linear predictor coefficients (LPCs) using covariance method, autocorrelation method and lattice (or Porcor analysis) method. We also discuss pitch period and formant estimation using LPCs.

6.2 ADVANTAGES AND DISADVANTAGES OF LINEAR PREDICTIVE ANALYSIS

Advantages

- LPC provides good model of speech signal.
- It is a production based method.
- Speech sample at time (n) can be represented as a linear combination of (p) previous samples.

- LPC represents the spectral envelope by low dimension feature vectors.
- Provides linear characteristics.
- LPC leads to a reasonable source-vocal tract separation.
- LPC is analytically tractable model.
- The method of LPC is mathematically precise and straightforward to implement in either software or in hardware.

Disadvantages

- The LP models the input signal with constant weighing for the whole frequency range. However, human perception does not have constant frequency perception in the whole frequency range.
- A serious problem with the LPCs is that they are highly correlated but are desirable to obtain less correlated features for acoustic modeling.
- An inherent drawback of conventional LP is its inability to include speech specific a priori information in the modelling process.

6.3 LINEAR PREDICTIVE ANALYSIS

In speech synthesis model it was shown that speech can be modelled as a output of linear-time varying system excited by either quasi-periodic pulses (during voiced speech) or random noise (during unvoiced speech) (Figure 6.1).

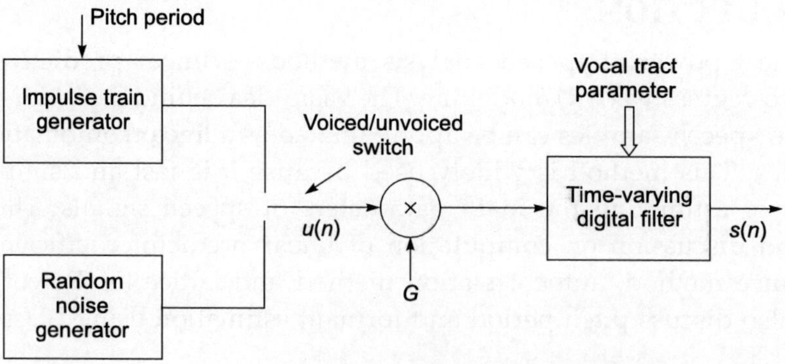

Figure 6.1 Simplified model for speech production

As discussed in chapter 2, if α_k are vocal tract parameters, then we could model the filter $V(Z)$ in Figure 6.1 as

$$V(Z) = \frac{S(Z)}{U(Z)} = \frac{G}{1 - \sum_{k=1}^{p} \alpha_k Z^{-k}} \tag{6.1}$$

$$S(Z)\left(1-\sum_{k=1}^{p}\alpha_k Z^{-k}\right)=GU(Z) \tag{6.2}$$

$$S(Z)=\sum_{k=1}^{p}\alpha_k Z^{-k}S(Z)+GU(Z) \tag{6.3}$$

Taking inverse Z-transform of Eq. (6.3), we obtain

$$s(n)=\sum_{k=1}^{p}\alpha_k s(n-k)+Gu(n) \tag{6.4}$$

The linear predictive coding gets its name from the fact that it predicts the current sample as a linear combination of its past p samples.

$$\tilde{s}(n)=\sum_{k=1}^{p}a_k s(n-k) \tag{6.5}$$

where, a_k are LPCs.

The prediction error when using this approximation is given by

$$e(n)=s(n)-\tilde{s}(n)=s(n)-\sum_{k=1}^{p}a_k s(n-k) \tag{6.6}$$

To estimate the predictor coefficients from a set of speech samples, we use the short-term analysis techniques. Let us define $s_n[m]$ as a segment of speech selected in the vicinity of sample n

$$s_n[m] = s(n + m) \tag{6.7}$$

We define short-term prediction error for that segment as

$$\varepsilon_n=\sum_m e_n^2(m) \tag{6.8}$$

$$\varepsilon_n=\sum_m (s_n[m]-\tilde{s}_n[m])^2 \tag{6.9}$$

$$\varepsilon_n=\sum_m\left[s_n[m]-\sum_{k=1}^{p}a_k s_n(m-k)\right]^2 \tag{6.10}$$

We can find the values of a_k that minimize ε_n in the above equation by setting

$\frac{\partial\varepsilon_n}{\partial a_i}=0,\ i = 1, 2,..p$, thereby obtaining the equation

$$\sum_m \left[s_n[m] - \sum_{k=1}^{p} a_k s_n(m-k) \right] s_n(m-i) = 0 \tag{6.11}$$

where, a_k are the values that minimize ε_n. The above equation can be written as

$$\sum_m s_n(m-i)s_n[m] = \left(\sum_{k=1}^{p} a_k \sum_m s_n(m-i)s_n(m-k) \right) \text{for } 1 \le i \le p \tag{6.12}$$

For convenience, we can define the correlation coefficients as

$$\phi_n(i, k) = \sum_m s_n(m-i)s_n(m-k) \tag{6.13}$$

Hence, Eq. 6.12 can be written more compactly as

$$\phi_n(i, 0) = \sum_{k=1}^{p} a_k \, \phi_n(i, k) \quad \text{for} \quad i = 1, 2, \ldots, p \tag{6.14}$$

Solution of the set of p linear equation results in the p LPCs that minimize the prediction error.

6.4 SOLUTION TO LPC EQUATIONS

The above LPC equation can be solved with any standard matrix inversion package. Because of the special form of the matrix, some efficient solutions are possible and each solution offers a different insight. The following three algorithms are used to obtain the solution of the LPC equation.

- Covariance method
- Autocorrelation method
- Lattice method

6.4.1 Covariance Method

The covariance method is derived by directly defining the interval over which the summation in Eq. (6.8) takes place

$$\varepsilon_n = \sum_{m=0}^{N-1} e_n^2(m) \tag{6.15}$$

and $\phi_n(i, k)$ in Eq. (6.13) becomes

$$\phi_n(i, k) = \sum_{m=0}^{N-1} s_n(m-i)s_n(m-k) \; 1 \le i \le p \text{ and } 0 \le k \le p \tag{6.16}$$

In this case if we change the index of summation, we can express $\phi_n(i, k)$ as either

$$\phi_n(i,k) = \sum_{m=-i}^{N-i-1} s_n(m)s_n(m+i-k) \quad 1 \le i \le p \text{ and } 0 \le k \le p \tag{6.17}$$

or

$$\phi_n(i,k) = \sum_{m=-k}^{N-k-1} s_n(m)s_n(m+k-i) \quad 1 \le i \le p \text{ and } 0 \le k \le p \tag{6.18}$$

Equation (6.14) becomes

$$\begin{bmatrix} \phi_n(1,1) & \phi_n(1,2) & \phi_n(1,3) & \dots & \phi_n(1,p) \\ \phi_n(2,1) & \phi_n(2,2) & \phi_n(2,3) & \dots & \phi_n(2,p) \\ \phi_n(3,1) & \phi_n(3,2) & \phi_n(3,3) & \dots & \phi_n(3,p) \\ \dots & \dots & \dots & \dots & \dots \\ \phi_n(p,1) & \phi_n(p,2) & \phi_n(p,3) & \dots & \phi_n(p,p) \end{bmatrix} \begin{bmatrix} a_1 \\ a_2 \\ a_3 \\ \dots \\ a_p \end{bmatrix} = \begin{bmatrix} \phi_n(1,0) \\ \phi_n(2,0) \\ \phi_n(3,0) \\ \dots \\ \phi_n(p,0) \end{bmatrix} \tag{6.19}$$

Cholesky decomposition solution for the covariance method

For the covariance method the set of equations which must be solved is of the form

$$\sum_{k=1}^{p} a_k \, \phi_n(i,k) = \phi_n(i,0) \text{ for } i = 1, 2, \dots. p \tag{6.20}$$

This can be expressed as the following in matrix notation

$$\phi a = \psi \tag{6.21}$$

where, the matrix ϕ in the above equation is positive definite symmetric matrix, for which efficient methods are available such as Cholesky decomposition. For this method matrix ϕ is expressed as

$$\phi = \mathrm{VDV}^t \tag{6.22}$$

here, V is the lower triangular matrix (whose main diagonal elements are all 1's) and D is the diagonal matrix. The subscript (t) denotes matrix transpose.

$$\mathrm{VDV}^t a = \psi \tag{6.23}$$

Let $$Y = DV^t a$$

or $$V^t a = D^{-1} Y \tag{6.24}$$

Equation 6.23 can be written as

$$VY = \psi \tag{6.25}$$

Consider an example with $p = 4$ and the matrix elements $\phi_n(i, j) = \phi_{ij}$, the above Eq. 16.22 can be written as

$$\begin{bmatrix} \phi_{11} & \phi_{12} & \phi_{13} & \phi_{14} \\ \phi_{21} & \phi_{22} & \phi_{23} & \phi_{24} \\ \phi_{31} & \phi_{32} & \phi_{33} & \phi_{34} \\ \phi_{41} & \phi_{42} & \phi_{43} & \phi_{44} \end{bmatrix} = \begin{bmatrix} 1 & 0 & 0 & 0 \\ V_{21} & 1 & 0 & 0 \\ V_{31} & V_{32} & 1 & 0 \\ V_{41} & V_{42} & V_{43} & 1 \end{bmatrix} \begin{bmatrix} d_1 & 0 & 0 & 0 \\ 0 & d_2 & 0 & 0 \\ 0 & 0 & d_3 & 0 \\ 0 & 0 & 0 & d_4 \end{bmatrix} \begin{bmatrix} 1 & V_{21} & V_{31} & V_{41} \\ 0 & 1 & V_{32} & V_{42} \\ 0 & 0 & 1 & V_{43} \\ 0 & 0 & 0 & 1 \end{bmatrix} \quad (6.26)$$

From Eq. 6.25 $VY = \psi$

$$\therefore \quad \begin{bmatrix} 1 & 0 & 0 & 0 \\ V_{21} & 1 & 0 & 0 \\ V_{31} & V_{32} & 1 & 0 \\ V_{41} & V_{42} & V_{43} & 1 \end{bmatrix} \begin{bmatrix} Y_1 \\ Y_2 \\ Y_3 \\ Y_4 \end{bmatrix} = \begin{bmatrix} \psi_1 \\ \psi_2 \\ \psi_3 \\ \psi_4 \end{bmatrix} \quad (6.27)$$

with $p = 4$, solve for Y_i assuming V and D are now known

$$\begin{aligned} Y_1 &= \psi_1 \\ Y_2 &= \psi_2 - V_{21}Y_1 \\ Y_3 &= \psi_3 - V_{31}Y_1 - V_{32}Y_2 \\ Y_4 &= \psi_4 - V_{41}Y_1 - V_{42}Y_2 - Y_{43}Y_3 \end{aligned} \quad (6.28)$$

By solving, Eq. (6.24), we get

$$\begin{bmatrix} 1 & V_{21} & V_{31} & V_{41} \\ 0 & 1 & V_{32} & V_{42} \\ 0 & 0 & 1 & V_{43} \\ 0 & 0 & 0 & 1 \end{bmatrix} \begin{bmatrix} a_1 \\ a_2 \\ a_3 \\ a_4 \end{bmatrix} = \begin{bmatrix} \frac{1}{d_1} & 0 & 0 & 0 \\ 0 & \frac{1}{d_2} & 0 & 0 \\ 0 & 0 & \frac{1}{d_3} & 0 \\ 0 & 0 & 0 & \frac{1}{d_4} \end{bmatrix} \begin{bmatrix} Y_1 \\ Y_2 \\ Y_3 \\ Y_4 \end{bmatrix} = \begin{bmatrix} Y_1/d_1 \\ Y_2/d_2 \\ Y_3/d_3 \\ Y_4/d_4 \end{bmatrix} \quad (6.29)$$

From Eq. (6.29), we get

$$a_4 = \frac{Y_4}{d_4}$$

$$a_3 = \frac{Y_3}{d_3} - V_{43}a_4$$

$$a_2 = \frac{Y_2}{d_2} - V_{32}a_3 - V_{42}a_4$$

$$a_1 = \frac{Y_1}{d_1} - V_{21}a_2 - V_{31}a_3 - V_{41}a_4 \tag{6.30}$$

Thus, completing the solution to the covariance equation for $p = 4$. Hence, the use of Cholesky decomposition procedure leads to a simple expression for the minimum error.

So each element of ϕ can be expressed as

$$\phi_n(i, j) = \sum_{k=1}^{j} V_{ik}\, d_k\, V_{jk} \quad \text{for } 1 \le j \le i-1 \tag{6.31}$$

or alternatively,

$$V_{ij}\, d_j = \phi_n(i, j) - \sum_{k=1}^{j-1} V_{ik}\, d_k\, V_{jk} \quad \text{for } 1 \le j \le i-1 \tag{6.32}$$

And for diagonal elements,

$$\phi_n(i, i) = \sum_{k=1}^{i} V_{ik}\, d_k\, V_{ik} \tag{6.33}$$

or alternatively,

$$d_i = \phi_n(i, i) - \sum_{k=1}^{i-1} V_{ik}^2\, d_k \quad \text{for } i \ge 2 \tag{6.34}$$

With

$$d_1 = \phi_n(1, 1) \tag{6.35}$$

Therefore, the given matrix (V) and Eq. (6.24), Y can be solved recursively as

$$Y_i = \psi_i - \sum_{j=1}^{i-1} V_{ij}\, Y_j \quad 2 \le i \le p \tag{6.36}$$

With initial conditions

$$Y_1 = \psi_1$$

Having determined Y, Eq. (6.24) can be solved recursively for a in a similar way

$$a_i = \frac{Y_i}{d_i} - \sum_{j=i+1}^{p} V_{ji}\, a_j \quad 1 \le i < p-1 \tag{6.37}$$

With the initial condition,

$$a_p = \frac{Y_p}{d_p}$$

6.4.2 Autocorrelation Method

The summation in Eq. 6.7 had no specific range. In the autocorrelation method, we assume $s_n(m)$ is zero outside the interval $0 \le m < N$.

$$s_n(m) = s[n+m]w(m) \tag{6.38}$$

With $w(m)$ being a window (e.g., Hamming window) which is 0 outside the interval $0 \le m < N$. With this assumption the corresponding prediction error, $e_n(m)$, is non-zero over the interval $0 \le m < N + p - 1$, and, therefore, the total prediction error takes the value,

$$E_n = \sum_{m=0}^{N+p-1} e_n^2(m) \tag{6.39}$$

With this range, Eq. (6.13) can be expressed as

$$\phi_n(i,k) = \sum_{m=0}^{N+p-1} s_n(m-i)s_n(m-k) \quad 1 \le i \le p \text{ and } 0 \le k \le p \tag{6.40}$$

This is equivalent to the form

$$\phi_n(i,k) = \sum_{m=0}^{N-1-(i-k)} s_n(m)s_n(m+i-k) \quad 1 \le i \le p \text{ and } 0 \le k \le p \tag{6.41}$$

or alternatively

$$\phi_n(i,k) = R_n(i-k) \tag{6.42}$$

With $R_n(i)$ being the autocorrelation sequence of $S_n(m)$

$$R_n(i) = \sum_{m=0}^{N-1-i} s_n(m)s_n(m+i) \tag{6.43}$$

combining Eqs. (6.42), (6.43) and (6.14) and using properties of autocorrelation, we obtain

$$\sum_{k=1}^{p} a_k R_n(|i-k|) = R_n(i) \text{ for } i = 1, 2, \ldots p \tag{6.44}$$

which corresponds to the following matrix equation

$$\begin{bmatrix} R_n(0) & R_n(1) & R_n(2) & \ldots & R_n(p-1) \\ R_n(1) & R_n(0) & R_n(p-1) & \ldots & R_n(p-2) \\ R_n(2) & R_n(1) & R_n(0) & \ldots & R_n(p-3) \\ \ldots & \ldots & \ldots & \ldots & \ldots \\ R_n(p-1) & R_n(p-2) & R_n(p-3) & \ldots & R_n(0) \end{bmatrix} \begin{bmatrix} a_1 \\ a_2 \\ a_3 \\ \ldots \\ a_p \end{bmatrix} = \begin{bmatrix} R_n(1) \\ R_n(2) \\ R_n(3) \\ \ldots \\ R_n(p) \end{bmatrix} \tag{6.45}$$

The matrix in Eq. (6.45) is symmetric and all the elements in its diagonal are identical, such matrices are called Toeplitz.

Durbin's recursive solution for the autocorrelation equation

By exploiting the toeplitz nature of the matrix of coefficients, Durbin's recursion results is an efficient algorithm (for convenience, we omit the subscript n of the autocorrelation function) which can be stated as follows:

1. Initialization

$$E^{(0)} = R(0) \tag{6.46}$$

2. Iteration, for $i = 1, 2, \ldots . p$ do the following recursion

$$k_i = \frac{\left[R(i) - \sum_{j=1}^{i-1} a_j^{i-1} R[i-j]\right]}{E^{i-1}} \tag{6.47}$$

$$a_i^i = k_i \tag{6.48}$$

$$a_j^i = a_j^{i-1} - k_i\, a_{i-j}^{i-1} \qquad 1 \le j < i-1 \tag{6.49}$$

$$E^i = \left(1 - k_i^2\right) E^{i-1} \tag{6.50}$$

Equations (6.47) – (6.50) are solved recursively for $i = 1, 2, \ldots . p$.

3. Final solution

$$a_j = a_j^p \qquad 1 \le j < p \tag{6.51}$$

where, the coefficients (k_i), called reflection coefficients, are bounded between –1 and 1.

In the process of computing the predictor coefficients of order p, the recursion finds the solution of the predictor coefficients for all orders less than p.

The above procedure can be illustrated with predictor order 2. The matrix Eq. (6.45) is of the form

$$\begin{bmatrix} R(0) & R(1) \\ R(1) & R(0) \end{bmatrix} \begin{bmatrix} a_1 \\ a_2 \end{bmatrix} = \begin{bmatrix} R(1) \\ R(2) \end{bmatrix}$$

Using Eq. (6.47) – (6.50), for $i = 1$, we obtain

$$E^{(0)} = R(0)$$

$$k_1 = \frac{R(1)}{R(0)}$$

$$a_1^1 = k_1 = \frac{R(1)}{R(0)}$$

$$E^{(1)} = (1-k_1^2)E^{(0)} = \left(1-\left(\frac{R(1)}{R(0)}\right)^2\right)R(0) = \frac{R^2(0)-R^2(1)}{R(0)} \tag{6.52}$$

For $i = 2$, we obtain

$$k_2 = \frac{\left[R(2)-\sum_{j=1}^{1} a_j^{(1)}\, R[2-j]\right]}{E^{(2-1)}} = \frac{\left[R(2)-a_1^{(1)}R[1]\right]}{E^{(1)}}$$

$$k_2 = \frac{\left[R(2)-\frac{R(1)}{R(0)}R[1]\right]}{\left[\frac{R^2(0)-R^2(1)}{R(0)}\right]} = \frac{R(2)R(0)-R^2(1)}{R^2(0)-R^2(1)}$$

$$a_2^2 = k_2 = \frac{R(2)R(0)-R^2(1)}{R^2(0)-R^2(1)} \tag{6.53}$$

From Eq. (6.49) $a_j^i = a_j^{i-1} - k_i\, a_{i-j}^{i-1} \qquad 1 \le j < i$

For $i = 2$, $j = 1$, we obtain

$$a_1^2 = a_1^{(1)} - k_2 a_1^{(1)} = \frac{R(1)}{R(0)} - \left[\frac{R(2)R(0)-R^2(1)}{R^2(0)-R^2(1)}\right]\frac{R(1)}{R(0)} = \frac{R(1)R(0)-R(1)R(2)}{R^2(0)-R^2(1)} \tag{6.54}$$

Final solution is obtained by using Eq. (6.51), $a_j = a_j^p \qquad 1 \le j < p$

$$a_1 = a_1^2$$

$$a_2 = a_2^2$$

6.4.3 Lattice Method (Parcor Analysis)

Parcor coefficients are computed by taking the correlation of forward prediction error and backward prediction error as shown in Figure 6.2.

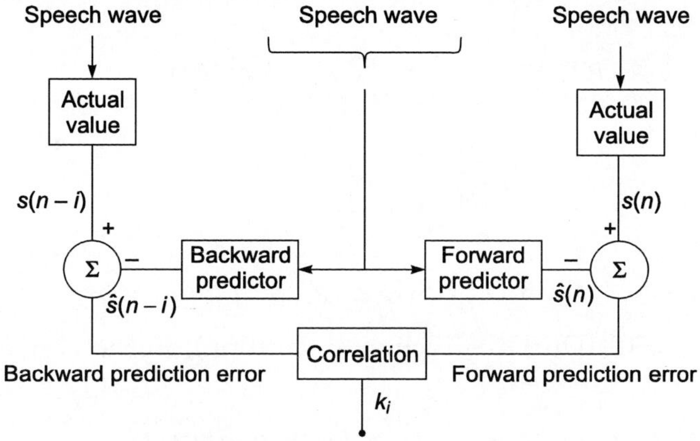

Figure 6.2 Calculation of Parcor coefficients

Let us define $s_n[m]$ as a segment of speech selected in the vicinity of sample n. For the sake of simplicity, we shall henceforth drop the subscript n. Error is the difference between the actual signal and synthetic signal. Forward prediction error at stage i is given by

$$e^i(m) = s(m) - \sum_{k=1}^{i} a_k^i \, s(m-k) \tag{6.55}$$

Taking Z transform on both the sides

$$E^i(Z) = S(Z) - \sum_{K=1}^{i} a_k^i \, Z^{-k} S(Z) \tag{6.56}$$

$$E^i(Z) = S(Z)\left[1 - \sum_{K=1}^{i} a_k^i \, Z^{-k}\right] \tag{6.57}$$

$$\text{Let } a^i(Z) = \left[1 - \sum_{K=1}^{i} a_k^i \, Z^{-k}\right] \tag{6.58}$$

$$E^i(Z) = S(Z) a^i(Z) \tag{6.59}$$

From Durbin's recursive solution

$$a^i(Z) = a^{i-1}(Z) - k_i \, Z^{-i} \, a^{i-1}(Z^{-1}) \tag{6.60}$$

Backward prediction error is given by

$$b^i(m) = s(m-i) - \sum_{k=1}^{i} a_k^i \, s(m-i+k) \tag{6.61}$$

$$B^i(Z) = Z^{-i}S(Z) - \sum_{K=1}^{i} a_k^i Z^{-i} Z^k S(Z) \tag{6.62}$$

$$B^i(Z) = S(Z)Z^{-i}\left[1 - \sum_{K=1}^{i} a_k^i Z^k\right] \tag{6.63}$$

Using Eq. (6.58), we obtain $B^i(Z) = S(Z)Z^{-i}a^i\left(Z^{-1}\right)$ (6.64)

Substituting Eq. (6.60) (Durbin's recursive solution), in the above equation, we obtain

$$B^i(Z) = S(Z)Z^{-i}\left[a^{i-1}\left(Z^{-1}\right) - k_i Z^i a^{i-1}(Z)\right] \tag{6.65}$$

$$B^i(Z) = S(Z)Z^{-i}a^{i-1}\left(Z^{-1}\right) - k_i a^{i-1}(Z)S(Z) \tag{6.66}$$

Multiplying the first term in the above equation by ZZ^{-1}, we obtain

$$B^i(Z) = Z^{-1}S(Z)Z^{-(i-1)}a^{i-1}\left(Z^{-1}\right) - k_i a^{i-1}(Z)S(Z) \tag{6.67}$$

Using Eqs. (6.64) and (6.59), we obtain

$$B^i(Z) = Z^{-1}B^{i-1}(Z) - k_i E^{i-1}(Z) \tag{6.68}$$

By taking inverse Z transform

$$b^i(m) = b^{i-1}(m-1) - k_i e^{i-1}(m) \tag{6.69}$$

From Eq. (6.59)

$$E^i(Z) = S(Z)a^i(Z) \tag{6.70}$$

Substituting $a^i(Z)$ from Eq. (6.60), we obtain

$$E^i(Z) = S(Z)\left[a^{i-1}(Z) - k_i Z^{-i}a^{i-1}(Z^{-1})\right] \tag{6.71}$$

$$E^i(Z) = S(Z)a^{i-1}(Z) - k_i Z^{-i}a^{i-1}(Z^{-1})S(Z) \tag{6.72}$$

Multiplying the second term in the above equation by ZZ^{-1}, we obtain

$$E^i(Z) = S(Z)a^{i-1}(Z) - Z^{-1}k_i Z^{-(i-1)}a^{i-1}(Z^{-1})S(Z) \tag{6.73}$$

By using Eqs. (6.59) and (6.64)

$$E^i(Z) = E^{i-1}(Z) - k_i Z^{-1}B^{i-1}(Z) \tag{6.74}$$

By taking inverse Z transform, we obtain

$$e^i(m) = e^{i-1}(m) - k_i b^{i-1}(m-1) \tag{6.75}$$

k_i coefficients are computed by minimizing sum of forward prediction error. Average forward prediction error is given by

$$E^i = \sum_{m=0}^{N-1} [e^i(m)]^2 \tag{6.76}$$

Substituting $e^i(m)$ from Eq. (6.75)

$$E^i = \sum_{m=0}^{N-1} [e^{i-1}(m) - k_i\, b^{i-1}(m-1)]^2 \tag{6.77}$$

Taking derivative with respect to k_i and equating to zero

$$\sum_{m=0}^{N-1} 2[e^{i-1}(m) - k_i\, b^{i-1}(m-1)]b^{i-1}(m-1) = 0$$

$$\sum_{m=0}^{N-1} e^{i-1}(m)b^{i-1}(m-1) = k_i \sum_{m=0}^{N-1} [b^{i-1}(m-1)]^2$$

$$k_i = \frac{\sum_{m=0}^{N-1} e^{i-1}(m)b^{i-1}(m-1)}{\sum_{m=0}^{N-1} [b^{i-1}(m-1)]^2} \tag{6.78}$$

Equation (6.78) has the form of a normalized cross-correlation function, and therefore, the reflection coefficients are also called partial correlation coefficients or Parcor coefficients. In any normalized cross-correlation function, the k_i coefficients are bounded by $-1 \le k_i \le 1$.

6.5 BURG'S ALGORITHM

In lattice method k_i coefficients are computed by minimizing the sum of forward prediction error. Burg's algorithm uses both forward and backward prediction errors and minimizes the sum of the squared error.

$$E^i = \sum_{m=0}^{N-1} [e^i(m)]^2 + [b^i(m)]^2 \tag{6.79}$$

Substituting $e^i(m)$ and $b^i(m)$ values using Eq. (6.75) and Eq. (6.69), we get

$$e^i(m) = e^{i-1}(m) - k_i b^{i-1}(m-1)$$

$$b^i(m) = b^{i-1}(m-1) - k_i e^{i-1}(m)$$

$$E^i = \sum_{m=0}^{N-1} [e^{i-1}(m) - k_i\, b^{i-1}(m-1)]^2 + [b^{i-1}(m-1) - k_i\, e^{i-1}(m)]^2 \tag{6.80}$$

Taking derivative with respect to k_i and equating to zero

$$\sum_{m=0}^{N-1} -2[e^{i-1}(m) - k_i\, b^{i-1}(m-1)]b^{i-1}(m-1) - 2[b^{i-1}(m-1) - k_i\, e^{i-1}(m)]e^{i-1}(m) = 0$$

$$k_i = \frac{2\sum_{m=0}^{N-1} e^{i-1}(m)b^{i-1}(m-1)}{\sum_{m=0}^{N-1} [e^{i-1}(m)]^2 + \sum_{m=0}^{N-1} [b^{i-1}(m-1)]^2} \tag{6.81}$$

6.6 RELATIONSHIP BETWEEN PARCOR AND LPCs

If either k_i or a_i^i is given, the other can be obtained by iterative computation using Durbin's solution.

$$a_i^i = k_i \tag{6.82}$$

$$a_j^i = a_j^{i-1} - k_i a_{i-j}^{i-1} \qquad 1 \le j < i \tag{6.83}$$

Final solution

$$a_j = a_j^p \qquad 1 \le j < p \tag{6.84}$$

6.7 LINE SPECTRUM PAIR (LSP) FREQUENCY ANALYSIS

Parcor coefficients are parameters operating in the time domain, whereas LSPs are parameters functioning in the frequency domain. Hence, LSP parameters are advantageous in that distortion they produce is smaller than that of Parcor coefficients. Consider all-pole transfer function of the vocal tract. Coefficients of the denominator polynomial of the all-pole transfer function are the linear prediction coefficients (LPCs). These pole locations are the formant frequencies. In LSPs, we decompose denominator polynomial into roots on the unit circle and roots are represented by angular positions. Roots occur in complex conjugate pairs and hence, they are called line spectrum pair (LSP) frequencies. Hence, Parcor analysis and LSP analysis is based on all-pole model. The forward and backward prediction errors

$$e^i(m) = e^{i-1}(m) - k_i b^{i-1}(m-1) \tag{6.85}$$

$$b^i(m) = b^{i-1}(m-1) - k_i e^{i-1}(m) \tag{6.86}$$

Transforming the equations in Z-domain, we get

$$E^i(Z) = E^{i-1}(Z) - Z^{-1} k_i b^{i-1}(Z) \tag{6.87}$$

$$B^i(Z) = Z^{-1}b^{i-1}(Z) - k_i E^{i-1}(Z) \tag{6.88}$$

Let us assume $E^p(Z)$ is given, and represents two $E^{P+1}(Z)$ types: $P(Z)$ and $Q(Z)$ under the condition $k_{P+1} = 1$ and $k_{P+1} = -1$ respectively. The condition $|k_{P+1}| = 1$ corresponds to the case where airflow is completely reflected at the glottis in the vocal tract model represented by Parcor coefficients.

$P(Z)$ and $Q(Z)$ can be represented as

For $k_{P+1} = 1$, $P(Z) = E^P(Z) - B^P(Z)$ – Difference filter

For $k_{P+1} = -1$, $Q(Z) = E^P(Z) + B^P(Z)$ – Sum filter

$$E^P(Z) = \frac{P(Z) + Q(Z)}{2} \tag{6.89}$$

$$E^1(Z) = 1 - Z^{-1}k_1 \text{ and } B^1(Z) = Z^{-1} - k_1 = Z^{-1}(1 - k_1 Z) = Z^{-1}E^1\left(Z^{-1}\right)$$

Continuing this transformation, we can derive general equation

$$B^p(Z) = Z^{-p}E^p\left(Z^{-1}\right) \tag{6.90}$$

If p is assumed to be even, $P(Z)$ and $Q(Z)$ are factorized as

$$P(Z) = (1 - Z^{-1}) \prod_{i=2,4,\ldots p} (1 - 2Z^{-1}\cos\omega_i + Z^{-2}) \tag{6.91}$$

$$Q(Z) = (1 + Z^{-1}) \prod_{i=1,3,\ldots p-1} (1 - 2Z^{-1}\cos\omega_i + Z^{-2}) \tag{6.92}$$

The coefficients (ω_i) which appear in the factorization are referred to as **LSP parameters**, ω_i are ordered as

$$0 < \omega_1 < \omega_2 \ldots\ldots\ldots\ldots < \omega_{p-1} < \omega_p < \pi$$

6.8 PREDICTION ERROR SIGNAL

From digital model, we know that

$$s(n) = Gu(n) - \sum_{k=1}^{p} \alpha_k s(n-k) \tag{6.93}$$

where, α_k are the vocal tract parameters. Prediction error signal is given by

$$e(n) = s(n) - \sum_{k=1}^{p} a_k s(n-k) \tag{6.94}$$

Substituting $s(n)$ values from Eq. (6.6), we obtain

$$e(n) = Gu(n) - \sum_{k=1}^{p} \alpha_k s(n-k) - \sum_{k=1}^{p} a_k s(n-k) = Gu(n) \text{ if } \alpha_k = a_k \tag{6.95}$$

Hence, $e(n)$ is good approximation of the excitation source. Prediction error will be large for voiced signal. Pitch period can be estimated by performing an autocorrelation analysis on $e(n)$ and detecting the largest peak.

6.9 LPC APPLICATIONS

The theory of linear prediction is highly developed. Based on this theory, a large variety of applications of linear predictive analysis to speech processing has evolved. Schemes have been developed for estimating all the basic parameters from linear predictive analysis. Finally, these techniques have been used in many speech processing systems for speaker verification and identification, speech recognition, speech classification, etc.

6.9.1 Pitch Detection using LPC Parameters

Pitch detection algorithm was proposed by Markel. This algorithm is called the SIFT (simple inverse filtering tracking) method. Figure 6.3 shows a block diagram of the SIFT algorithm. As shown in block diagram the speech signal, $s(n)$, is low-pass filtered with a cut-off frequency of about 900 Hz and then by decimation process sampling rate is reduced to 2 kHz from 10 kHz, i.e., 4 out of every 5 samples are dropped at the output of low-pass filter. The signal, $x(n)$, is then inverse filtered to give $y(n)$, a signal with an approximately flat spectrum. The output, $y(n)$, is simply the prediction error for the fourth order predictor. The short-time autocorrelation of the inverse filtered signal is computed and the largest peak in the appropriate range is chosen as the pitch period.

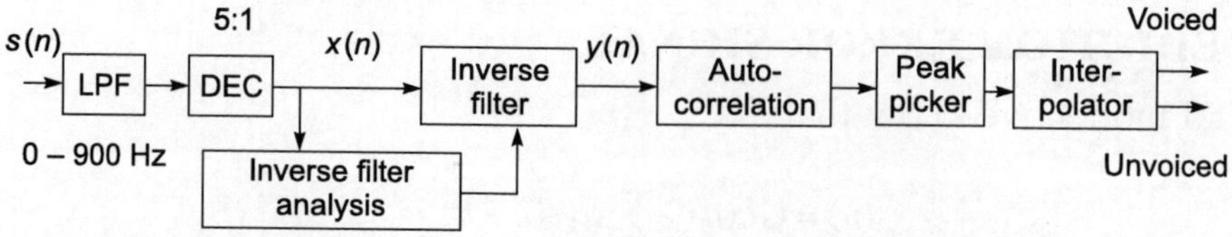

Figure 6.3 Pitch detection using SIFT algorithm

6.9.2 Formant Analysis using LPC Parameters

A very powerful method for speech analysis is based on linear predictive coding (LPC), also known as auto-regressive (AR) modelling. Formants can be estimated from the predictor parameters. This method is widely used because it is fast and simple, yet an effective way of estimating the main parameters of speech signals.

6.9.3 LPC Vocoder

One of the important applications of linear predictive analysis has been in the area of transmission and storage. Figure 6.4 shows the block diagram of an LPC vocoder it consists of a transmitter and receiver. The transmitters perform the LPC analysis and pitch detection and then code the parameters for transmission. Then through a channel the parameters are sent. The receiver decodes the parameters and synthesizes the speech.

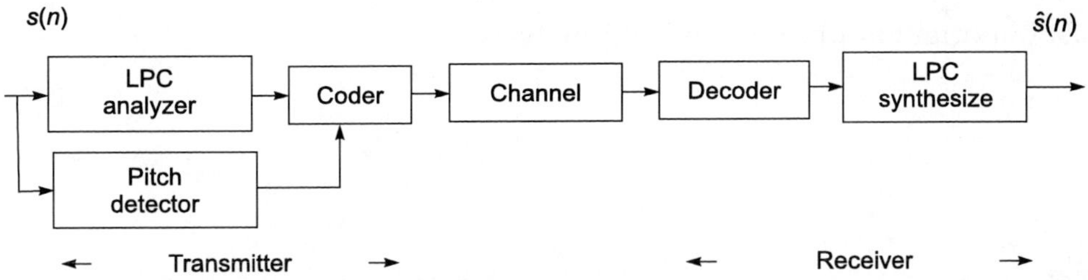

Figure 6.4 LPC vocoder

Linear predictive coding is an analysis/synthesis technique to lossy speech compression that attempts to model the human production of sound instead of transmitting an estimate of the sound wave. Linear predictive coding achieves a bit rate of 2400 bits/second which makes it ideal for use in secure telephone systems. Secure telephone systems are more concerned that the content and meaning of speech, rather than the quality of speech, be preserved.

6.10 SUMMARY

This chapter presents the versatility of LPC algorithms for estimating and modelling of speech signals using various techniques.

EXERCISES

Review Questions

1. With necessary diagram, explain LPC analysis, list its advantages and disadvantages.
2. Explain covariance method for calculating LPCs.
3. Explain Cholesky decomposition solution for the covariance method.

4. Explain autocorrelation method for calculation of LPCs.
5. Explain Durbin's recursive algorithm for the autocorrelation equation
6. Explain Durbin's recursive solution for the autocorrelation equation with predictor order 2.
7. With necessary block diagram, explain computations of LPC coefficients by taking the correlation of forward prediction error and backward prediction error.
8. Explain Burg's algorithm for calculation of linear predictor coefficients.
9. With necessary block diagram, explain computations of LPCs by LSP frequency analysis.
10. List important application of LPC analysis.

7

CHAPTER

Speech Coding

This chapter provides insights on

- Mid-tread and mid-riser uniform quantizers
- Logarithmic and companded quantizers
- Other types of quantizers—feedforward, feedback, differential and Jayant.
- Coding in time and frequency domains
- Vector quantization
- Parametric vocoders—channel, formant, cepstral (homomorphic)
- Hybrid coders—multi-pulse excited linear prediction coder (MPC) and code excited linear prediction coder (CELP)

7.1 INTRODUCTION

Speech coding is a technique that represents speech data in a compressed form. A major application of speech signal processing is speech coding, used for efficient storage and transmission. Speech is transformed by a quantizer and encoder into a sequence of bits (i.e., 0's and 1's), then it is temporarily stored or transmitted over a channel and then reconverted by the decoder. For efficient transmission whatever done by the encoder must be inverted by the decoder, to reproduce the output speech. Recent advances in speech compression allow a wide variety of applications for speech coding. Two important applications of speech coding are mobile telephony and digital voice transmission. Techniques developed for speech coding have also been applied to other application areas such as speech synthesis, speech recognition and speaker recognition.

7.2 OVERVIEW

The goal of all speech coding systems is to transmit speech with the highest possible quality using the least possible channel capacity. In speech coding, the analog speech signal is converted to a sequence of binary digits or bits by a quantizer and encoder. These bits are then transmitted on the channel. A speech decoder receives the sequence of binary digits or bits and synthesizes speech.

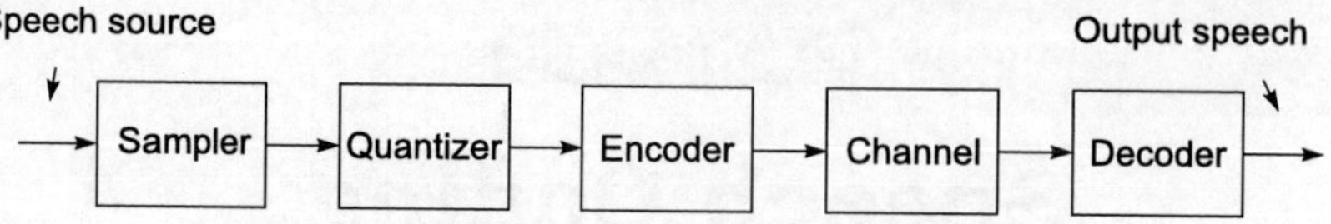

Figure 7.1 Speech coding and decoding system

Figure 7.1 shows a typical setup for coding speech waveform and decoding the sample. Speech coding and decoding process consists of following stages:

- Sampling the input signal.
- Converting the sample amplitude into a set of amplitudes and then it is represented by bit streams (a string of 0's and 1's). This process is called *quantization*.
- Digitized speech enters the encoder to produce the encoded bitstream or compressed speech data.
- After transmission the decoder takes the encoded bitstream (code words) and converts the code words to quantized signal (same rate as the input to the encoder).

7.3 DESIRABLE PROPERTIES IN SPEECH CODING

Following are the desirable properties of speech in coding techniques.

- Low bit rate: The lower the bit rate of the encoded bitstream, the less bandwidth is required for transmission, leading to a more efficient system.
- Good performance on non-speech signals: In telecommunication systems, besides speech other signaling tones, dual tone multi-frequency (DTMF) in keypad dialing and music, etc. are often encountered. It should not generate annoying artifacts for alternate signals.
- Low coding delay: In the process of speech encoding and decoding, delay is introduced. Efficient algorithms must be used such that delay can be shortened.
- Low memory size: In order for the speech coder to be practible, it must have low amount of memory needed to support its operation and hence cost associated with its implementation will be low.

7.4 DIGITAL REPRESENTATION OF SPEECH SIGNALS

Speech signals are analog in nature because they originate as sound pressure waves. The digital conversion process can be split sampling, quantization and encoding. Sampling converts a continuous time signal into a discrete time signal by measuring the signal value at regular intervals of time. Quantization converts a continuous amplitude signal into a discrete amplitude signal. Encoding stage represents each quantized sample by a code word.

7.4.1 Sampling Theorem

Sampling is a process by which a continuous time signal is converted into a discrete time signal. The process of sampling is illustrated in Fig. 7.2. A discrete time sinusoid, $x(n)$, is obtained by sampling a continuous time sinusoid $x(t)$ with sampling interval. As shown in Fig. 7.2 sampling unit can be simple electronic component, switch, and input to the sampling unit. The signal is available at the output only when switch is closed. Hence, we get discrete time sinusoid signal, $x(n)$, at the output.

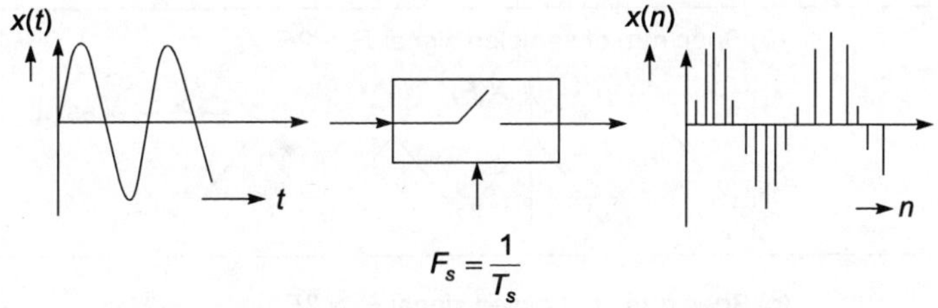

Figure 7.2 Sampling of continuous to discrete time signal

If the highest frequency component in a digital signal is $F_{\max}$, then sampling theorem or Nyquist theorem states that continuous time signal can be completely represented by its samples and can be recovered, if the sampling frequency,

$$F_s \leq 2F_{\max} \tag{7.1}$$

where, F_s is the sampling frequency and $F_{\max}$ is the maximum frequency component in an analog signal. Most speech coding systems were designed to support telecommunication applications, with the frequency contents limited between 300 and 3400 Hz. According to Nyquist theorem, sampling frequency must be at least twice the bandwidth of the continuous time signal in order to avoid aliasing. A value of 8 kHz is commonly selected as the standard sampling frequency for speech signals. If sampling is less than the rate specified by the sampling theorem, it results in the aliasing effect and original signal cannot be recovered. Figure 7.3 shows the effect of sampling rate. Figure 7.3 (c) shows the spectrum of an under-sampled signal and in the fold over region the signals are not recoverable. The overlap or aliasing occurs about half the sampling frequency point and this frequency is called Nyquist frequency, F_N.

$$F_N = \frac{F_s}{2} \tag{7.2}$$

Before sampling, the signal is always passed through an anti-aliasing filter to remove the noise or signal outside the band of interest.

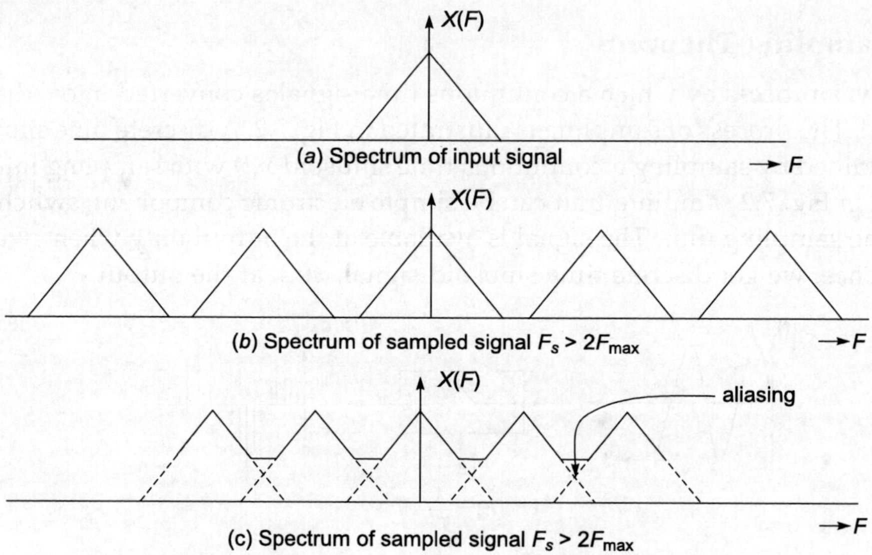

Figure 7.3 Sampling of an analog band limited signal

7.4.2 Quantization and Encoding

In order to transmit the sequence of samples over a digital communication channel or to store them in digital memory, or to use them as input to a digital signal processing algorithm, the sample values must be quantized to a finite set of amplitudes so that they can be represented by a bitstream (a string of 0's and 1's). It consists of two stages: quantization stage and encoding stage as shown in Figure 7.4.

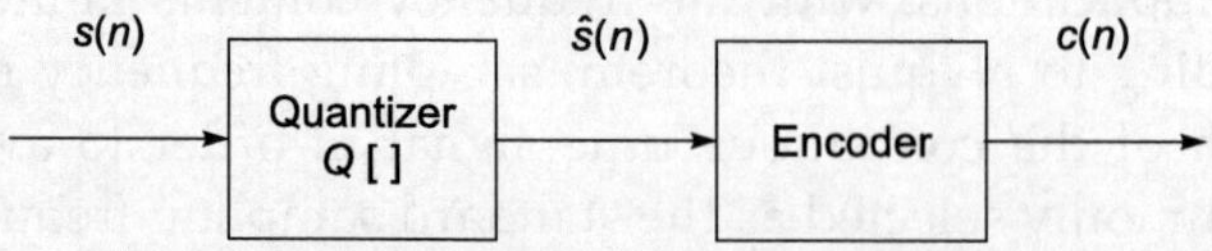

Figure 7.4 Quantization and coding

Quantization converts a continuous amplitude signal into a discrete amplitude signal, $\{S(n) = \{Q[S(n)]\}$, and the encoding stage represents each quantized sample by a code word, $c(n)$. The distance between the finite set of amplitude levels is called the quantize step size, Δ.

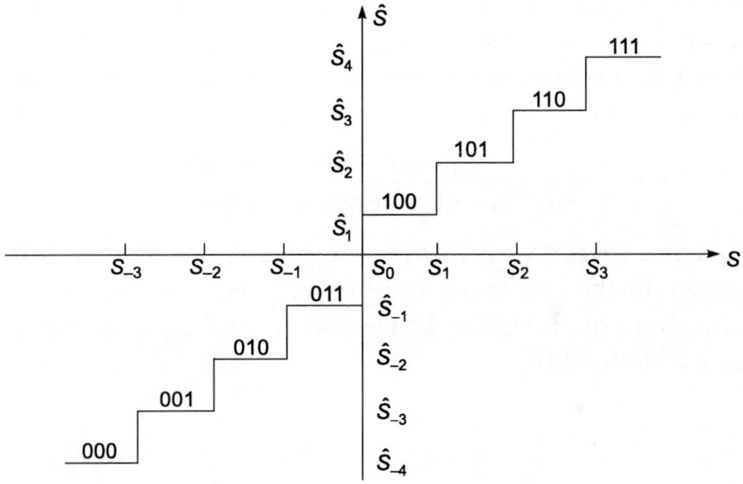

Figure 7.5 Input-output characteristic of a 3-bit quantizer

In this coding mechanism all values of S between S_i and S_{i+1} and $\hat{S}_i$ and $\hat{S}_{i+1}$ are represented by Δ. Figure 7.5 shows the input and output characteristics of a 3-bit quantizer.

$$S_i - S_{i+1} = \Delta$$

$$\hat{S}_i - \hat{S}_{i+1} = \Delta$$

Each discrete amplitude level of the encoder is represented by a code word, $c(n)$, for transmission purposes. After transmission, the decoder converts the code word to quantized signal. If the code word is represented by B bits, then it is possible to represent 2^B different quantization levels. Hence, the

$$\text{Number of quantization levels} = 2^B \tag{7.3}$$

Transmission bit rate or information capacity, $I = B \times F_s$ bits per second or bit rate (7.4)

where, F_s is the sampling rate (samples/sec) and B is the number of bits per sample.

If sampling frequency, F_s = 8 kHz and number of bits per sample, (B) = 16, then transmission bit rate or information capacity is given by

$$I = B \times F_s = 8 \times 16 = 128 \text{ Kbits per second (Kbps)}$$

The design of quantzier has a significant impact on the conversion process, hence we will discuss different quantization techniques such as uniform quantization, nonuniform quantization, companded quantizer, adaptive quantizer, differential quantizer, and jayant quantizer.

7.4.2.1 *Uniform quantizer*

A quantizer can be specified by its input partitions and output levels. If the input range is divided into levels of equal spacing, then the quantizer is termed a uniform quantizer, and if not, it is termed a non-uniform quantizer. Uniform quantizer uses constant step size and is given by

$$\text{Step size} = \frac{\text{Dyamic range of signal}}{\text{Number of quantizer levels}} \tag{7.5}$$

In this technique the speech waveform is sampled regularly at uniform intervals and then quantized and the quantized samples are encoded and then transmitted. In uniform quantization, quantization levels and ranges are distributed uniformly. Uniform quantizer is defined by

$$S_i - S_{i-1} = \Delta \tag{7.6}$$

$$\hat{S}_i - \hat{S}_{i-1} = \Delta \tag{7.7}$$

where, Δ is the quantization step size.

Two common uniform quantizer characteristics are shown in Figure 7.6 with eight quantization levels. In Figure 7.6(a) origin contains the rising part of the staircase function. This class of quantizer is called the mid-riser. For B bit mid-riser quantizer, there are 2^B quantization levels. The number of levels is a power of 2. Positive voltage is coded using a sign bit of logic 0 and negative voltage is coded using a sign bit of logic 1. In case of mid-raser quantizer, the zero input is coded as 0.5Δ. Mid-riser quantizer has the same number of +ve and –ve levels and they are symmetrical. In Figure 7.6 (a) we interpret the binary code as a sign magnitude representation with the leftmost bit being the sign bit.

If input range is –5 V to +5 V, quantizer step size is Δ and if B is the number of bits used for quantiztion, then the number of quantizer level is 2^B. If $B = 2$, then step size is given by,

$$\Delta = \frac{10}{2^B} = \frac{10}{2^2} = 2.5 \text{ volts}$$

Figure 7.6 (b) shows the mid-tread class quantizer. Positive voltage is coded using a sign bit of logic 0 and negative voltage is coded using a sign bit of logic 1. In case of mid-tread quantizer, any input $-0.5\Delta \le s \le 0$ is coded as 100. $0 \le s \le 0.5\Delta$ is coded as 000. For B bit mid-tread quantizer, there are $(2^B - 1)$ quantization levels.

If input range is –5 V to +5 V, quantizer step size is Δ and if B is the number of bits used for quantiztion, then the number of quantizer level is 2^B. If $B = 2$, then step size is given by

$$\Delta = \frac{10}{2^B - 1} = \frac{10}{2^2 - 1} = 1.4286 \text{ volts}$$

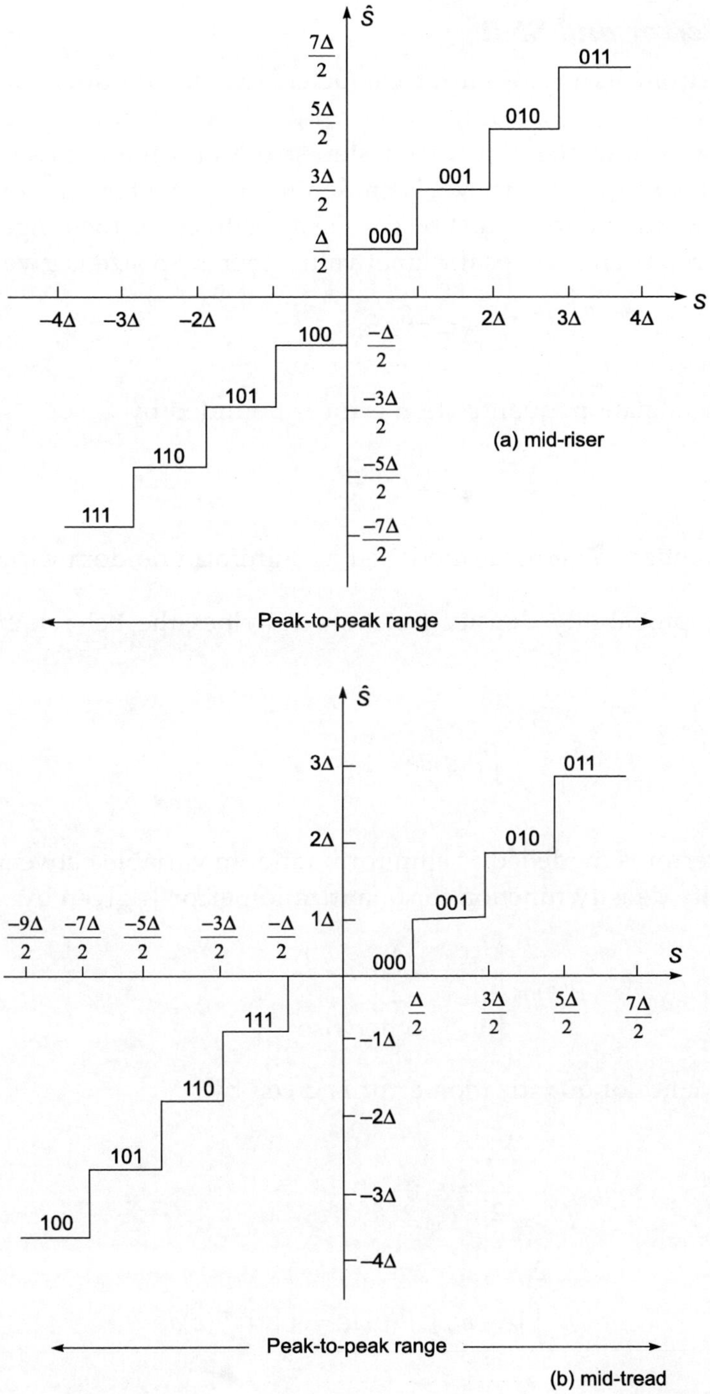

Figure 7.6 Input-output characteristic of a 3-bit mid-tread and mid-riser quantizer

Quantization error and SNR

As can be seen from its input-output characteristics, in uniform quantization all the quantizer levels (steps) are of the same width. Hence, a uniform quantizer can be defined by two parameters: quantizer step size Δ and the number of quantizer levels. The number of quantizer level is chosen to be of the form 2^B, where, B bit is the binary code word. Δ and B must be chosen so as to cover the range of the input samples. If $S_{\max}$ represents the peak signal value, then step size is given by

$$\Delta = \frac{2S_{\max}}{2^B} \tag{7.8}$$

In uniform quantization, quantization error is bounded by

$$-\frac{\Delta}{2} \leq \varepsilon \leq \frac{\Delta}{2} \tag{7.9}$$

Hence, the quantization error is modelled as a uniform random variable between $-\frac{\Delta}{2}$ and $\frac{\Delta}{2}$. The probability density function describes the behaviour of random variable and is defined by

$$\int_{-\infty}^{\infty} P(\varepsilon) d\varepsilon = 1 \tag{7.10}$$

Quantization error is modelled as a uniform random variable between $-\frac{\Delta}{2}$ and $\frac{\Delta}{2}$. Hence, probability density function for quantization error is given by

$$P(\varepsilon) = \begin{cases} \frac{1}{\Delta} & -\frac{\Delta}{2} < \varepsilon < \frac{\Delta}{2} \\ 0 & \text{otherwise} \end{cases} \tag{7.11}$$

Mean and variance of quantization error is given by

$$m_\varepsilon = \frac{-\frac{\Delta}{2} + \frac{\Delta}{2}}{2} = 0 \tag{7.12}$$

$$v_\varepsilon = \int_{-\frac{\Delta}{2}}^{\frac{\Delta}{2}} [\varepsilon - m_\varepsilon]^2 P(\varepsilon) d\varepsilon = \int_{-\frac{\Delta}{2}}^{\frac{\Delta}{2}} [\varepsilon]^2 P(\varepsilon) d\varepsilon \tag{7.13}$$

$$v_\varepsilon = \left[\frac{[\varepsilon]^3}{3} \frac{1}{\Delta} \right]_{-\frac{\Delta}{2}}^{\frac{\Delta}{2}} = \frac{\Delta^2}{12} \tag{7.14}$$

If standard deviation of the signal is σ_S (sigma) and standard deviation of the noise σ_ε, then signal-to-noise ratio (SNR) is defined as

$$\mathrm{SNR} = \frac{\sigma_S^2}{\sigma_\varepsilon^2} \tag{7.15}$$

Since standard deviation of the noise $\sigma_\varepsilon^2 = v_\varepsilon$ hence when step size decreases, quantization noise is reduced. If noise is uniformly distributed over the quantizer dynamic range, then standard deviation of the noise is given by

$$\sigma_\varepsilon^2 = v_\varepsilon = \frac{\Delta^2}{12} = \frac{\left[\frac{2S_{\max}}{2^B}\right]^2}{12} = \frac{\left[\frac{4[S_{\max}]^2}{2^{2B}}\right]}{12} = \left[\frac{[S_{\max}]^2}{3\times 2^{2B}}\right]$$

$$\mathrm{SNR} = \frac{\sigma_S^2}{\sigma_\varepsilon^2} = \frac{\sigma_S^2}{\left[\frac{[S_{\max}]^2}{3\times 2^{2B}}\right]} = \frac{3\times 2^{2B}}{\left[\frac{[S_{\max}]^2}{\sigma_S^2}\right]} \tag{7.16}$$

$$\text{SNR in decibels} = 10\log_{10}\left[\frac{\sigma_S^2}{\sigma_\varepsilon^2}\right] = 6B + 4.77 - 20\log_{10}\left[\frac{S_{\max}}{\sigma_S}\right] \tag{7.17}$$

σ_S is the variance of the signal, normally the probability distribution for speech signal is Laplacian. It can be shown that only 0.35% of the speech samples would fall outside the range $-4\sigma_S \le S(n) \le 4\sigma_S$. If we assume $S_{\max} = 4\sigma_S$

$$\text{SNR in decibels} = 6B + 4.77 - 20\log_{10}\left[\frac{4\sigma_S}{\sigma_S}\right] = 6B - 7.2 \tag{7.18}$$

Hence, every additional bit used in the quantizer will increase SNR by 6 dB.

Example 7.1 For a 3-bit mid-tread with a range of –5 to 5 volts analog signal, determine the quantization step size, binary codes, recovered voltages, and quantization errors when the input is –3.65 volt and 0.49 volt, respectively.

Solution:

Mid-tread quantizer

1. Quantization step size is

$$\text{Step size} = \frac{\text{Dyamic range of signal}}{\text{Number of quantizer levels}}$$

$$\Delta = \frac{10}{2^3 - 1} = \frac{10}{2^3 - 1} = 1.4286 \text{ volts}$$

2. Quantization table for the 3-bit mid-tread quantizer

TABLE 7.1 Quantization table for the 3-bit mid-tread quantizer

Binary code	Quantization level	Input signal sub range
011	3Δ	$2.5\Delta \le s \le 3.5\Delta$
010	2Δ	$1.5\Delta \le s \le 2.5\Delta$
001	Δ	$0.5\Delta \le s \le 1.5\Delta$
000	0	$0 \le s \le 0.5\Delta$
100	0	$-0.5\Delta \le s \le 0$
101	$-\Delta$	$-1.5\Delta \le s \le -0.5\Delta$
110	-2Δ	$-2.5\Delta \le s \le -1.5\Delta$
111	-3Δ	$-3.5\Delta \le s \le -2.5\Delta$

3. For volt, we have $S = \dfrac{-3.65\Delta}{1.4286} = -2.5549\,\Delta$, hence the binary code 111 and the recovered voltage, $\hat{S} = -3\Delta = -4.2858$ volts. Thus, the quantization error,

$$\varepsilon = \hat{S} - S = -4.2858 - (-3.65) = -0.6358 \text{ volt}$$

For volt, we have $S = \dfrac{0.49\Delta}{1.4286} = 0.343\Delta$, hence, the binary code 000 and the recovered voltage is $\hat{S} = 0$ volts. Thus, the quantization error,

$$\varepsilon = \hat{S} - S = 0 - (0.49) = 0.51 \text{ volt}$$

Example 7.2 For the 3-bit mid-riser quantizer, with a range of –5 to 5 volts analog signal, determine the quantization step size, binary codes, recovered voltages and quantization errors when the input is –3.65 volt and 0.49 volt, respectively.

Solution:

Mid-riser quantizer

1. Quantization step size is

$$\text{Step size} = \frac{\text{Dyamic range of signal}}{\text{Number of quantizer levels}}$$

$$\Delta = \frac{10}{2^3} = \frac{10}{2^3} = 1.25 \text{ volts}$$

2. Quantization table for the 3-bit mid-riser quantizer

TABLE 7.2 Quantization table for the 3-bit mid-riser quantizer

Binary code	Quantization level	Input signal sub range
011	3.5Δ	$3\Delta \le s \le 4\Delta$
010	2.5Δ	$2\Delta \le s \le 3\Delta$
001	1.5Δ	$\Delta \le s \le 2\Delta$
000	0	$0 \le s \le \Delta$
100	0	$-\Delta \le s \le 0$
101	-1.5Δ	$-2\Delta \le s \le -\Delta$
110	-2.5Δ	$-3\Delta \le s \le -2\Delta$
111	-3.5Δ	$-4\Delta \le s \le -3\Delta$

3. For –3.65 volt, we have $S = \frac{-3.65\Delta}{1.25} = -2.92\Delta$, hence the binary code 010 and the recovered voltage, $\hat{S} = -2.5\Delta = -3.125$ volts. Thus, the quantization error,

$\varepsilon = \hat{S} - S = -3.125 - (-3.65) = 0.525$ volt

For 0.49 volt, we have $S = \frac{0.49\Delta}{1.25} = 0.392\Delta$, hence, the binary code 000 and the recovered voltage, $\hat{S} = 0$ volt. Thus, the quantization error is

$$\varepsilon = \hat{S} - S = 0 - (0.49) = 0.51 \text{ volt}$$

Example 7.3 Speech signal is coded using 5 bits and has a root mean square amplitude of 5 V. Find the step size, variance of quantization noise or quantizatization noise power, and SNR if $S_{max} = 4\sigma_S$.

Solution: Signal has rms value of 5 V, $S_{max} = 5 \times \sqrt{2}$

$$\text{Step size} = \frac{\text{Dyamic range of signal}}{\text{Number of quantizer levels}} \quad \therefore \Delta = \frac{2 \times S_{max}}{2^5} = \frac{10}{2^5} = 10 \times 2^{-5} \text{ volts}$$

Variance of quantization noise or quantization noise power is given by

$$v_\varepsilon = \sigma_\varepsilon^2 = \frac{\Delta^2}{12} = \frac{[10\times 2^{-5}]^2}{12}$$

If $S_{\max} = 4\sigma_S$, then $\text{SNR} = 6B - 7.2 = 6\times 5 - 7.2 = 22.8$ dB

Example 7.4 Speech signal is coded using 6 bits and dynamic range of the signal is 2 V. Signal density is assumed to be Laplacian, find the step size, variance of quantization noise or quantizatization noise power and SNR if $S_{\max} = 4\sigma_S$.

Solution:

$$\text{Step size} = \frac{\text{Dyamic range of signal}}{\text{Number of quantizer levels}} \quad \therefore \Delta = \frac{2V}{2^5} = \frac{1}{2^4} = 2^{-4} \text{ volts}$$

Variance of quantization noise or quantization noise power is given by

$$v_\varepsilon = \sigma_\varepsilon^2 = \frac{\Delta^2}{12} = \frac{2^{-8}}{12}$$

If $S_{\max} = 4\sigma_S$, then $\text{SNR} = 6B - 7.2 = 6\times 6 - 7.2 = 28.8$ dB

7.4.2.2 *Non-uniform quantization or logarithmic quantization*

In uniform quantization, the quantization error is bounded by $\frac{\Delta}{2}$. As discussed in chapter 2 speech signals are classified into two types as voiced and unvoiced. Voiced speech signal has maximum energy but contains less information, whereas unvoiced speech signal has low energy but contains maximum information. Therefore, small signals have large relative quantization error, while large signals have small relative error. As discussed (section 7.4.2.1), when the step size decreases, quantization noise is reduced. In speech and video quantization perceptual quality is better when the quantization error is proportional to the size of the signal.

For a voiced signal, signal level is large, we use larger step size and for a unvoiced signal, signal level is small, smaller step size is used to maintain constant SNR irrespective of signal strength. In non-uniform quantizer or logarithmic quantizer, amplitude is compressed by logarithmic transformation before linear quantization and coding. This is shown in Figure 7.7. Figure 7.7 shows the input amplitudes are compressed by logarithmic function prior to quantization and is expanded by the exponential function after decoding.

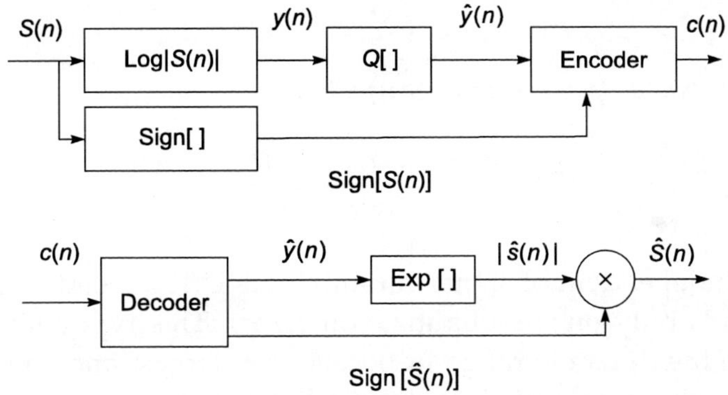

Figure 7.7 Logarithmic encoder and decoder

In encoder

$$y(n) = \log|S(n)| \tag{7.19}$$

$$\hat{y}(n) = Q[\log|S(n)|] \tag{7.20}$$

$$\hat{y}(n) = \log|S(n)| + \varepsilon(n) \tag{7.21}$$

In decoder

$$\hat{S}(n) = |S(n)|\,\text{sign}[S(n)] \tag{7.22}$$

where, sign$[S(n)]$ is +1 if $S(n)$ is positive and –1 if $S(n)$ is negative

But
$$|\hat{S}(n)| = \exp[\hat{y}(n)] \tag{7.23}$$

Hence,

$$\hat{S}(n) = \exp[\hat{y}(n)]\text{sgn}[S(n)] \tag{7.24}$$

$$\hat{S}(n) = \exp[\log|S(n)| + \varepsilon(n)]\text{sgn}[S(n)] \tag{7.25}$$

$$\hat{S}(n) = \log|S(n)|\exp\varepsilon(n)]\text{sgn}[S(n)] \tag{7.26}$$

$$\hat{S}(n) = S(n)\exp[\varepsilon(n)] \tag{7.27}$$

If $\varepsilon(n)$ is very small, $\exp[\varepsilon(n)] = 1 + \varepsilon(n)$

$$\hat{S}(n) = S(n)[1 + \varepsilon(n)] = S(n) + \varepsilon(n)S(n) \tag{7.28}$$

Let

$$f(n) = \varepsilon(n)S(n) \tag{7.29}$$

Then

$$\hat{S}(n) = S(n) + f(n) \tag{7.30}$$

$$\text{SNR} = \frac{\sigma_S^2}{\sigma_f^2} = \frac{\sigma_S^2}{\sigma_S^2 \sigma_\varepsilon^2} = \frac{1}{\sigma_\varepsilon^2} \tag{7.31}$$

Hence, SNR is independent of signal variance and SNR depends only on the step size (i.e., it depends only on the quantization error). This type of quantization is not practical since the dynamic range (ratio between largest and smallest value) is infinite. Hence, the above analysis is impractical. An alternative method is to mimic human ear. They are quasi-logarithmic, they are linear for small signal values and logarithmic for large signal values.

7.4.2.3 Companded quantizer

In companding quantization, lower signals are given more sensitivity and the higher level signals have the least sensitivity. In this method, signal is multiplied with a nonlinear curve known as compressor curve which is roughly logarithmic and on the receiving end, expanders are used. The use of compressor-expander is called companding technique. The block diagram of companded quantizer is shown in Figure 7.8 (a) and the transfer curves for the compressor and expander are shown in Figure 7.8 (b).

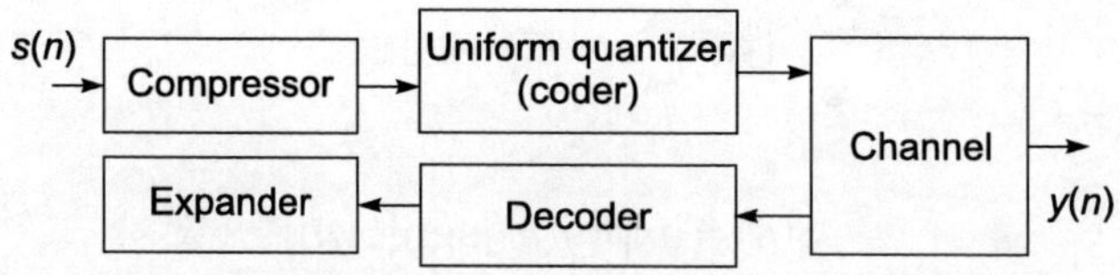

Figure 7.8 (a) Logarithmic encoder and decoder

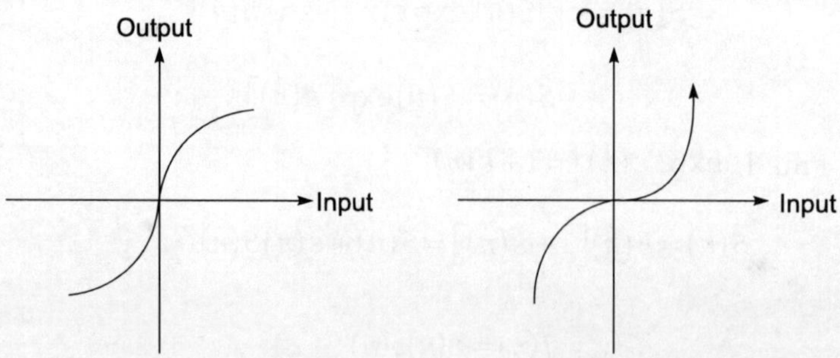

Figure 7.8 (b) Characteristic of compressor and expander

The use of compressor and expander allows us to use a uniform quantizer. Compressor enhances the low-level signal and compresses high-level signal. Characteristics of expander is opposite to compressor, it compresses low-level signal and enhances high-level signal. The speech signal is passed through a compressor and then uniformly quantized and encoded in binary form. The binary output is transmitted through the channel. Decoder receives the binary bitstream and converts it to quantized signal and then passed through the expander. The output of the expander is the speech signal. Compression characteristics are obtained using A law.

Formula for A law companding is

$$y = \begin{cases} \dfrac{A|S|}{\log_e A}; & 0 \le |S| \le \dfrac{1}{A} \\ \dfrac{1+\log_e A|S|}{1+\log_e A} & \dfrac{1}{A} < |S| \le 1 \end{cases} \tag{7.32}$$

A law companding is used in European telephony with $A = 87.56$. The curve with A law and $A = 87.56$ is approximated with 13 segments.

Formula for μ law compressor is given by

$$y = \text{sign}(S) \frac{\log_e\left(1 + \dfrac{\mu|S|}{|S|_{\max}}\right)}{\log_e(1+\mu)} \quad \mu \ge 0 \tag{7.33}$$

where $|S|_{\max}$ is the maximum amplitude of the inputs, while μ is the positive parameter to control the degree of the compression. $\mu = 0$ corresponds to no compression, in North American telephony $\mu = 255$ is adopted.

$$\text{sign}(S) = \begin{cases} 1 & s \ge 0 \\ -1 & s < 0 \end{cases} \tag{7.34}$$

Solving Eq. (7.33) by substituting the quantized value, i.e., $y = y_q$, we get expander equation as

$$\hat{S} = |S|_{\max} \,\text{sign}(y_q) \frac{\log_e (1+\mu)^{|y_q|} - 1}{\mu} \tag{7.35}$$

North American telephony employs μ law companding with $\mu = 255$. The curve with μ law and $\mu = 255$ is approximated with 17 segments. Characteristics of μ law compressor and expander curve are shown in Figure 7.9.

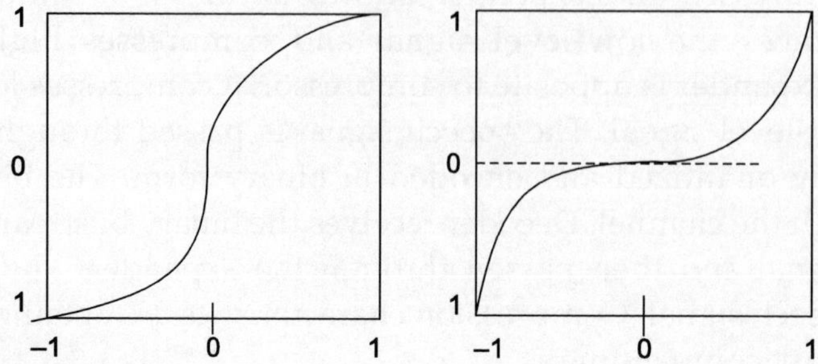

Figure 7.9 (a) Characteristics for the μ-law compander
(b) Characteristics for the μ-law expander

Example 7.5 For μ law companding technique with 3-bit mid-tread quantizer, with a range of –5 to 5 volts analog signal and $\mu = 255$, determine binary codes, recovered voltages and quantization errors for input –3.6 volt and 0.5 volt.

Solution:

(a) For input $S = -3.6$ volt

Input to mid-tread qunatizer, can be determined by μ law compression

$$y = \text{sign}(S)\frac{\log_e\left(1+\frac{\mu|S|}{|S|_{\max}}\right)}{\log_e(1+\mu)} = \text{sign}(-3.6)\frac{\log_e\left(1+\frac{255|-3.6|}{5}\right)}{\log_e(1+255)} = -0.94$$

For a range of –5 to 5 volts, range of y is –1 to 1 volt, quantization step size is

$$\text{Step size} = \frac{\text{Dyamic range of signal}}{\text{Number of quantizer levels}}$$

$$\Delta = \frac{2}{2^3-1} = \frac{2}{2^3-1} = 0.286 \text{ volt}$$

For –0.94 volt, we have $y = \frac{-0.94\Delta}{0.286} = -3.28\Delta$. Based on Table 7.1, the binary code 111 and the recovered voltage is $y_q = -3\Delta = -0.858$ volt.

Applying μ law expander, we get

$$\hat{S} = |S|_{\max}\,\text{sign}(y_q)\frac{\log_e(1+\mu)^{|y_q|}-1}{\mu} = 5\,\text{sign}(-0.858)\frac{\log_e(1+255)^{|-0.858|}-1}{255} = -2.264$$

Thus, the quantization error,

$$\varepsilon = \hat{S} - S = -2.264 - (-3.6) = 1.336 \text{ volts}$$

(b) For input $S = 0.5$ volt

Input to mid-tread qunatizer, can be determined by μ law compression

$$y = \text{sign}(S)\frac{\log_e\left(1 + \frac{\mu|S|}{|S|_{\max}}\right)}{\log_e(1+\mu)} = \text{sign}(0.5)\frac{\log_e\left(1 + \frac{255|0.5|}{5}\right)}{\log_e(1+255)} = 0.591$$

For 0.591 volt, we have $y = \frac{0.591\Delta}{0.286} = 2.1\Delta$, based on Table 7.1, the binary code 010 and the recovered voltage is $y_q = 2\Delta = 0.572$ volt.

Applying μ law expander, we get

$$\hat{S} = |S|_{\max} \text{sign}(y_q)\frac{\log_e(1+\mu)^{|y_q|} - 1}{\mu} = 5\,\text{sign}(0.572)\frac{\log_e(1+255)^{|0.572|} - 1}{255} = 0.448 \text{ volt}$$

Thus, the quantization error is

$$\varepsilon = \hat{S} - S = 0.448 - (0.5) = -0.052 \text{ volt}$$

7.4.2.4 Adaptive quantization

In uniform quantization small signals have large quantization error, while large signals have small relative error. In logarithmic quantizer, dynamic range is infinite (ratio between largest and smallest values), hence quasi-logarithmic quantizers (*A* law and μ law) are used in telephone systems. Adaptive quantization method reduces the quantization error and it gives 4-8 dB SNR improvement over μ law for the same bit rate. In adaptive quantizer step size Δ or gain vary according to the input signal. In the first case step size increases or decreases according to the variance of the input signal. In the second method gain changes inversely, with changes in the variance of the input signal. There are two methods for varying step size or gain. They are:

- Feedforward adaptation
- Feedback adaptation

In the first method, i.e., feedforward adaptation method, the step size or gain is changed for every block. The principles of these two methods are shown in Figures 7.10 and 7.11. In the first case the step size should increase and decrease with increase and decrease of the variance or the standard deviation of the input.

$$\Delta(n) = \Delta_0 \sigma(n) \tag{7.36}$$

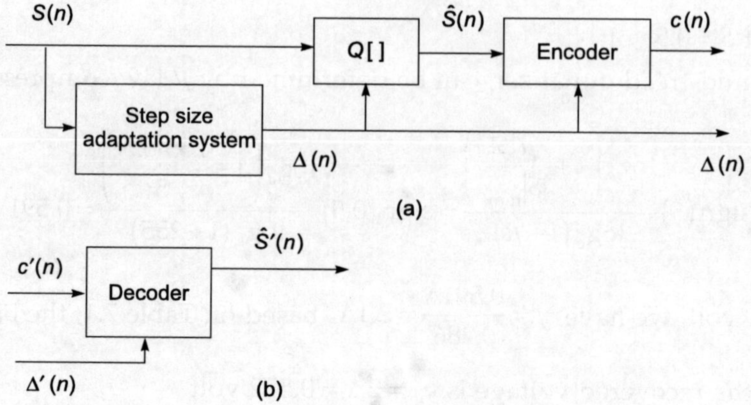

Figure 7.10 Feedforward quantizer (a) coder (b) decoder

In the second case time varying gain is inversely proportional to the variance or standard deviation of the input.

$$G(n) = \frac{G_0}{\sigma(n)} \tag{7.37}$$

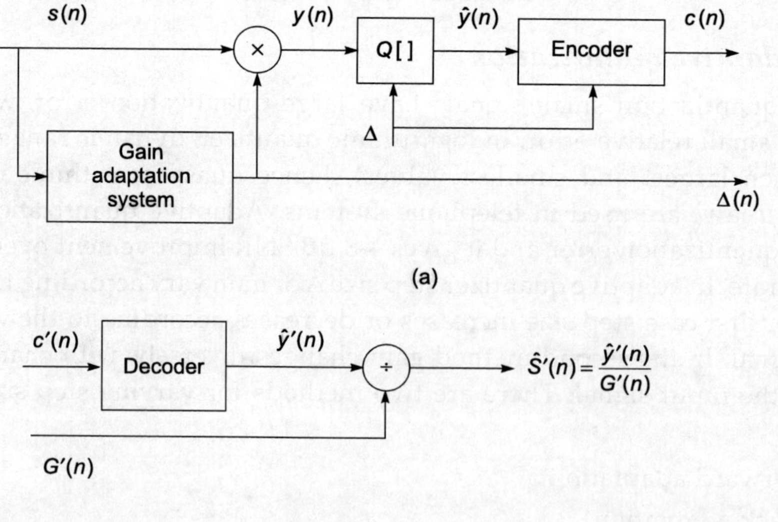

Figure 7.11 Feed-forward adaptive quantizer with a time varying gain (a) coder, (b) decoder

$\sigma^2(n)$ can be estimated from the past samples

$$\sigma^2(n) = \frac{1}{M} \sum_{m=n-M}^{n-1} S^2(m) \tag{7.38}$$

In the second method, i.e., feedback adaptation method, the step size or gain is changed on a sample-by sample basis according to the decoded samples. In both the cases it is necessary to obtain the time varying properties of the input signal. The principles of these two methods are shown in Figures 7.12 and 7.13.

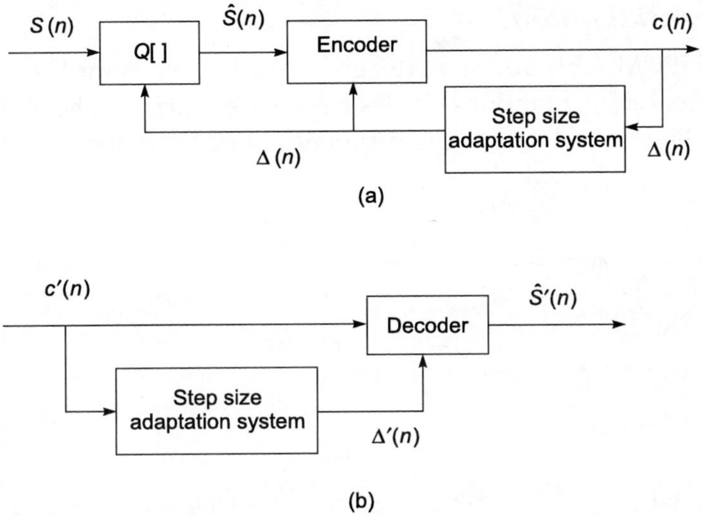

Figure 7.12 Feedback adaptation of the step size (a) coder, and (b) decoder

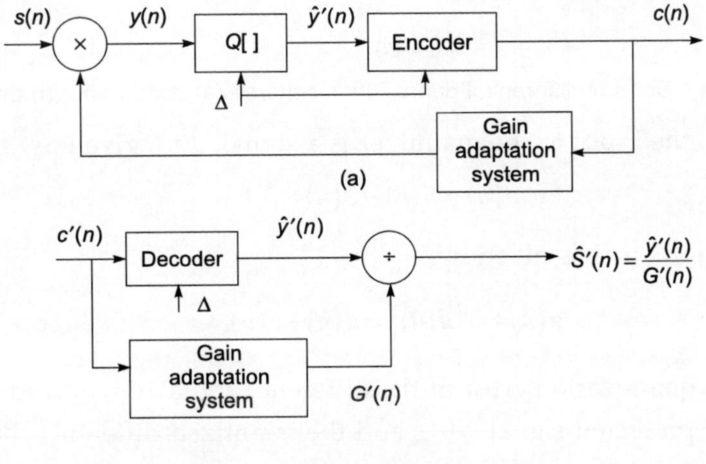

Figure 7.13 Feedback adaptation of the time varying gains (a) coder, and (b) decoder

Optimum step size or gain is decided according to the variance calculated for every block and is then transmitted as side information. Hence, feedforward adaptation scheme can accurately control gain level or step size, but the side information must be transmitted to the decoder. Hence, this method has a higher bit rate because of the side information. In the feedback adaptation method the step size and gain are estimated from the quantizer output. Hence, in the feedback adaptation method, the step size or gain does not need to be transmitted, since it can be automatically generated sample by sample by using reconstructed samples at both the ends.

7.4.2.5 *Differential quantization*

In a differential PCM, difference between the adjacent samples is quantized. Differences between adjacent input samples have lower variance, hence quantization error is reduced. Figure 7.14 shows the differential PCM scheme.

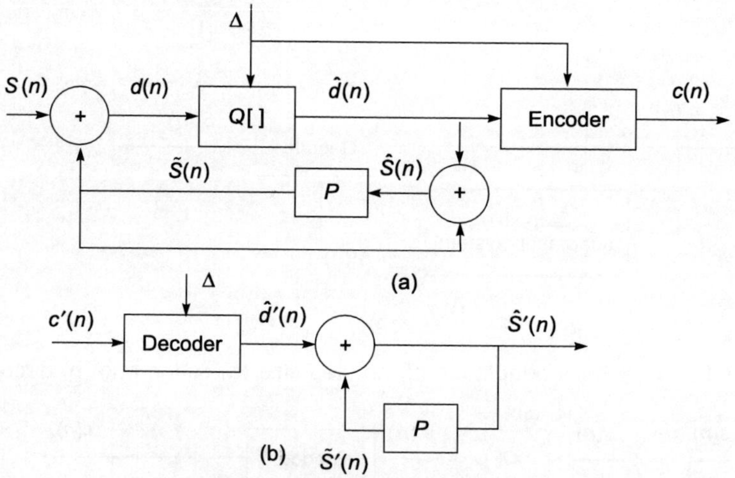

Figure 7.14 General differential quantization scheme (a) coder, and (b) decoder

In this system, the input to the quantizer is a signal, $d(n)$, given by

$$d(n) = S(n) - \tilde{S}(n) \tag{7.39}$$

Quantizer output, $\hat{d}(n)$, is given by

$$\hat{d}(n) = q\{d(n)\} = d(n) + e(n) \tag{7.40}$$

where $e(n)$ is the quantization error of the difference signal. Quantized signal, $\hat{S}(n)$, is the sum of the predicted signal, $\tilde{S}(n)$, and the quantized difference, $\hat{d}(n)$.

$$\hat{S}(n) = \tilde{S}(n) + \hat{d}(n) \tag{7.41}$$

where, the prediction value, $\tilde{S}(n)$, is a linear combination of the past, p, quantized values.

$$\tilde{S}(n) = \sum_{k=1}^{p} a_k \hat{s}(n-k) \tag{7.42}$$

Substituting Eqs. (7.39) and (7.40) in Eq.(7.41), we obtain

$$\hat{S}(n) = S(n) - d(n) + d(n) + e(n) = S(n) + e(n) \tag{7.43}$$

The quantized speech sample differs from the input only by the quantization error of the difference signal. Thus, if the prediction is good, the variance, $d(n)$, will be smaller than the variance of $S(n)$.

Signal-to-noise ratio (SNR)

SNR of a system is given by

$$\text{SNR} = \frac{E\left[S^2(n)\right]}{E\left[e^2(n)\right]} = \frac{\sigma_S^2}{\sigma_e^2} \tag{7.44}$$

$$= \frac{\sigma_S^2}{\sigma_d^2}\frac{\sigma_d^2}{\sigma_e^2} = G_P\,\text{SNR}_Q \tag{7.45}$$

where, $G_P = \frac{\sigma_S^2}{\sigma_d^2}$ and $\text{SNR}_Q = \frac{\sigma_d^2}{\sigma_e^2}$

where, G_P is defined as the gain due to differential configuration. SNR can be improved by maximizing G_P, G_P can be maximized by maximizing the denominator. If $d(n)$ is small, SNR is improved.

7.4.2.6 Jayanth quantizer

N.S. Jayanth at Bell Labs showed that step size of the quantizer can be adjusted by observing a single output instead of tracking N output quantized samples. In this method:

- Quantizer tracks only a single output quantized sample and takes a decision on the step size.
- If the output falls in the outer or inner level of the quantizer, then it increases or decreases the step size, respectively.
- The increasing and decreasing of the step size is done by assigning a multiplier, M_i, to each interval. Update Δ acoording to

$$\Delta(n) = M_i(n-1)\Delta(n-1)$$

where, $\Delta(n)$ are new Δ values, $M_i(n-1)$ last samples level index and $\Delta(n-1)$ are old Δ values.

- Multipliers are symmetric, multipliers for outer levels are > 1 and multipliers for outer levels are < 1 and specify Δ_{min} and Δ_{max} to avoid going out of the range.

The block diagram of Jayant quantizer is shown in Figure 7.15

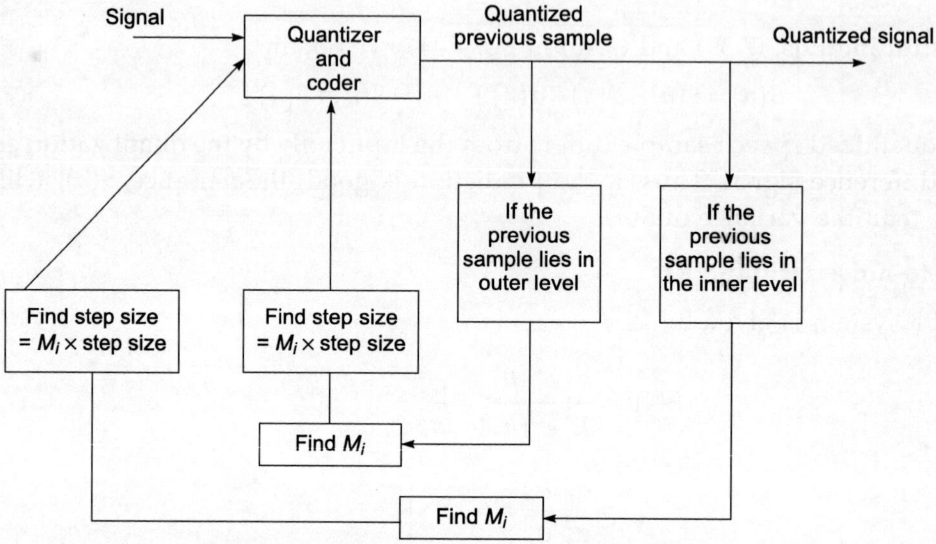

Figure 7.15 Block diagram of Jayant quantizers

Input and output characteristics of the quantizer with multiplier values, $M_1 = 0.8$, $M_2 = 0.9$, $M_3 = 1.0$ and $M_4 = 1.2$, are shown in Figure 7.16. Jayanth quantizer works fine for the speech signals, since values of the speech signal do not change drastically from sample to sample.

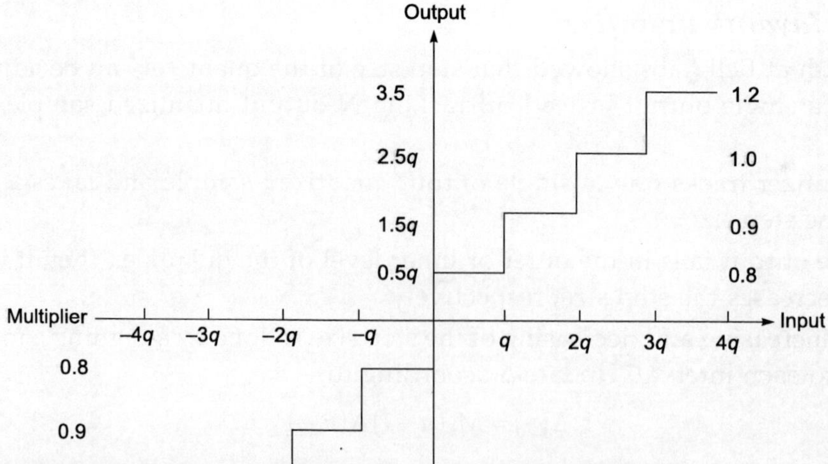

Figure 7.16 Multipliers for different levels of quantizer

7.5 CLASSIFICATION OF SPEECH CODERS

Speech coders are typically categorized in three groups.

- Waveform coders
- Parametric coders
- Hybrid coders

7.5.1 Waveform Coders

The simplest approach to encoding the speech signal is to use the time waveform for the digital representation of speech. This type of coding is called waveform coding. These coders preserve the original shape of the signal waveform and suited for high bit-rate coding. In practice, these coders work best at rates greater than 8 kbps and produce good quality of speech. The process, employed in waveform coding, consists of first sampling the analog signal at the Nyquist rate and then using an appropriate method sampled signal is quantized and encoded. Then each sample of the encoded signal is transmitted. Some example of this class includes various kinds pulse code modulation, delta modulation. Sub-band coding, adaptive transform coding, vector quantization, etc. Waveform coding is used for a faithful reproduction of speech waveform. Hence, waveform coded speech is useful for message transmission from one place to another. It is not useful for applications such as voice conversion and text-to-speech synthesis. Generally, signal-to-noise ratio can be used to measure the quality of waveform coders.

7.5.2 Parametric Coders

These coding schemes are based on speech production model which is controlled by some parameters and this class of coders works well for low bit-rate. In this model, the excitation signal is separated from the spectral envelope of the speech signal and each of them coded separately leading to low bit-rate saving. They use explicit model of speech production as shown in Figure 7.17.

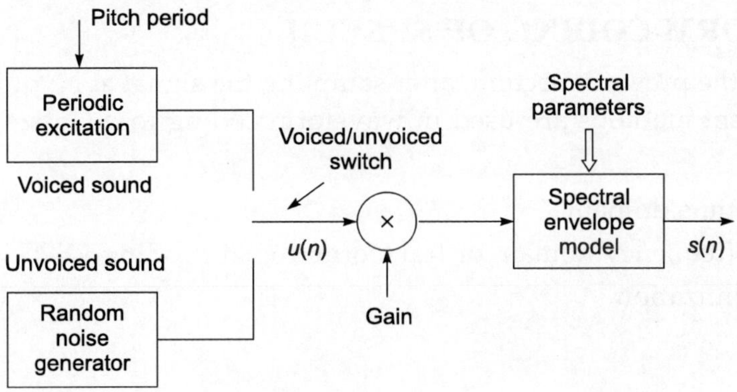

Figure 7.17 Speech production model for parametric coders

During encoding parameters of the model are estimated from the input speech signal (analysis stage), then these parameters are transmitted as the encoded bit-stream. Then these parameters are used at the receiver to reconstruct the speech signal (synthesis stage). These techniques are popular because they can afford representation of the speech signal at low-bit rates (2.4 -9.6 Kbits per sec). Waveform coders cannot preserve such good quality of speech at such low rates. But disadvantage of this method is speech quality cannot be improved beyond certain level even if the data rate grows to the level of waveform coders. Increasing the bit rate normally does not translate into better quality, since it is restricted by the chosen model. The analysis /synthesis methods have also been traditionally called vocoders (for voice coders). In this chapter, we will discuss four types of parametric coders, they are:

- Channel vocoder
- Formant vocoder
- Linear predictive vocoder
- Cepstral (homomorphic) vocoder

The above techniques differ only in the way in which the information about the spectral envelope is supplied. In the above mentioned techniques, the channel vocoder is the oldest (developed in 1939) and the linear predictive coder is the most successful and most popular one. In this approach the human speech production mechanism is summarized using a time varying filter, with the coefficients of the filter found using the linear prediction analysis procedure (Chapter 6).
Hybrid coders are partly waveform based partly parametric based and operate in the 2.4 -16 Kbps range. Hence, hybrid coders combine the strength of waveform coders and parametric coders. Hybrid coders like a parametric coders, it relies on a speech production model. Hybrid coders tend to behave like a waveform coder for high bit rate and like a parametric coder at a low bit rate, with fair to good quality for medium bit rate.

7.6 WAVEFORM CODING OF SPEECH

As discussed in the previous section, after sampling the signal at Nyquist rate, then following different methods are used in waveform coding to quantize and encode each sample:

- Coding in time domain
- Coding in frequency domain or transform domain coding
- Vector quantization

7.6.1 Coding in Time Domain

The simplest approach to encoding the speech signal is to use the time waveform for digital representation of the speech signal. Some examples of this class include various kinds of pulse code modulation (PCM), companded PCM, adaptive PCM, differential PCM, delta modulation, and adaptive PCM. A quantizer can be specified by its input partitions and output levels (also called reproduction points). If the input range is divided into levels of equal spacing, then the quantizer is termed a *uniform quantizer*, and if not, it is termed a *non-uniform quantizer*.

7.6.1.1 Pulse code modulation (PCM)

Pulse code modulation is the simplest form of coding the speech waveform. PCM coding is a basic coding scheme used for a telephone speech signal. In this technique, the speech waveform is sampled regularly at uniform intervals and then quantized using either a mid-riser or a mid-tread quantizer and the quantized samples are encoded and then transmitted. In 8-bit PCM every sample is encoded using 8-bit binary code and is transmitted in the pulse coded form. If one is represented by +5 V and zero is represented by 0 V, PCM coded waveform is shown in Figure 7.18.

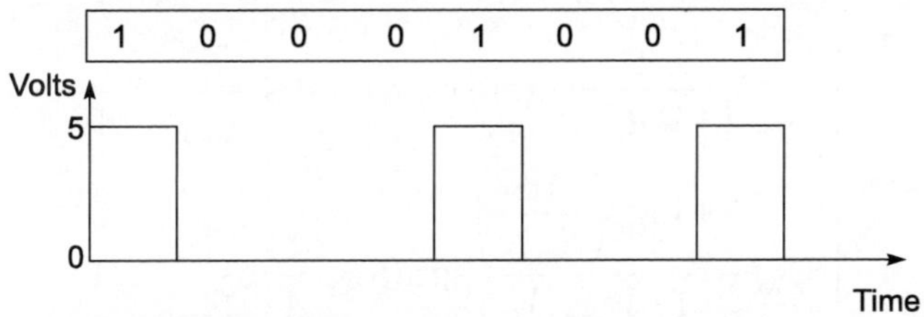

Figure 7.18 PCM coded waveform for 10001001

Manchester coding is also used in PCM. In this coding one is represented as +5 V for half the bit duration and as –5 V for the next half of the bit duration, zero is represented as –5 V for half the bit duration and as +5 V for the next half of bit duration as is shown in Figure 7.19.

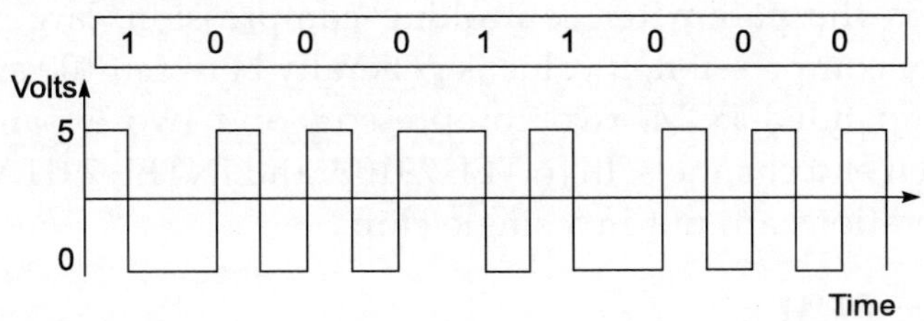

Figure 7.19 Manchester coding waveform for 10001001

In case of PCM, step size is calculated

$$\text{Step size} = \frac{\text{Dyamic range of signal}}{\text{Number of quantizer levels}} \tag{7.46}$$

If input range is 0 V to +1 V, quantizer step size is Δ and if *B* bits are used for quantization, then the number of quantizer level is 2^B. If two bits are used for quantization, then the step size (Δ) is given by

$$\Delta = \frac{1}{2^B} = \frac{1}{2^2} = 0.25$$

7.6.1.2 Companded PCM

In companded PCM, companded quantizers are used to maintain SNR constant by enhancing the low-level signals and compressing the high-level signals as discussed in section 7.4.2.3. Two kinds of transformation formulae *A*-law and *μ*-law are used in actual PCM system.

Formula for *A*-law companding is

$$c(x) = \begin{cases} \dfrac{A|S|}{1+\log_e A}\,\text{sgn}(S); & 0 \le \dfrac{|S|}{S_{\max}} \le \dfrac{1}{A} \\[2ex] S_{\max}\dfrac{1+\log_e\left(\dfrac{A|S|}{S_{\max}}\right)}{1+\log_e A}\,\text{sgn}(S) & \dfrac{1}{A} < \dfrac{|S|}{S_{\max}} \le 1 \end{cases} \tag{7.47}$$

Formula for *μ* law companding is

$$c(x) = S_{\max}\frac{\log_e\left(1+\mu|S_t|/S_{\max}\right)}{\log_e(1+\mu)}\,\text{sgn}(S_t)\ \mu \ge 0 \tag{7.48}$$

Here S_t is a sample value of the speech signal, $S_{\max}$ is the maximum permissible input level and *μ* is the parameter controlling compression, larger the *μ* value, larger the amount of compression, *μ* value is generally between 100 and 500. In case of multiplexed companded PCM, two compressors and two expanders are used alternatively for even end channels. In INTEL 2910A and INTEL 2911A, compressing and expanding operations are put in a single chip.

7.6.1.3 Adaptive PCM

For improving the SNR of quantized speech, in adaptive quantizer step size or gain vary according to the input signal, as discussed in section 7.4.2.5. In the first case step size increases or decreases according to the variance of the input signal and in the second case gain changes inversely, with changes in the variance of the input signal. There are two methods for varying step size or gain. They are:

- Feedforward adaptation
- Feedback adaptation

In the first method, i.e., feedforward adaptation method, the step size or gain is changed for every block. In the first case, the step size should increase and decrease with increase and decrease of the variance or the standard deviation of the input. In the second method, known as feedback adaptation method, step size or gain is changed on a sample-by-sample basis. In the second case, to hold the uniform quantizer at a fixed level, code word and AGC signals are transmitted. We have to calculate the variance of the block of data and divide each sample value by the tracked variance of the block. This normalizes the signal by the variance and provides automatic gain control for the signal. In the receiver, normalized signal is again denormalized by multiplying it with the variance received.

7.6.1.4 Differential pulse code modulation (DPCM)

In DPCM, instead of coding signal directly, difference between adjacent samples or the difference between the actual sample value and the predicted value calculated is quantized (section 7.4.2.5). Since the speech signals have high correlation between successive samples, the difference between the actual sample value and the predicted value is very small. The signal has a smaller dynamic range and hence it increases the SNR as discussed in section 7.4.2.5. Number of bits used for quantization will be reduced; this will reduce the bit rate for transmission and improve the speech coding efficiency.

Example 7.6 Perform DPCM encoding to produce the binary code for 5-bit input data $S(0)=6, S(1)=8$ and $S(2)=13$, which is compressed to 3 bits and performs decoding to recover the data using the binary code. Assume the first order predictor.

Solution:

TABLE 7.3 Quantization table for the 3-bit mid-tread quantizer

Binary code	Quantization level	Input signal sub-range
011	11	$7 \le d(n) \le 15$
010	5	$3 \le d(n) \le 7$
001	2	$1 \le d(n) \le 3$
000	0	$0 \le d(n) \le 1$
100	0	$-1 \le d(n) \le 0$
101	–2	$-3 \le d(n) \le -1$
110	–5	$-7 \le d(n) \le -3$
111	–11	$-15 \le d(n) \le -7$

DPCM encoding with the first order predictor

$$\tilde{S}(n) = \hat{s}(n-1)$$

For $n = 0$, we get

$$\tilde{S}(0) = \hat{s}(-1) = 0$$

$$d(0) = S(0) - \tilde{S}(0) = 6 - 0 = 0$$

From Table 7.3, $\hat{d}(0) = q\{d(0)\} = 5$

$$\hat{S}(0) = \tilde{S}(0) + \hat{d}(0) = 0 + 5 = 5$$

From Table 7.3, Binary code is 010

For $n = 1$, we get

$$\tilde{S}(1) = \hat{s}(0) = 5$$

$$d(1) = S(1) - \tilde{S}(1) = 8 - 5 = 3$$

From Table 7.3, $\hat{d}(1) = q\{d(1)\} = 2$

$$\hat{S}(1) = \tilde{S}(1) + \hat{d}(1) = 5 + 2 = 7$$

From Table 7.3, Binary code is 001

For $n = 2$, we get

$$\tilde{S}(2) = \hat{s}(1) = 7$$

$$d(2) = S(2) - \tilde{S}(2) = 13 - 7 = 6$$

From Table 7.3, $\hat{d}(2) = q\{d(2)\} = 5$

$$\hat{S}(2) = \tilde{S}(2) + \hat{d}(2) = 7 + 5 = 12$$

From Table 7.3, Binary code is 010

DPCM decoding with the first order predictor

For $n = 0$, binary code is 010

From Table 7.3, $\hat{d}(0) = 5$

$$\tilde{S}(0) = \hat{s}(-1) = 0$$

$$\hat{S}(0) = \tilde{S}(0) + \hat{d}(0) = 0 + 5 = 5$$

For n = 1, binary code is 001

From Table 7.3, $\hat{d}(0)=2$

$$\tilde{S}(1)=\hat{s}(0)=5$$

$$\hat{S}(1)=\tilde{S}(1)+\hat{d}(1)=5+2=7$$

For n = 0, binary code is 010

From Table 7.3, $\hat{d}(2)=5$

$$\tilde{S}(2)=\hat{s}(1)=7$$

$$\hat{S}(2)=\tilde{S}(2)+\hat{d}(2)=7+5=12$$

Hence, the output of the decoder is (i.e., recovered data) $\hat{S}(0)=5, \hat{S}(1)=7, \hat{S}(2)=12$. Recovered data has a quantization error. Hence, DPCM is a lossy data compression scheme.

7.6.1.5 Adaptive differential PCM (ADPCM)

The differential pulse code modulation (DPCM) described in the previous section has a fixed predictor and a fixed quantizer. Adaptation can be performed on the quantizer and predictor and the resulting system is called adaptive differential quantizer (ADPCM). Figure 7.20 shows the ADPCM with forward adaptation. In the forward ADPCM scheme the side information, step size or gain is transmitted.

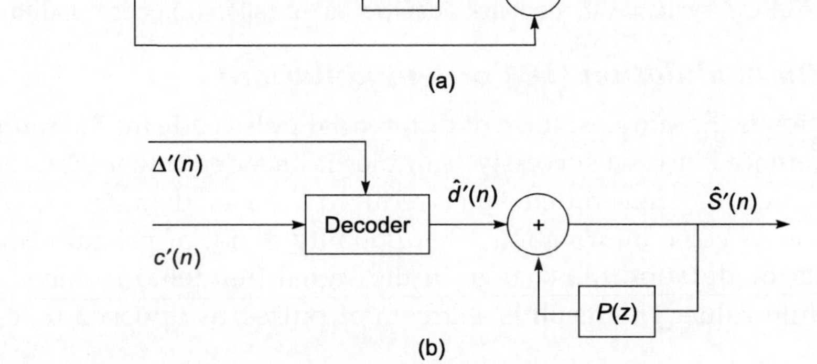

Figure 7.20 ADPCM system with feed-forward adaptive quantization (a) coder, and (b) decoder

Figure 7.21 shows an alternative ADPCM scheme with backward prediction. In this predictor gain and step size are derived from the quantized normalized prediction error samples: hence, there is no need to transmit any additional parameters. In feed-forward and feedback ADPCM will achieve SNR that is 10-12 dB greater than the fixed quantizer with the same number of levels. Both methods produce an SNR of roughly 22 dB at 32 Kbps (8 kHz sampling and 4-bit quantization), which is roughly 8 dB higher than that of Log PCM for the same data rate. It covers wider amplitude range than log PCM with the same bit rate. One shortcoming of the forward adaption scheme is the delay introduced due to the necessity of collecting a given number of samples before processing can start. Hence, backward adaptation is often preferred in those applications where delay is critical.

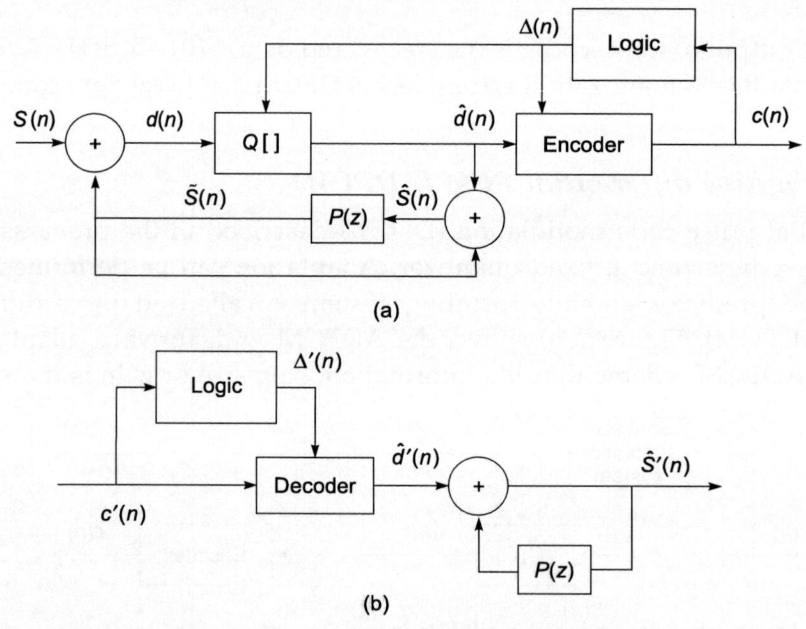

Figure 7.21 ADPCM system with feedback adaptive quantization (a) coder, and (b) decoder

7.6.1.6 *Delta modulation (DM or Δ-modulation)*

Delta modulation is the simplest form of differential pulse code modulation (DPCM) where the difference between successive samples is encoded into n-bit data streams. In delta modulation, the transmitted data is reduced to a 1-bit data stream. DM is used for transmission of voice information where quality is not of primary importance. Hence, in delta modulation the change in the signal (its delta) is encoded, rather than the absolute value. The result is a stream of pulses, as opposed to a stream of numbers.

Its main features are:

- The analog signal is approximated with a series of segments and each segment of the approximated signal is compared with the previous signal to determine whether there is a increase or decrease in relative amplitude.
- The increase or decrease of the signal amplitude from the previous sample is sent as 0 or 1. Whereas a no-change condition causes the modulated signal to remain at the same 0 or 1 state of the previous sample.

Selection of step size (Δ) and sampling time (T_s):

In delta modulation, step size (Δ) should not be too small, nor too large.

(1) • small Δ+ signal changes rapidly $\Rightarrow$ underestimation

(2) • large Δ+ signal changes slowly $\Rightarrow$ overestimation

Sampling time (T_s)

- smaller T_s increase overall accuracy
- but, small T_s increases output data rate, i.e., # of bps

Hence, in delta modulation rule is smaller $\Delta \Rightarrow$ smaller T_s, larger $\Delta \Rightarrow$ larger T_s.

The simplest form of differential coding is the first order, one-bit linear delta modulator. It has a single predictor. Figure 7.22 shows the delta modulation scheme.

$$d(n) = S(n) - \tilde{S}(n)$$

where, the prediction value $\tilde{S}(n)$ is a linear combination of the past quantized values.

$$\tilde{S}(n) = \sum_{k=1}^{p} a_k S(n-k)$$

In delta modulation, since it has single predictor, if $a_1 = 1$, then

$$\tilde{S}(n) = \hat{S}(n-1)$$

Using Eq. (7.40) and substituting the value of $\hat{S}(n-1)$, $d(n)$ is given by

$$d(n) = S(n) - \hat{S}(n) = S(n) - S(n-1) - e(n-1)$$

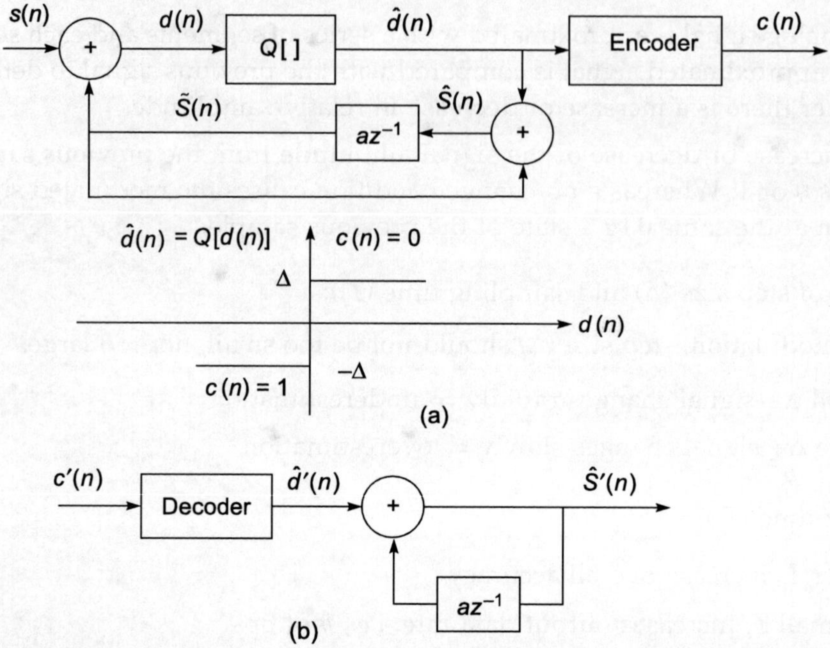

Figure 7.22 Delta modulation system (a) coder, and (b) decoder

where, $e(n-1)$ is the quantization error of the difference signal. In this case, the quantizer has only two levels and the step size is fixed. In this method, we transmit whether the current sample is above or below the previous sample.

$$\hat{d}(n) = \begin{cases} \Delta & \text{if } S(n) > S(n-1) \\ -\Delta & \text{if } S(n) \le S(n-1) \end{cases}$$

The main advantage of LDM is its simplicity. The system can be implemented with simple analog and digital integrated circuits. Since only one bit code is required, no synchronization of bit patterns is required between the transmitter and receiver. Figure 7.23 shows the delta modulation output.

Delta modulation offers higher signal-to-noise ratio (SNR) than other conventional modulation techniques. SNR is the strength ratio between data and noise in a transmitted signal, and is calculated in decibels. In data communications, signal power is always inversely proportional to the noise power, which means that higher signal power results in higher SNR. Other common modulation techniques such as pulse code modulation and differential pulse code modulation have lower SNR compared to delta modulation.

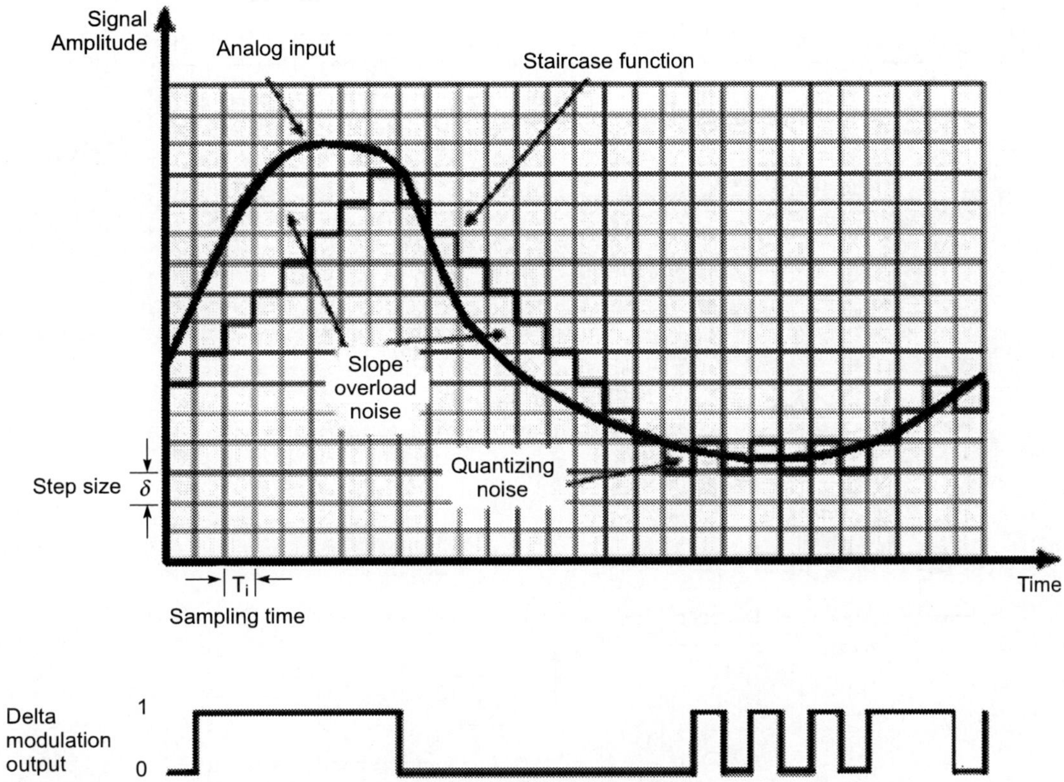

Figure 7.23 Delta modulation output

7.6.1.7 *Adaptive delta modulation*

Adaptive delta modulation (ADM) or continuously variable slope delta modulation (CVSD) is a modification of DM in which the step size is not fixed. Most of these schemes are of the feedback types in which the step size for the quantizer is adapted from the output code words. Figure 7.24 shows the feedback type adaptive delta modulation scheme.

In such schemes no synchronization bit pattern is required, since step size information is derived from the code word sequence at both the transmitter and the receiver.

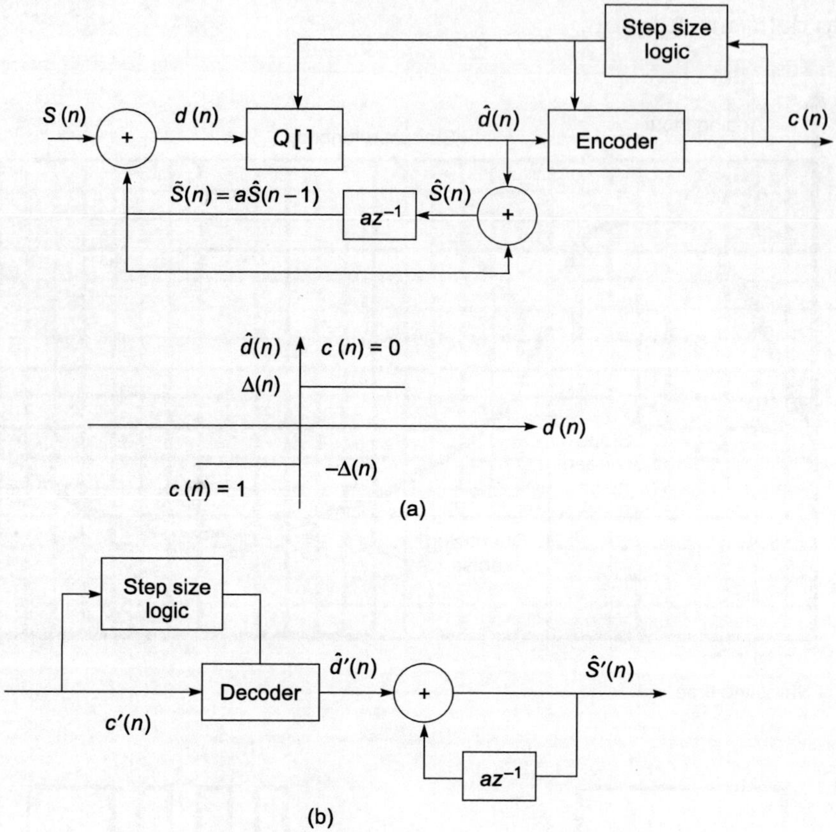

Figure 7.24 Feedback type ADM with step size (a) coder, and (b) decoder

7.6.1.8 Second-order and higher-order delta modulators

In delta modulation systems, greater SNR is possible with a second order predictor, i.e.,

$$\tilde{S}(n) = a_1\hat{S}(n-1) + a_2\hat{S}(n-2)$$

$$d(n) = S(n) - \hat{S}(n) = S(n) - a_1 s(n-1) + a_2 s(n-2) - e(n-1) - e(n-2)$$

where, $e(n-1)$ and $e(n-2)$ are the quantization errors of the difference signal. Hence, second order prediction gives improved performance over the first order predictor, may be as high as 4 dB depending on the speaker and speech material. But higher order predictor in adaptive delta modulation systems is not just a simple matter since algorithm interacts with prediction algorithm.

7.6.2 Coding in Frequency Domain or Transform Domain Coding

Frequency domain techniques for speech coding have recently received considerable attention. The basic concept of these methods is to divide the speech into frequency components by a filter bank (sub-band coding), or by a suitable transform (transform coding), and then encode them using adaptive PCM. Three basic factors are involved in the design of these coders: 1) the type of the filter bank or transform, 2) the choice of bit allocation, and 3) the control of the step size of the encoders. In this section two types of coders are discussed, they are:

- Sub-band coders (SBC)
- Adaptive transform coding (ATC)

7.6.2.1 Sub-band coders (SBC)

The coding method, in which a speech band is divided into several contiguous bands by a bank of band-pass filters (BPFs) and a specific coding strategy is employed for each band signal, is called sub-band coding. The block diagram of SBC is depicted in Figure 7.25.

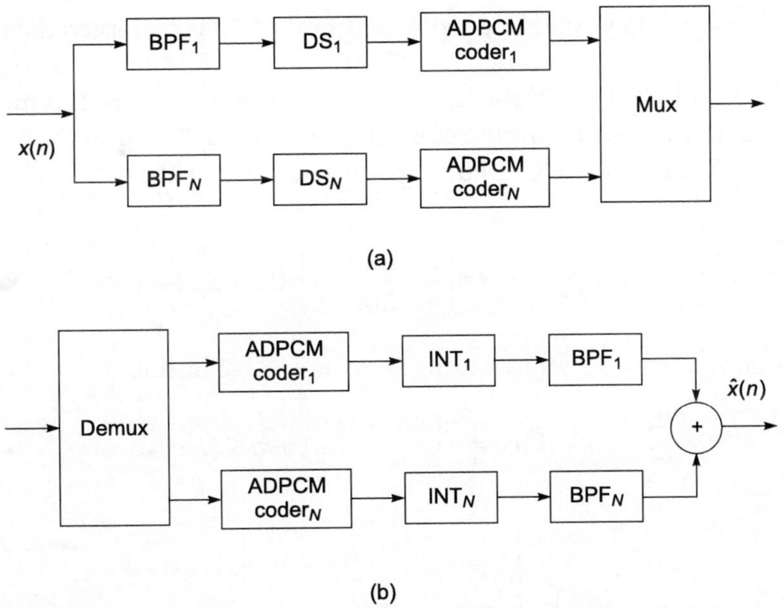

Figure 7.25 (a) coder (b) decoder SBC DS = Downsampling, INT = interpolation

In voiced speech, the sub-band usually contains at most one formant, however the formant can differ in amplitude by 30-40 dB (F_1 is having most amplitude). As shown in Figure 7.25, the SBC divides the spectrum into 4-8 bands. Decimation process reduces the bands sampling rate below its corresponding Nyquist rate. After decimation it codes each signal using adaptive pulse code modulation (APCM).

Low frequency bands are sent with more bits than higher frequency bands. Since the low frequency bands contain pitch and formant information of voiced speech and high frequency bands contain random waveforms. The inverse procedure Figure 7.25 reproduces the original signal. The method is advantageous for two reasons. One is that human auditory characteristics can be easily applied. The other is that a higher bit rate can be allocated to low frequency bands (voiced speech) to preserve good quality of speech.

7.6.2.2 *Adaptive transform coding (ATC)*

Adaptive transform coding (ATC) is a method in which a speech signal is divided into several frequency bands in a way similar to that with SBC. Speech wave of around 20 ms is considered stationary, is extracted as a block or as frames. Hence, in this method speech wave is not buffered and divided into frames. The speech wave of every block is transferred to frequency domain components, which are subsequently encoded and quantized by adaptive quantization. Although various kinds of orthogonal transforms can be used such as FFT, DCT, etc. The most popular choice in ATC is the discrete cosine transform. MPEG was established in 1988, where the DCT plays a key role. The block diagram of ATC is depicted as shown in Figure 7.26. On the basis of periodicity, DCT can be classified into four types: DCT-1, DCT-2, DCT-3 and DCT-4. Of these, DCT-2 representations are the most widely used transforms and is often simply referred to as " DCT ". Throughout this book we have used DCT-2 has just DCT. For a given sequence $s(n)\, n = 0, 1 \ldots N-1$ DCT-2 is defined as

$$S_c(k) = c(k) \sum_{n=0}^{N-1} s(n) \cos\left(\frac{\pi k(2n+1)}{2N}\right) \quad 0 \le k \le N-1 \tag{7.49}$$

$x(n)$ can be obtained from $X_c(k)$ by taking inverse DCT-2 and is defined as

$$s(n) = \sum_{k=0}^{N-1} c(k) S_c(k) \cos\left(\frac{\pi k(2n+1)}{2N}\right) \quad 0 \le n \le N-1 \tag{7.50}$$

where
$$c(k) = \begin{cases} \dfrac{1}{\sqrt{N}} & \text{if } k = 0 \\ \sqrt{\dfrac{2}{N}} & \text{if } 1 \le k \le N-1 \end{cases}$$

Since Matlab supports $n = 1 \ldots N$ DCT-2 and IDCT-2 is also defined by

$$S_c(k) = c(k) \sum_{n=1}^{N} s(n) \cos\left(\frac{\pi(k-1)(2n-1)}{2N}\right) \quad 1 \le k \le N \tag{7.51}$$

$$s(n)=\sum_{k=1}^{N}c(k)S_c(k)\cos\left(\frac{\pi(k-1)(2n-1)}{2N}\right)\quad 1\le n\le N \tag{7.52}$$

where $$c(k)=\begin{cases}\frac{1}{N} & \text{if } k=1\\ \sqrt{\frac{2}{N}} & \text{if } 2\le k\le N\end{cases} \tag{7.53}$$

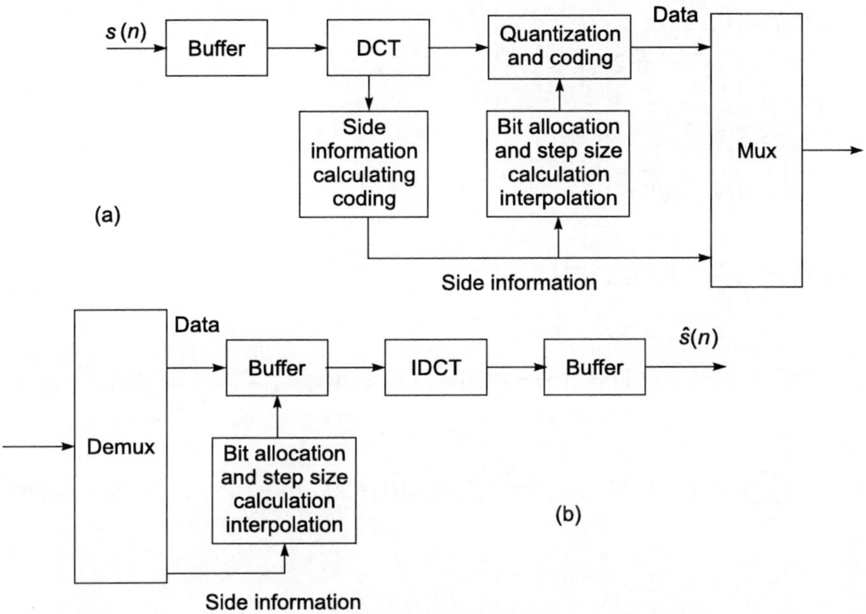

Figure 7.26 ATC (a) coder, and (b) decoder

The basic structure of ATC is depicted in Figure 7.26. The speech wave in each block is transformed to frequency domain using DCT and the resultant DCT coefficients are roughly divided into 20 sub-bands. The mean energy of each sub-band is calculated and coded as side information. A spectral envelope is obtained by interpolating the mean energy values of adjacent sub-bands. The number of bit allocation and corresponding quantization steps are optimally allocated to each DCT coefficient. This allocation produces a uniform level of quantization noise along the frequency axis. The inverse procedure (Figure 7.26) produces high good quality of speech as original speech signal.

Example 7.7 If $s(n)=\begin{Bmatrix}2, & 4, & 3, & 6, & 0, & 1, & 5, & 2\\ \uparrow & & & & & & & \end{Bmatrix}$, compute signal in frequency domain using DCT algorithm.

Solution:

Signal in frequency domain using DCT is given by

$$S_c(k) = c(k)\sum_{n=1}^{N} s(n)\cos\left(\frac{\pi(k-1)(2n-1)}{2N}\right) \quad 1 \le k \le N$$

where, $$c(k) = \begin{cases} \dfrac{1}{\sqrt{N}} & \text{if } k = 1 \\ \sqrt{\dfrac{2}{N}} & \text{if } 2 \le k \le N \end{cases}$$

$$c(1) = \frac{1}{\sqrt{N}} = \frac{1}{\sqrt{8}} = 0.3535$$

$$c(k) = \sqrt{\frac{2}{N}} \quad \text{if } 2 \le k \le N$$

$$c(2) = c(3) = c(4) = c(5) = c(6) = c(7) = c(8) = \sqrt{\frac{2}{8}} = \sqrt{\frac{1}{4}} = 0.5$$

$$S_c(1) = c(1)\sum_{n=1}^{N} s(n)\cos(0) = 0.3535\left[s(1)\cos(0) + s(2)\cos(0) + \ldots + s(8)\cos(0)\right]$$

$$S_c(1) = 8.05$$

$$S_c(k) = 0.5\sum_{n=1}^{N} s(n)\cos\left(\frac{\pi(k-1)(2n-1)}{2N}\right) \quad 2 \le k \le 8$$

$$S_c(2) = 0.5\sum_{n=1}^{N} s(n)\cos\left(\frac{\pi(2n-1)}{16}\right) = 0.7251$$

$$S_c(3) = 0.5\sum_{n=1}^{N} s(n)\cos\left(\frac{2\pi(2n-1)}{16}\right) = 0.0327$$

$$S_c(4) = 0.5\sum_{n=1}^{N} s(n)\cos\left(\frac{3\pi(2n-1)}{16}\right) = -2.5499$$

$$S_c(5) = 0.5\sum_{n=1}^{N} s(n)\cos\left(\frac{4\pi(2n-1)}{16}\right) = -1.0605$$

$$S_c(6) = 0.5\sum_{n=1}^{N} s(n)\cos\left(\frac{5\pi(2n-1)}{16}\right) = 3.1799$$

$$S_c(7) = 0.5\sum_{n=1}^{N} s(n)\cos\left(\frac{6\pi(2n-1)}{16}\right) = -2.6924$$

$$S_c(8) = 0.5\sum_{n=1}^{N} s(n)\cos\left(\frac{7\pi(2n-1)}{16}\right) = -1.833$$

Hence, $S_c(k) = \left\{\begin{matrix} 8.05, & 0.7251, & 0.0327, & -2.5499, & -1.0605, & 3.1799, & -2.6924, & -1.833 \\ \uparrow & & & & & & & \end{matrix}\right\}$

Example 7.8 If $S_c(k) = \left\{\begin{matrix} 8.05, & 0.7251, & 0.0327, & -2.5499, & -1.0605, & 3.1799, & -2.6924, & -1.833 \\ \uparrow & & & & & & & \end{matrix}\right\}$

compute signal in time domain using IDCT algorithm.

Signal in time domain using IDCT is given by

$$s(n) = \sum_{k=1}^{N} c(k) S_c(k)\cos\left(\frac{\pi(k-1)(2n-1)}{2N}\right) \quad 1 \le n \le N$$

where, $$c(k) = \begin{cases} \dfrac{1}{\sqrt{N}} & \text{if } k = 1 \\ \sqrt{\dfrac{2}{N}} & \text{if } 2 \le k \le N \end{cases}$$

$$c(1) = \frac{1}{\sqrt{N}} = \frac{1}{\sqrt{8}} = 0.3535$$

$$c(k) = \sqrt{\frac{2}{N}} \text{ if } 2 \le k \le N$$

$$c(2) = c(3) = c(4) = c(5) = c(6) = c(7) = c(8) = \sqrt{\frac{2}{8}} = \sqrt{\frac{1}{4}} = 0.5$$

$$s(1) = \sum_{k=1}^{N} c(k) S_c(k)\cos\left(\frac{\pi(k-1)}{16}\right)$$

$$s(1)=\Big[c(1)S_c(1)\cos 0+c(2)S_c(2)\cos\left(\frac{\pi}{16}\right)+c(3)S_c(3)\cos\left(\frac{2\pi}{16}\right)$$

$$+c(4)S_c(4)\cos\left(\frac{3\pi}{16}\right)+c(5)S_c(5)\cos\left(\frac{4\pi}{16}\right)+c(6)S_c(6)\cos\left(\frac{5\pi}{16}\right)$$

$$+c(7)S_c(7)\cos\left(\frac{6\pi}{16}\right)+c(8)S_c(8)\cos\left(\frac{6\pi}{16}\right)\Big]$$

$$s(2)=\sum_{k=1}^{N}c(k)S_c(k)\cos\left(\frac{3\pi(k-1)}{16}\right)=4$$

$$s(3)=\sum_{k=1}^{N}c(k)S_c(k)\cos\left(\frac{5\pi(k-1)}{16}\right)=3$$

$$s(4)=\sum_{k=1}^{N}c(k)S_c(k)\cos\left(\frac{7\pi(k-1)}{16}\right)=6$$

$$s(5)=\sum_{k=1}^{N}c(k)S_c(k)\cos\left(\frac{9\pi(k-1)}{16}\right)=0$$

$$s(6)=\sum_{k=1}^{N}c(k)S_c(k)\cos\left(\frac{11\pi(k-1)}{16}\right)=1$$

$$s(7)=\sum_{k=1}^{N}c(k)S_c(k)\cos\left(\frac{13\pi(k-1)}{16}\right)=5$$

$$s(8)=\sum_{k=1}^{N}c(k)S_c(k)\cos\left(\frac{15\pi(k-1)}{16}\right)=2$$

$$s(n)=\left\{\begin{matrix}2, & 4, & 3, & 6, & 0, & 1, & 5, & 2\\ \uparrow & & & & & & & \end{matrix}\right\}$$

Example 7.9 If $s(n)=\left\{\begin{matrix}0, & 1, & 1, & 2, & 3, & 1, & 4, & 2\\ \uparrow & & & & & & & \end{matrix}\right\}$, compute signal in frequency domain using DCT algorithm.

Solution:

Signal in frequency domain using DCT is given by

$$S_c(k) = c(k)\sum_{n=1}^{N} s(n)\cos\left(\frac{\pi(k-1)(2n-1)}{2N}\right) \quad 1 \le k \le N$$

where, $$c(k) = \begin{cases} \dfrac{1}{\sqrt{N}} & \text{if } k = 1 \\ \sqrt{\dfrac{2}{N}} & \text{if } 2 \le k \le N \end{cases}$$

$$c(1) = \frac{1}{\sqrt{N}} = \frac{1}{\sqrt{8}} = 0.35$$

$$c(k) = \sqrt{\frac{2}{N}} \quad \text{if } 2 \le k \le N$$

$$c(2) = c(3) = c(4) = c(5) = c(6) = c(7) = c(8) = \sqrt{\frac{2}{8}} = \sqrt{\frac{1}{4}} = 0.5$$

$$S_c(1) = c(1)\sum_{n=1}^{N} s(n)\cos(0) = 0.35\left[s(1)\cos(0) + s(2)\cos(0) + \ldots + s(8)\cos(0)\right]$$

$$S_c(1) = 4.949$$

$$S_c(k) = 0.5\sum_{n=1}^{N} s(n)\cos\left(\frac{\pi(k-1)(2n-1)}{2N}\right) \quad 2 \le k \le 8$$

$$S_c(2) = 0.5\sum_{n=1}^{N} s(n)\cos\left(\frac{\pi(2n-1)}{16}\right) = -2.3255$$

$$S_c(3) = 0.5\sum_{n=1}^{N} s(n)\cos\left(\frac{2\pi(2n-1)}{16}\right) = -0.8118$$

$$S_c(4) = 0.5\sum_{n=1}^{N} s(n)\cos\left(\frac{3\pi(2n-1)}{16}\right) = -0.2610$$

$$S_c(5) = 0.5\sum_{n=1}^{N} s(n)\cos\left(\frac{4\pi(2n-1)}{16}\right) = 0$$

$$S_c(6) = 0.5\sum_{n=1}^{N} s(n)\cos\left(\frac{5\pi(2n-1)}{16}\right) = 0.4998$$

$$S_c(7) = 0.5\sum_{n=1}^{N} s(n)\cos\left(\frac{6\pi(2n-1)}{16}\right) = -1.9598$$

$$S_c(8) = 0.5\sum_{n=1}^{N} s(n)\cos\left(\frac{7\pi(2n-1)}{16}\right) = 1.1286$$

$$\text{Hence, } S_c(k) = \left\{\begin{matrix} 4.949, & -2.3255, & -0.8118, & -2.261, & 0, & 0.4998, & 1.1286 \\ \uparrow & & & & & & \end{matrix}\right\}$$

Example 7.10 If $S_c(k) = \left\{\begin{matrix} 4.949, & -2.3255, & -0.8118, & -2.261, & 0, & 0.4998, & 1.1286 \\ \uparrow & & & & & & \end{matrix}\right\}$

compute signal in time domain using IDCT algorithm.

Signal in time domain using IDCT is given by

$$s(n) = \sum_{k=1}^{N} c(k)S_c(k)\cos\left(\frac{\pi(k-1)(2n-1)}{2N}\right) \quad 1 \le n \le N$$

where,
$$c(k) = \begin{cases} \frac{1}{\sqrt{N}} & \text{if } k = 1 \\ \sqrt{\frac{2}{N}} & \text{if } 2 \le k \le N \end{cases}$$

$$c(1) = \frac{1}{\sqrt{N}} = \frac{1}{\sqrt{8}} = 0.35$$

$$c(k) = \sqrt{\frac{2}{N}} \quad \text{if } 2 \le k \le N$$

$$c(2) = c(3) = c(4) = c(5) = c(6) = c(7) = c(8) = \sqrt{\frac{2}{8}} = \sqrt{\frac{1}{4}} = 0.5$$

$$s(1)=\sum_{k=1}^{N}c(k)S_c(k)\cos\left(\frac{\pi(k-1)}{16}\right)$$

$$s(1)=\left[c(1)S_c(1)\cos 0+c(2)S_c(2)\cos\left(\frac{\pi}{16}\right)+c(3)S_c(3)\cos\left(\frac{2\pi}{16}\right)\right.$$

$$+c(4)S_c(4)\cos\left(\frac{3\pi}{16}\right)+c(5)S_c(5)\cos\left(\frac{4\pi}{16}\right)+c(6)S_c(6)\cos\left(\frac{5\pi}{16}\right)$$

$$\left.+c(7)S_c(7)\cos\left(\frac{6\pi}{16}\right)+c(8)S_c(8)\cos\left(\frac{6\pi}{16}\right)\right]=0$$

$$s(2)=\sum_{k=1}^{N}c(k)S_c(k)\cos\left(\frac{3\pi(k-1)}{16}\right)=1$$

$$s(3)=\sum_{k=1}^{N}c(k)S_c(k)\cos\left(\frac{5\pi(k-1)}{16}\right)=1$$

$$s(4)=\sum_{k=1}^{N}c(k)S_c(k)\cos\left(\frac{7\pi(k-1)}{16}\right)=2$$

$$s(5)=\sum_{k=1}^{N}c(k)S_c(k)\cos\left(\frac{9\pi(k-1)}{16}\right)=3$$

$$s(6)=\sum_{k=1}^{N}c(k)S_c(k)\cos\left(\frac{11\pi(k-1)}{16}\right)=1$$

$$s(7)=\sum_{k=1}^{N}c(k)S_c(k)\cos\left(\frac{13\pi(k-1)}{16}\right)=4$$

$$s(8)=\sum_{k=1}^{N}c(k)S_c(k)\cos\left(\frac{15\pi(k-1)}{16}\right)=2$$

$$s(n)=\left\{\begin{matrix}0, & 1, & 1, & 2, & 3, & 1, & 4, & 2\\ \uparrow & & & & & & & \end{matrix}\right\}$$

7.6.2.3 Wavelet transform coding

The wavelet transform Haar, Daubechies, etc. (discussed in chapter 3) are used for speech compression. For many signals, the low-frequency content is the most important part. It is what gives the signal its identity. The high-frequency content, on the other hand, imparts flavour. Consider the human voice, if we remove the high-frequency components, the voice sounds different, but you can still tell what's being said. Speech compression using wavelet transform is performed with the following steps:

1. From the standard speech file, read speech samples frame by frame. Each frame consists of 256 samples.
2. Apply wavelet transform on each frame of 256 samples.
3. Apply thresholding (replace small wavelet coefficients by zero).
4. Encode wavelet coefficients.
5. Store encoded wavelet coefficients.
6. Read next frame and repeat steps 2 to 5 for entire speech file.

Speech is reconstructed using inverse wavelet transform with the following steps:

1. From the compressed file, read the first frame of wavelet coefficients.
2. Decode wavelet coefficients (restore zeros).
3. Apply inverse wavelet transform on the frame.
4. Store the frame in the standard speech file.
5. Read the next frame of wavelet coefficient and repeat steps 2 to 4 for the entire compressed file.

7.6.3 Vector Quantization

Vector quantization considers k samples of a speech signal or set of scalars of waveform coding or in the analysis and synthesis method are grouped into blocks or vectors. This scheme is therefore, also sometimes called *block quantization*. During transmission, blocks of vectors are represented by a single vector code, rather than values independently.

Figure 7.27 indicates the principle of vector quantization. In vector quantization, in waveform coding a certain period of sampled waveform is extracted and the waveform in this period is represented by a single code. The procedure is achieved by storing typical waveform patterns and giving a code to each pattern. The correspondence between patterns and codes is termed codebook. It provides an appropriate set of patterns in waveform coding. Selecting the most appropriate pattern from the codebook as efficiently as possible is an important issue in vector quantization.

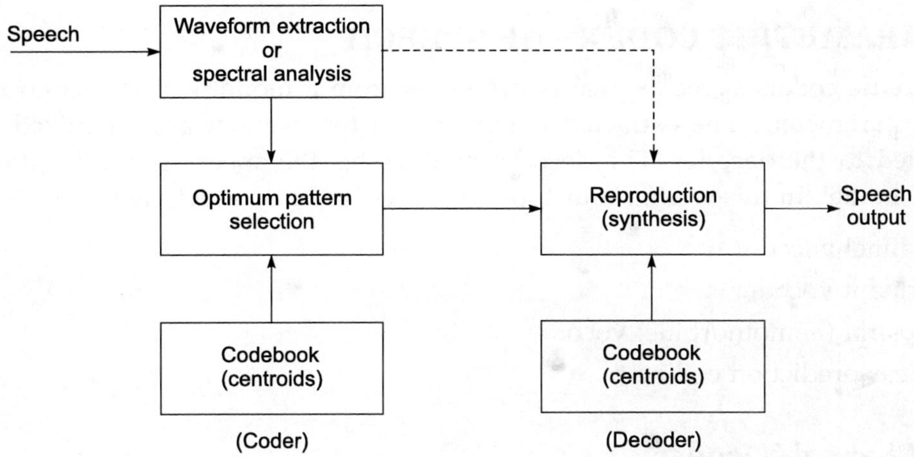

Figure 7.27 Vector quantizer

One of the most practical selection methods is binary tree coding (BTC). In binary tree coding, N-dimensional space is first divided into two regions, then each region is further divided in two sub-regions, process is repeated until the space is divided into 8 regions. Here, N is restricted to be a power of 2 and $N = 2^B$, where, B is an integer representing the number of bits. The number of binary decisions is $B = \log_2 N$. Figure 7.28 shows the division of the space into $N = 8$ regions.

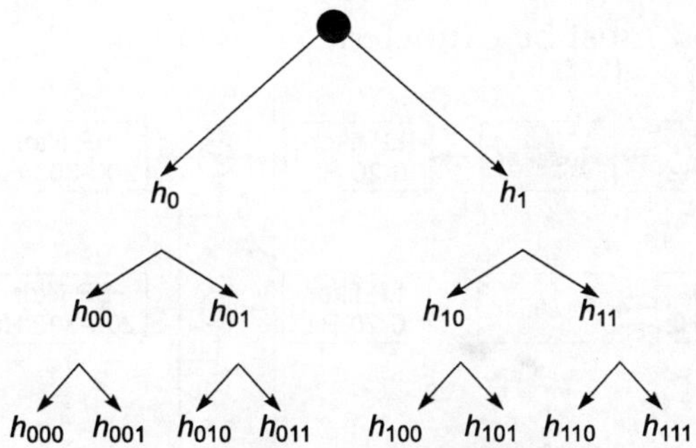

Figure 7.28 Principle of BTC

Hence, N-dimensional space is successively divided, and the input vector is compared with two code vectors at the last decision stage. Binary tree search replaces 2^B calculations with only N comparisons. Hence, binary search codebook results in low computation cost with higher storage. Whereas full search code requires high computation cost and low storage requirements.

7.7 PARAMETRIC CODERS OF SPEECH

In parametric coders speech signal is generated from a model which is controlled by some parameters. The extracted parameters at the encoder are quantized and transmitted to the decoder. The decoder synthesizes the speech according to the specified model. In this section four types of coders are discussed, they are:

- Channel vocoder
- Formant vocoder
- Cepstral (homomorphic) vocoder
- Linear prediction coding

7.7.1 Channel Vocoder

The oldest form of speech coding device is the channel vocoder which was invented by Dudley. This was the first practical proof that speech could be presented by analysis-synthesis systems, where the vocal source and the vocal tract characteristics are separated and presented separately and was demonstrated in New York in 1939. The vocal source and the vocal tract information are separated in the analyzer and then they are coded and transmitted to the synthesizer. At the synthesizer, the speech signal is reconstructed from the transmitted information. In this technique, determination of excitation signal for voiced/unvoiced signal is very important, if voiced signal pitch period must be extracted.

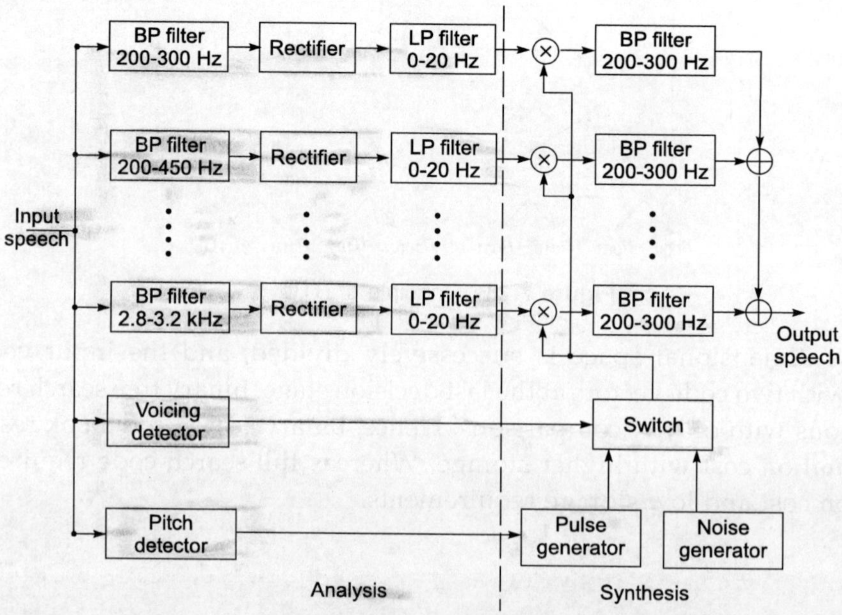

Figure 7.29 Channel vocoder

As shown in Figure 7.29, in the analyzer the speech signal is passed through a bank of band-pass filters covering the portion of the spectrum (0-4 kHz). Generally, 16-19 channels are chosen. Another consideration is the choice of bandwidth of the band-pass filters is according to the response of the human ear. The response of the human ear is linear on a logarithmic scale of the frequencies. The bandwidth is thus chosen to increase proportionally with frequency. The output of each filter is followed by a full-wave rectifier (magnitude block) which, in turn, followed by a low-pass filter. The full-wave rectifier and low-pass filter serve as an approximate envelope detector and determine the envelope of the signal in that bandwidth. The envelope is then sampled at a low bit rate and then encoded. Channel vocoder has an additional analysis component for determining the mode of excitation, i.e., voiced or unvoiced. If voiced, fundamental frequency or pitch period is estimated. The resulting excitation information together with the channel signals (output of low-pass filter) forms the representation for the speech signal.

The envelope is sampled at a low bit rate, encoded, multiplexed with outputs from the other filters and transmitted to the receiver. At the receiver, the inverse process takes place and a replica of input is produced.

7.7.2 Formant Vocoder

The most important elements of the speech spectrum are the vocal tract resonances which are called formants. Figure 7.30 shows an example where the formants have been marked. From the formant information speech can be synthesized as shown in Figure 7.30.

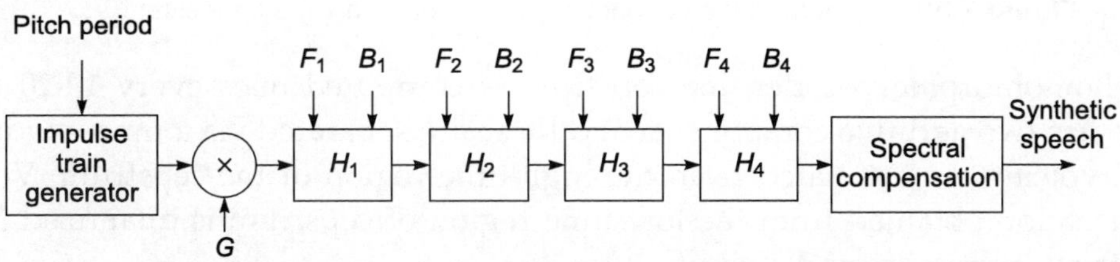

Figure 7.30 Speech production model using a formant synthesizer

Each formant is realized as a second order all pole filter according to the equation

$$H_i(Z) = \frac{1 - 2e^{-B_iT} \cos(2\pi F_i T) + e^{-2B_iT}}{1 - 2e^{-B_iT} \cos(2\pi F_i T) Z^{-1} + e^{-2B_iT} Z^{-2}} \tag{7.54}$$

Formant vocoder identifies and transmits formant frequencies and bandwidth of the first 3-4 formant frequencies of speech energies in a set of fixed bands as channel vocoders. Hence, it uses same model as that of the channel vocoder, but the vocal tract is represented by 3-4 formant frequencies and bandwidth. However, the difficulty of accurate formant tracking has hindered the use of formant vocoders.

7.7.3 Cepstral (Homomorphic) Vocoder

In cepstal processing, as discussed in section 5.9 leads to a convenient representation in which basic speech parameters: excitation information and vocal tract information are clearly displayed and isolated from one another. Excitation information is present in the high-time region of the cepstrum and vocal tract information in the low time region of the cepstrum. The homomorphic vocoder is depicted in Figure 7.31.

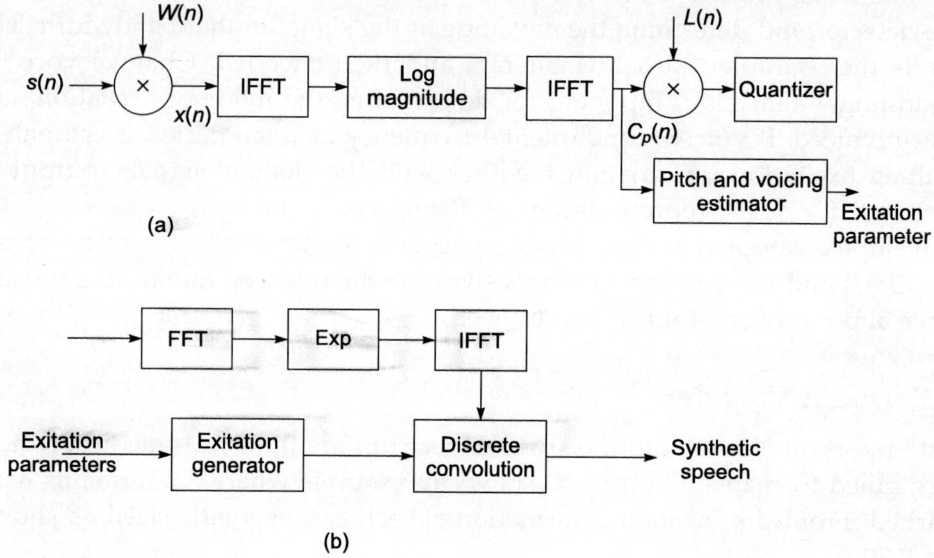

Figure 7.31 Homomorphic vocoder (a) analyser, and (b) synthesiser

In the homomorphic vocoder, the cepstrum is computed once every 10-20 ms. Cepstrum has two additive components and these are separated via a time window. Pitch and voicing are estimated from the high-time region of the cepstrum. Vocal tract information obtained from the low-time region of cepstrum is quantized and encoded for transmission and storage.

At the synthesizer the Fast Fourier transform of the low quefrency elements produces the logarithmic spectral envelope. Linear spectral envelope is obtained by taking the exponential transform of the logarithmic spectral envelope. At the synthesizer, the impulse response of voiced or unvoiced signal is computed by taking inverse FFT of linear spectral envelope. The impulse response is then convolved with the excitation function created at the synthesizer from pitch, voicing, and amplitude information. The main drawback of the cepstral vocoder has been the computation: two DFTs are required at the transmitter and also two DFTs are required at the receiver.

7.7.4 Linear Prediction Coding

One of the most powerful speech coding techniques is linear predictive coding (LPC) method for encoding good quality speech at a low bit rate and provides extremely accurate estimates of speech parameters. As discussed in section 6.3 the basic idea behind linear predictive analysis is that a speech sample can be approximated as a linear combination of past speech samples. The speech production model for LPC is shown in Figure 7.32.

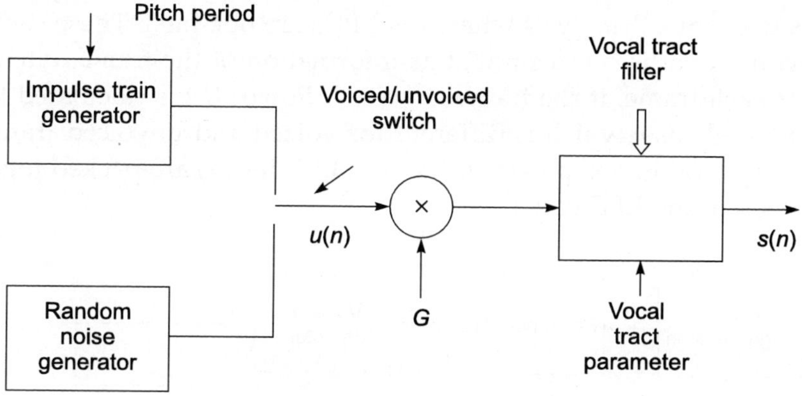

Figure 7.32 Speech production model for LPC

The vocal tract filter is defined by an all-pole model and is expressed as

$$H(Z) = \frac{G}{1+\alpha_1 Z^{-1} + \alpha_2 Z^{-2} + \ldots + \alpha_p Z^{-p}} = \frac{S(Z)}{E(Z)} \tag{7.55}$$

where, p is the order of the model. If $s(n)$ is the speech output of the model and $e(n)$ is the excitation input, the above equation in the time domain is expressed as

$$s(n) = Ge(n) - \alpha_1 s(n-1) - \ldots - \alpha_p s(n-p) \tag{7.56}$$

In other words every speech sample is computed as a linear combination of the previous speech sample with a contribution from the excitation.

The LPC method is accurate when it is applied to stationary signals whose behaviour does not change with time. However, this is not the case in speech. To be able to apply the LPC method, we segment the speech signal into small segments called frames, which are quasistationary. The segmentation is done by multiplying the speech signal, $s(n)$, by a window, $w(n)$, which is zero outside the interval we want to extract. The most popular choice of window is Hamming window and is given by

$$w(n) = \left.\begin{matrix} 0.54 - 0.46 \cos \dfrac{2\pi n}{N} & 0 \leq n \leq N-1 \\ 0 & \text{otherwise} \end{matrix}\right\} \tag{7.57}$$

Here, N is the window length in samples and is generally in the range 20-40 ms. Figure 7.33 shows the block diagram of the LPC encoder. The input speech is first segmented into non-overlapping frames. The voicing detector classifies the current frame as voiced or unvoiced and outputs one bit indicating the voicing state. Each frame signal is used for LP analysis where ten LPCs are obtained. These coefficients are encoded with the indices transmitted as information of the frame. Pitch period is estimated for each frame, if the frame is voiced. Power (P) is calculated for both voiced and unvoiced case, which is different for voiced and unvoiced frames. The voicing bit, pitch period index, power index, and LPC index, are packed together to form the bitstream of the LPC coder.

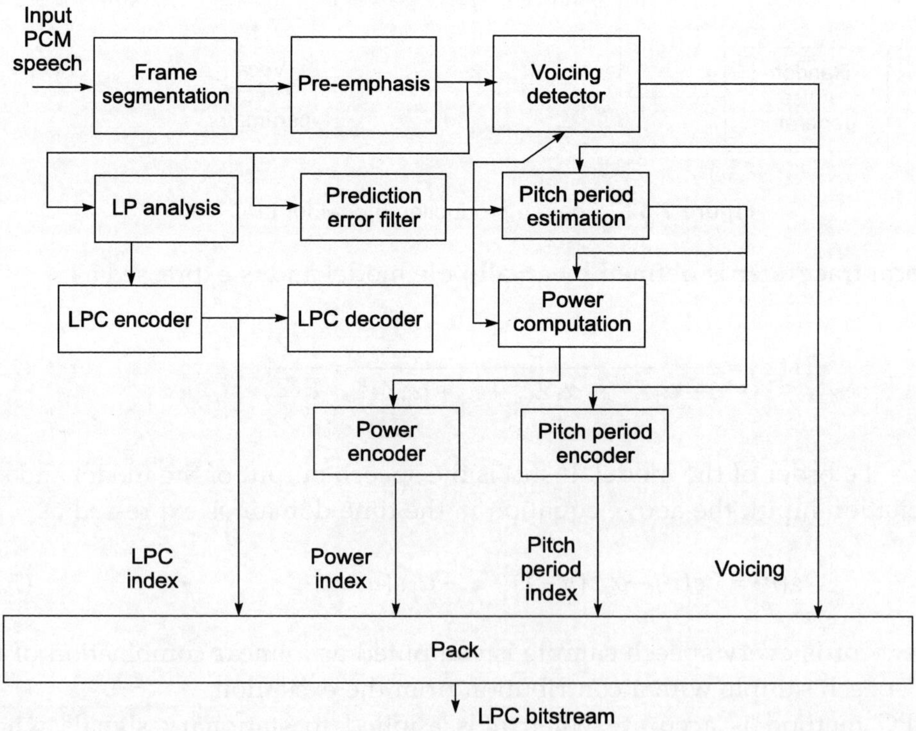

Figure 7.33 LPC encoder

Figure 7.34 shows the block diagram of the decoder and is essentially the LPC model of speech production with parameters controlled by the bitstream. It basically decodes various parameters from the bitstream which are directed to the corresponding frame.

Gain computation is performed as follows, for the unvoiced case, gain is given by

$$G = \sqrt{P}$$

For voiced case

$$G = \sqrt{TP}$$

where, T is the period and P is the power of impulse train. Finally, the output of the synthesis filter is de-emphasized using the above equation and yields synthesis speech.

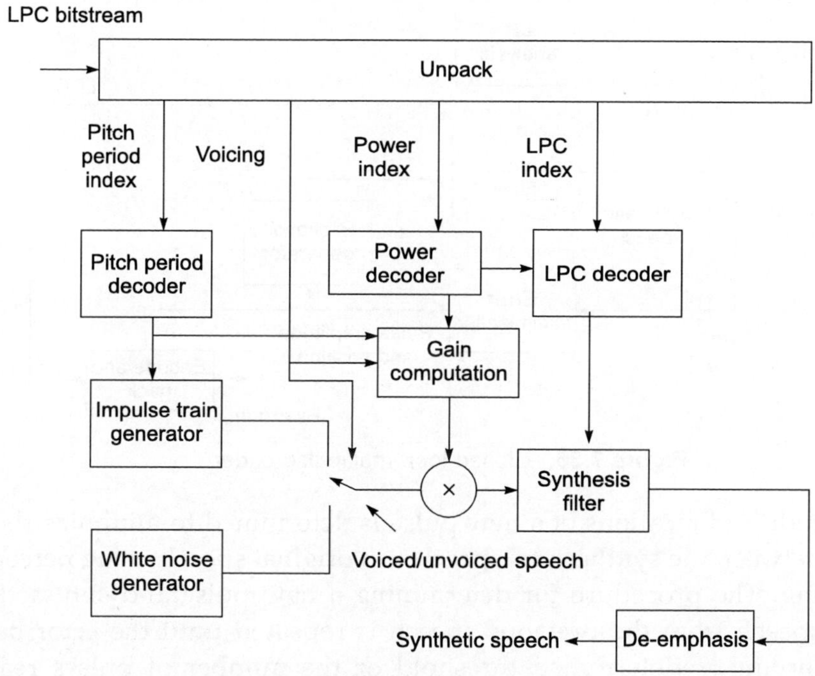

Figure 7.34 LPC decoder

7.8 HYBRID CODERS

After studying the shortcomings of waveform and parametric coding techniques, engineers combined the best features of each to produce hybrid coding techniques. In this section, two types of coders are discussed, they are:

- Multi-pulse excited linear prediction coder (MPC)
- Code-excited linear prediction coder (CELP)

7.8.1 Multi-Pulse Excited Linear Prediction Coder (MPC)

Multi-pulse excited LPC (MPC) is a system in which an LPC synthesis filter is excited by multiple pulses, regardless of whether the source is voiced or unvoiced. Figure 7.35 shows the encoder of a closed loop system. In this approach, the excitation sequence is generated through feedback sequence, where the difference between input speech and synthetic speech is minimized. If we assume the amplitude and locations of a certain number of pulses are already determined, the multiple pulses are transformed into synthesized speech through the LPC synthesis filter.

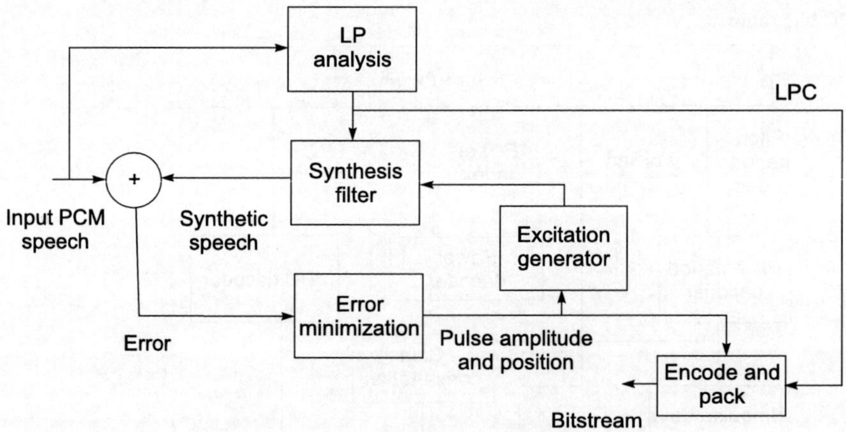

Figure 7.35 Closed loop multipulse coder

The amplitude and locations of a new pulse is determined to minimize the mean square error between the synthesized speech and original speech using perceptually based weighing. The procedure for determining a new pulse and subtracting the synthesized speech from the previous speech is repeated until the error becomes less than a certain predetermined threshold or the number of pulses reaches a predetermined number. The LPC index, pulse amplitude and position are encoded to form the bitstream. The decoder recovers the information, regenerates the excitation sequence, and passes it through a synthesis filter to obtain synthetic speech.

7.8.2 Code-Excited Linear Prediction Coder (CELP)

The most popular speech coders in the 4-8 Kbps range is code-excited linear prediction coder (CELP), which uses standard LP filters but with code words from a VQ-generated codebook, rather than by multi-pulses. The CELP strategy has been widely used in a variety of government and international standard coders. A block diagram of a CELP encoder is shown in Figure 7.36. The CELP coder relies on the long-term and short-term linear prediction models.

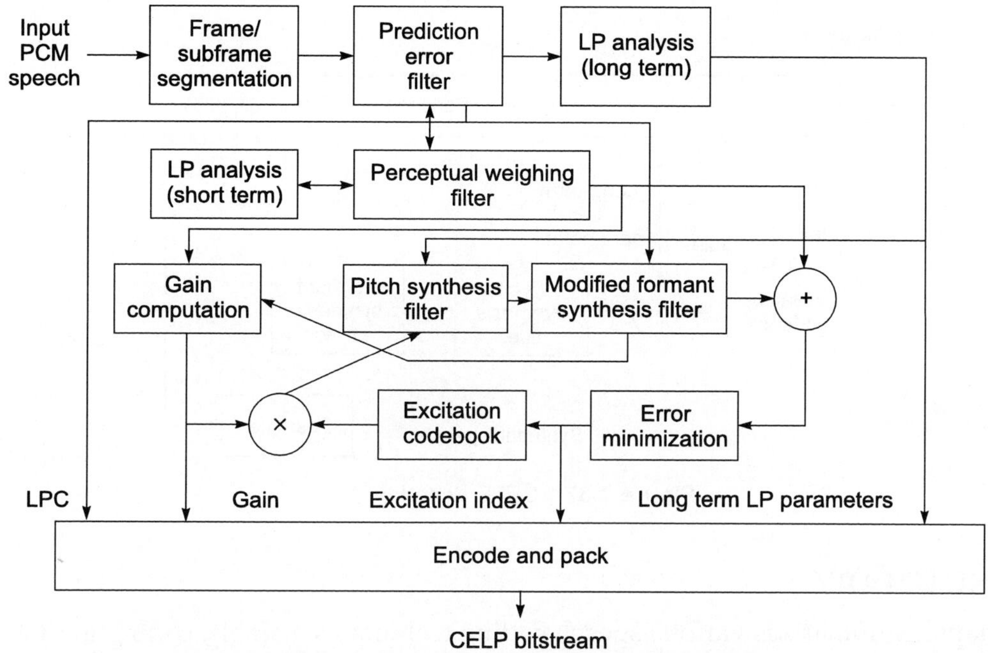

Figure 7.36 Block diagram of CELP encoder

Input speech signal is segmented into frames and subframes. Generally, one frame contains four subframes. Lengths of the frame and subframe are usually around 20 to 30 ms and 5 to 7.5 ms, respectively. Short-term LP analysis is performed on each frame to yield the LPC. (Short-term predictor recreates the correlation between nearby samples and the long-term predictor is responsible for generating correlation between samples that are one pitch period apart.) Input to short term LP analysis is the original speech, whereas input to long-term LP analysis is prediction error. Coefficients of the perceptual weighing filter, pitch synthesis filter and formant synthesis filters are computed.

The excitation sequence is now determined and the length of the excitation code vector is equal to that of the subframe. Hence, excitation codebook search is performed once every subframe. The search procedure begins with the generation of excitation sequences with the corresponding gains. Mean square error is computed for each excitation sequence. The code vector and gain are selected for the lowest error.

The block diagram of the CELP decoder is shown in Figure 7.37. It unpacks and decodes various parameters of the bitstream and synthesizes the speech signal.

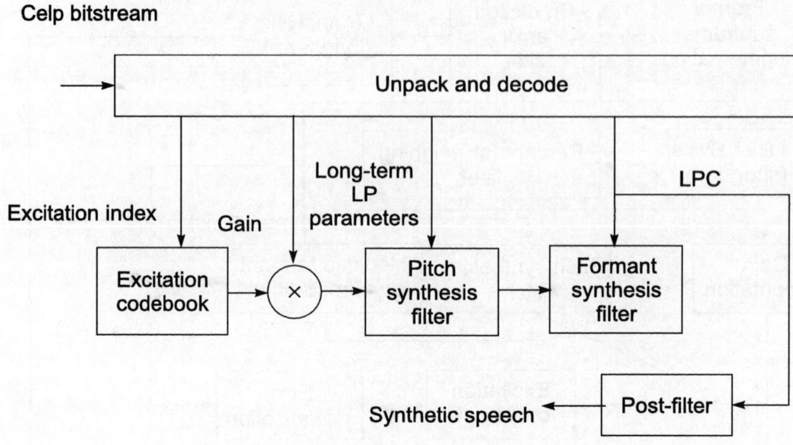

Figure 7.37 CELP decoder

7.9 SUMMARY

This chapter summarizes various speech coding techniques, namely, coding in time, frequency and transformation domains. Addition to this design of hybrid coders is also briefed.

EXERCISES

Review Questions

1. Discuss different types of uniform quantizer with input and output characteristics.
2. Show that in uniform quantizer, SNR in decibels $= 6B + 4.77 - 20\log_{10}\left[\frac{4\sigma_S}{\sigma_S}\right]$
3. Show that in uniform quantizer, if $S_{max} = 4\sigma_S$, then SNR in decibels $= 6B - 7.2$.
4. With necessary block diagram, explain functions of logarithmic encoder and decoder and show that signal to noise ratio is independent of signal variance.
5. Explain A-law and μ-law quantizer and its applications.
6. What is companding? Explain A-law and μ-law companders.
7. With block diagram, explain functions of encoder and decoder of differential quantization scheme and show that signal-to-noise ratio (SNR) of the quantizer is proportional to the gain of the differential configuration.
8. With necessary block schematic, explain feedforward and feedback adaptive quantizer with variable step size and gain representation.

9. Explain feedforward and feedback adaptive quantizer with necessary block diagram. List advantages and disadvantages of both the quantizers.
10. With block schematic, explain the operation of backward adaptive quantizer with variable step size and gain representation.
11. With block schematic, explain the operation of Jayanth quantizer.
12. Explain the working and selection of multiplier in Jayanth quantizer with the help of block schematic.
13. Discuss classification of speech coders.
14. What is waveform coding? Explain different types waveform coding techniques.
15. What is PCM coding? Explain why Manchester coding formant is preferred in PCM.
16. What is companded PCM? Explain how companded PCM will improve the SNR.
17. With block schematic, explain the operation of DPCM and adaptive PCM.
18. With diagram and example, explain the function of coder and decoder of delta modulation system.
19. With block schematic, explain the function of coder and decoder of adaptive delta modulation system.
20. Explain different types of frequency domain coding techniques.
21. With block schematic, explain vector quantization techniques.
22. What is parametric coding? Explain different parametric coding techniques.
23. With block schematic, explain the function of sub-band coders.
24. Explain how formant patterns are used in formant based coders.
25. What is sub-band coding? With block schematic, explain sub-band coding.
26. With block schematic and DCT equations, explain the function of adaptive transform coding.
27. What is DCT? Explain the use of DCT for speech coding.
28. With a block schematic, explain functions of cepstrum based speech encoder and decoder.
29. With a block schematic, explain functions of LPC based speech encoder and decoder.
30. What is hybrid coding? Explain different types of hybrid coding techniques.
31. With block schematic, explain the working of multi-pulse excited linear prediction coder.
32. With block schematic, explain the working of code-excited linear prediction.

8
CHAPTER

Speech Synthesis

This chapter provides insights on

- Challenges and types of speech synthesis
- Architecture of text-to-speech synthesis
- Speech quality and evaluation methods
- Applications of TTS system

8.1 INTRODUCTION

Speech synthesis is a process which artificially produces speech for various applications. Electronic system used for this purpose is called *speech synthesizer*. Several synthesis techniques have been introduced. In this chapter synthesis based on waveform coding, on speech production mechanism, and on analysis-synthesis methods such as linear predictive methods are covered. These methods enable a machine to pass an instruction or information to the user through "speaking". The application of speech synthesis includes banking services, public announcements such as those at train stations, reading emails, faxes and web pages over telephone, special equipment for physically challenged people, such as word processors with reading-out capability and book reading aids for visually handicapped people and speaking aids for vocally handicapped people. The progresses in VLSI, computer technology have collectively helped to advance speech synthesis research.

Voice response technology to convey messages has several advantages:

- This type of messaging is paper-free communication form.
- Without training, anybody can easily understand the message.
- Message can be received even when the listener is involved in other activities such as walking, etc.

8.2 HISTORY OF SPEECH SYNTHESIZERS

History of first speech synthesizers is said to have been constructed in 1779, more than 200 years ago. In 1779, C.G. Kratzenstein attempted to artificially produce and explain the differences among the five vowels. He constructed acoustic resonator which is similar in shape to the human vocal tract and excited them with a vibrating reed, which works like vocal folds, interrupting the airstream.

In 1791, Von Kempelen published a book describing his experiments and observations during the two decades he had been working on a speaking machine. This synthesizer was capable of producing both vowels and consonants. Fricative sounds were produced through 's' and 'sh' whistles. This synthesizer was able to produce words consisting of 19 consonants and 5 vowels. Von Kempelen's speech synthesizer machine used bellows to supply air to the reed, which, in turn, excited the resonator cavity. The resonator cavity was leather and the cross-sectional area varied. Consonant sounds were simulated using four separate constriction mechanisms controlled by the fingers of operators.

In 1939, one of the first of electrical speech synthesizers known as the "Voder", was introduced by H. Dudley. It produced continuous speech by controlling the fundamental period and band-pass filter characteristics, respectively using a foot pedal and 10 finger keys. The Voder structure became a principal foundation block for speech synthesis research. The Voder structure is still used in present speech synthesizers.

In 1950, H.K. Dunn achieved better quality than that of the Voder with an electrical vocal tract. The speech synthesizer was based on a vibrating energy source that replaced the vocal folds and transmission model using inductors and capacitors that provided the delay experienced by sound wave travelling along the vocal tract. Further developments in speech synthesis continue with the developments of VLSI and computers.

8.3 SPEECH SYNTHESIS METHODS

Speech synthesis methods can be divided into four types:

- Synthesis based on waveform coding: In this type speech waves of human voice are stored after recoding or after waveform coding. The stored information is used to produce desired messages.
- Synthesis by rule: In this method speech is produced based on phonetic and linguistic rules from sequences of phoneme symbols and prosodic features.

- Synthesis based on speech production mechanism: This method is based on the idea of generating periodic and non-periodic source signals and to feed them through a resonator circuit or a filter that models the vocal tract.
- Synthesis based on the analysis-synthesis method: In this method speech waves of recorded human voice are transformed into parameter sequences by the analysis-synthesis and speech synthesizers uses concatenated parameters to produce messages.

8.4 SYNTHESIS BASED ON WAVEFORM CODING

Synthesis based on waveform coding is the method by which short segmental units of human voice, typically words or phrases are stored. The speech is synthesized by selecting and connecting the appropriate units. Connecting pre-recorded natural utterances is probably the easiest way to produce intelligible and natural sounding synthetic speech. This is also called *cut-and-paste synthesis* in which short segments of speech are selected from a pre-recorded database and joined one after another to produce the desired utterances. In this method, the quality of synthesized speech is influenced by the quality of acoustic features at the connection between units. Acoustic features include the spectral envelope, amplitude, fundamental frequency and speaking rate. However, the drawback is that the systems are usually limited to one voice and often require more capacity than other methods. In order to reduce requirements for memory size, the units are sometimes compressed by waveform coding methods such as ADPCM.

One of the most important aspects of this method is to find the correct sound unit length. The selection is usually trade-off between longer and shorter units. With longer units high naturalness, less concatenation point and good control of coarticulation can be achieved but the memory requirement is increased. With shorter units, less memory is needed but the sample collecting and labelling procedure become more difficult and complex.

Two major problems exist is concatenating words to produce sentences.

- Spoken sentence is very different from a sequence of words uttered in isolation.
- The sentence stress pattern, rhythm, and intonation which are dependent on syntactic and semantic factors are unnatural when words are simply concatenated.

In order to solve the above problems, synthesis methods concatenating phonemes units have recently been widely employed. In these methods a large number of phoneme units or sub-phoneme (shorter than phonemes) units and pitch variations are stored and the most appropriate units are selected based on rules and evaluation procedure and are concatenated to synthesize the speech.

8.5 SYNTHESIS BASED BY RULE

Synthesis by rule is a method for producing any words or sentences based on sequences of phonetic symbols. In these methods small units of speech such as syllables, phonemes or one pitch period speech are stored and connected by rules. At the same time, prosodic features such as pitch and amplitude are controlled by rules. English language has more than 3500 syllables, and Japanese language has 100 syllables. When syllables or phonemes are taken as the fundamental units for speech production, the memory capacity can be greatly reduced. However, the rules for connecting phonemes are complicated to get high-quality speech. Control parameter for intonation, accent, stress, pause and duration were manually fed into the system to synthesize high-quality sentence speech.

8.6 SYNTHESIS BASED ON SPEECH PRODUCTION MECHANISM

Two methods are capable of producing speech by electroacoustically replicating the speech production mechanism. One is the modelling of vocal tract, which simulates the acoustic wave propagation in the vocal tract. The other is the formant synthesis method which simulates the frequency spectrum structure, that is, the resonance and anti-resonance characteristics, which reproduce articulation. In the early years, these methods were realized by analog circuits, but now most of the recent systems have digital circuits and digital signal processor.

8.6.1 Formant Synthesis

Formant synthesis was the first synthesis technique and was the dominant technique until the early 1980s. Formant synthesis is often called synthesis by rule, which models the pole frequencies of speech signal or transfer function of vocal tract based on source-filter model. That is, the speech is synthesized by simulating the formant frequencies and amplitudes and not the physical characteristics of the vocal tract. The human vocal tract has certain major resonant frequencies which are knows as formants.

At least three formants are generally required to produce intelligible speech and up to five formants to produce high-quality speech. Each formant is usually modelled with a two-pole resonator which enables both the formant frequency and its bandwidth to be specified.

The formant synthesis based on the source filter model of speech production is described in Figure 8.1.

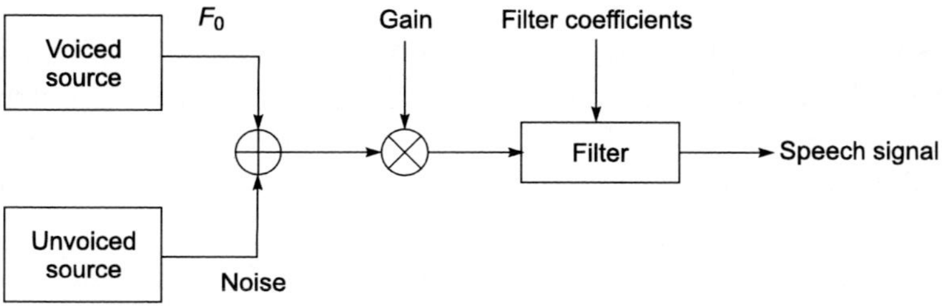

Figure 8.1 Formant synthesis based on the filter model

The excitation signal could be either voiced with fundamental frequency (F_0) or unvoiced noise. The excitation is then gained and filtered with a vocal tract filter which is constructed of resonators similar to formants of natural speech. For formant synthesis there are two basic structures in general, parallel and cascade. The cascade formant synthesizer (Figure 8.2) consists of band-pass resonators connected in series and the output of each formant resonator is applied to the input of the following one. The cascade structure needs only formant frequencies as control information. The main advantage of the cascade structure is that the relative formant amplitudes for vowels do not need individual controls.

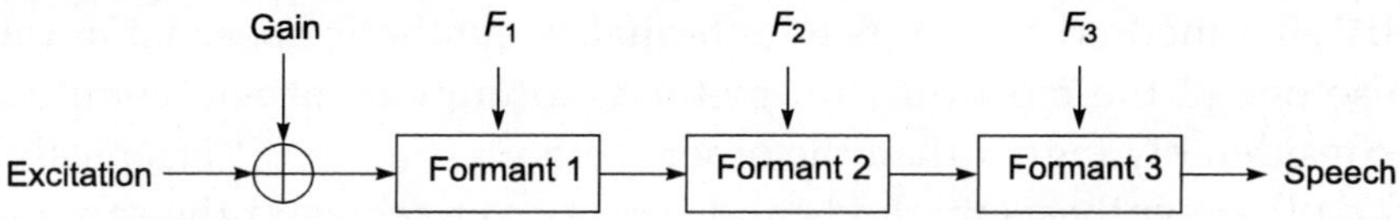

Figure 8.2 Basic structure of cascade formant synthesizer

The cascade structures have been found better for non-nasal voiced sounds and because it needs less control information than parallel structure, it is simpler to implement. However, with cascade model the generation of fricatives and plosives is a problem.

A parallel formant synthesizer (Figure 8.3) consists of resonators connected in parallel. The excitation signal is applied to all formants simultaneously and their outputs are summed.

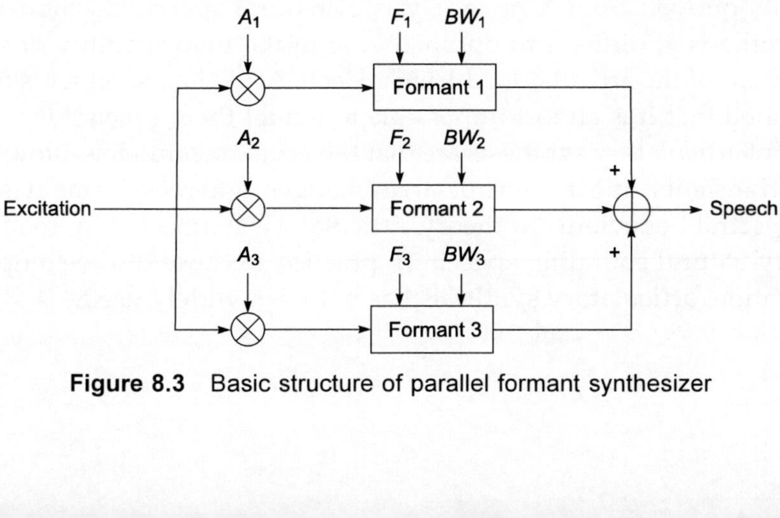

Figure 8.3 Basic structure of parallel formant synthesizer

The parallel structure enables control of bandwidth and gain for each formant individually and thus needs more control information. The parallel structure has been found to be better for nasals, fricatives and stop consonants but some vowels cannot be modelled with parallel formant synthesizer as well as with the cascade one.

Formant synthesizer can be constantly intelligible and it does not require any database of speech samples. However, many formant based speech synthesis systems generate speech which is artificial and robotic.

8.6.2 Articulatory Synthesis or Modelling of Vocal Tract

Speech generation is done by modelling of human articulatory system directly. In this case, we consider the shapes of the vocal tract, oral cavity, nasal cavity, size of the articulators, etc. This means that a mathematical model is defined for every organ of the human articulatory system. Thus, different models exist, for example, for the lungs, vocal cords, tongue, lips, etc. Hence, this method is inspired by the term articulators which imply speech organs like jaw, tongue, lips, etc. articulatory synthesis tries to model the human organs as perfectly as possible, so it is potentially the most satisfying method to produce high-quality synthetic speech. On the other hand, it is also one of the most difficult methods to implement and computational load is also considerably higher than the other methods. Further, it has received less attention than other synthesis methods and has not yet achieved the same level of success.

Articulatory synthesis typically involves models of the human articulators and vocal cords. The articulators are usually modelled with a set of area functions between the glottis and the mouth. For rule based synthesis the articulatory control parameter may be, for example, lip aperture, lip protrusion, tongue tip height, tongue tip position, tongue height, tongue position and velic aperture.

When speaking the vocal tract muscles cause articulators to move and change shape of the vocal tract which causes different sounds. The data for articulatory model is usually derived from X-ray analysis of natural speech. So the rule based articulatory synthesis is difficult to optimize due to the unavailability of sufficient data of the motions of the articulators during speech. Also, the movements of tongue are so complicated that it is almost impossible to model them precisely.

Advantages of articulatory synthesis are that the vocal tract models allow accurate modelling of transients due to abrupt area changes whereas formant synthesis models only spectral behaviour. In theory, articulatory synthesis is a good method to produce very natural sounding speech. In practice, because of the complexity of the implementation, articulatory synthesis has not been widely used.

8.7 ANALYSIS AND SYNTHESIS METHOD

In analysis and synthesis methods, words or phrases of human speech are analyzed based on speech production models and stored as time sequences of feature parameters. These parameter sequences are connected by a speech synthesizer to produce the desired spoken message. The amount of information stored in this method is less than the other methods but the naturalness of synthesized speech is slightly degraded. Speech synthesis based on LPC analysis method or cepstral analysis methods are used for this purpose.

In linear predictive coding (LPC) is based on the source-filter model of speech and filter coefficients are estimated automatically from a frame of natural speech. The linear predictor order (P) and the linear predictor coefficients are found by minimizing the sum of squared errors over a frame. As discussed in chapter 7, three methods, namely, the covariance method, autocorrelation method and lattice method are commonly used to calculate these coefficients. Then the speech sample $y(n)$ can be estimated from a finite number (P) of previous samples, $y(n-1)$ to $y(n-p)$, by a linear combination with a small error, $e(n)$. Output, $y(n)$, is given by

$$y(n) = \sum_{k=1}^{p} a(k)y(n-k) + e(n)$$

The main deficiency of the LPC method is that nasals and short plosives are poorly modelled. Several other variations of linear prediction have been developed such as multi-pulse linear prediction (MPC), residual excited linear prediction (RELP), and code excited linear prediction (CELP) to increase the quality compared to the basic model.

8.8 TEXT-TO-SPEECH SYSTEM

The text-to-speech (TTS) is a form of speech synthesis that converts text into spoken voice output. TTS system was first developed to aid the visually impaired by offering a computer-generated spoken voice. In 1971, Bell lab researcher Cecil H. Coker demonstrated a new computerized system for converting printed text into synthetic speech. He is a member of Bell laboratories acoustic research department, where techniques for "text" to "talk" conversion are being devised.

The important challenge that needs to be addressed in any TTS system is how to produce natural sounding speech for a plain text given as its input. It is not feasible to solve this by recording and storing all the words of a language and then concatenating the words in the given text to produce the corresponding speech.

A text-to-speech system finds a wide range of applications in a number of fields like accessing emails, various kinds of databases by voice over telephone, reading for the blind, etc.

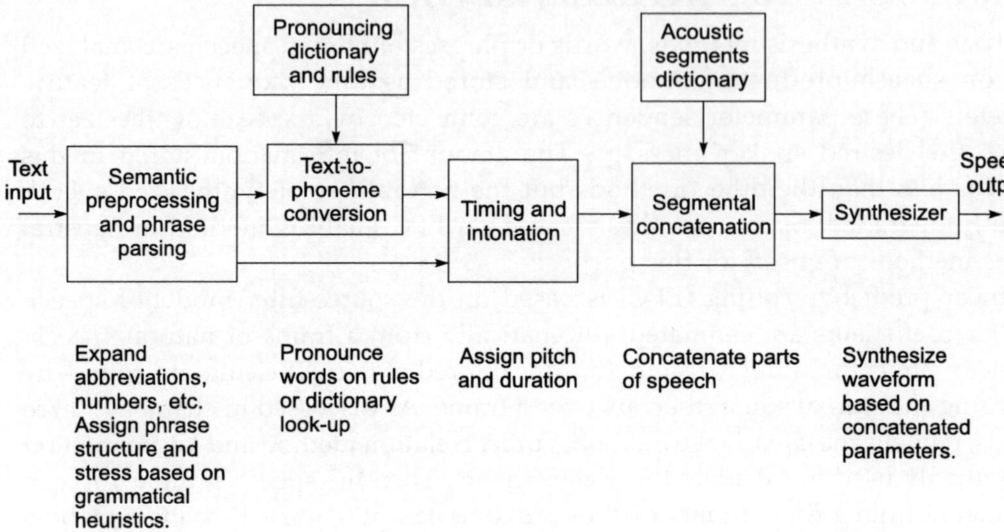

Figure 8.4 Principal elements of text-to-speech conversion system

Figure 8.4 shows the chief elements of text-to-speech conversion. The text preprocessing transforms the input text into a regularized format that can be processed to the rest of the system. These include breaking the input text into sentences, tokenizing them into words, expand abbreviations, numbers, etc. The pronunciation generation component is responsible for generating the acoustic sequence needed to synthesize the input text by finding the pronunciation of individual words in the input text. The function of intonation generation component is to generate fundamental frequency or pitch for the input text to be synthesized, duration value for each segment of speech is determined by the segmental duration generation component. Segmental concatenation concatenates parts of speech with acoustic segments dictionary. The synthesizer takes the input information generated by the various components described above and generates the speech output.

8.9 ARCHITECTURE OF THE TTS SYSTEM

A text-to-speech synthesis (TTS) is a process which automatically converts the orthographic text into artificial human speech. A computer based system which serves this purpose is called a *speech synthesizer*.

This section starts with a brief overview about the architecture of the text-to-speech synthesis system. Later the methods available for evaluating the naturalness and intelligibility of the synthesized speech are presented. Further, the applications of TTS system extending from visually impaired, deaf, and vocally challenged to educational applications are covered.

Text-to-speech synthesis processes are notably distinct from live human speech production. The live human speech production depends on the variations in the lung pressure and the vocal tract constrictions.

The objective of the TTS system is to transform an arbitrary given text into a corresponding spoken waveform. The TTS synthesis procedure mainly consists of two important phases, namely,

- text processing, and
- speech generation.

The objective of the text processing component is to process the input text and produce appropriate sequence of phonetic units or to provide some other linguistic representation. The speech generation component produces the acoustic output using the phonetic units produced by the text processing component employing either synthesis from parameters or by selection of a unit from a large speech corpus. For natural sounding speech synthesis it is essential that the text processing phase produces an appropriate sequence of phonetic units corresponding to the arbitrary input text. A simplified version of the TTS procedure is presented in Figure 8.5. The two phases are usually called the front end and backend of the TTS procedure.

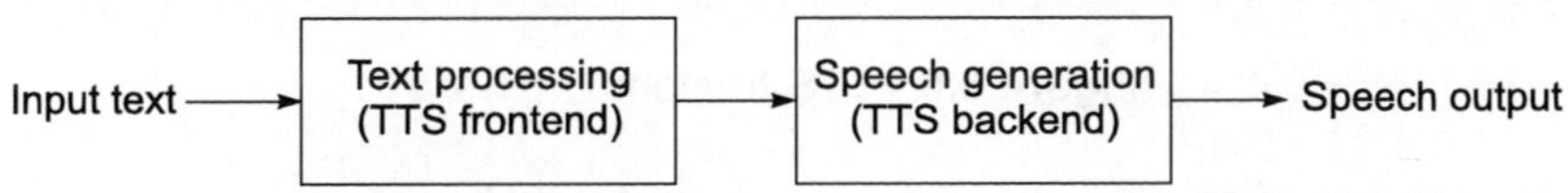

Figure 8.5 Simplified version of TTS procedure

8.9.1 TTS Frontend

The TTS frontend works on symbolic level, taking text as input and producing control information as output.

Frontend of the TTS system reveals about document structure detection, text normalization, interprets text marking and performs a linguistic analysis to produce what amounts to tagged text that then undergoes phonetic analysis to create phone based information and prosody analysis to determine pitch, duration as well as assigning stresses and pauses.

The TTS frontend comprises three fundamental components, they are:

- Text processing
- Phonetic analysis
- Prosody analysis

Figure 8.6 shows a block diagram of the frontend module.

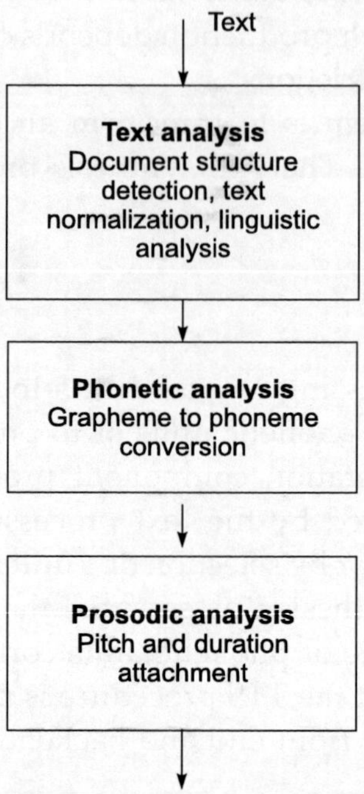

Figure 8.6 TTS frontend.

8.9.1.1 Text processing

Text processing is the fundamental step in TTS system which involves low-level processing issues such as sentence segmentation and word segmentation. It deals about document structure detection, text normalization and linguistic analysis.

Document structure detection: The document structure detection is a preprocessing part which analyses the input text and organizes into manageable list of words. The document structure can be detected by interpreting punctuation mark and paragraph formatting into account. This is also called *text analysis*.

Text normalization: Text normalization is used to transform the input text to transformable form. Usually, this part handles abbreviations and acronyms with the goal of matching how the educated human reader would render the input text. Proper normalization makes good output.

The vital roles of *text normalization* are in:

(i) Number converter: Number is pronounced differently in different situations like date, phone number, currency denomination, decimal and fractional numbers.

(ii) Abbreviation converter: Abbreviations may be expanded into full words, pronounced as written or pronounced letter by letter. For example, Dr could be rendered as doctor, Mrs. could be rendered as misses, etc.

(iii) Acronym converter: Acronym may be pronounced as written or pronounced letter by letter. For example: ISRO, NATO, RAM, etc., are pronounced as written and SAS or ADP letter by letter.

On the other end "Henry IV" may be pronounced as Henry the fourth and "chapter v" may be pronounced as chapter five.

(iv) Linguistic analysis: Linguistic analysis refers to the scientific analysis of a language sample. It involves at least one of the five main branches of linguistics, which are phonology, morphology, syntax, semantics, and pragmatics. Linguistic analysis can be used to describe the unconscious rules and processes that speakers of a language use to create spoken or written language, and this can be useful to those who want to learn a language or translate from one language to another. Some argue that it can also provide insight into the minds of the speakers of a given language, although this idea is controversial.

8.9.1.2 Phonetic analysis

The phonetic analysis, basically known as *"grapheme-to-phoneme conversion"* transforms the orthographic into phonological ones using a phonetic alphabet. Phone is a sound that has definite shape as a sound wave. Phone is the smallest sound unit and a collection of phones that consists of minimal distinctive phonetic units are called phoneme. Phonetic analysis focuses on the phone level within each word and tagging each phone with information about what sound to produce and how to produce it. (i.e., speaking style and emphasis). In regard to this process, it performs grapheme-to-phoneme conversion and homograph disambiguation.

Grapheme-to-phoneme conversion: The role of this is the determination of exact pronunciation of each word of the input sentence based on its spelling.

Homograph disambiguation: This process involves figuring out whether an input sentence uses the present tense or past tense of the word read.

For achieving grapheme to phoneme conversion, there are two main approaches.

- Dictionary based approach
- Rule based approach

Dictionary-based Approach: In this approach, a large dictionary containing all the words of a language and their correct pronunciations is stored in the database, now it is a matter of looking into the dictionary for each word and spell out the correct pronunciation. The advantage of this approach is quick and accurate. On the other hand, the major drawback is that it requires a huge database to store all the words and the approach completely fails if the word is not found in the dictionary.

Rule-based Approach: In this approach, correct pronunciation of words is determined based on some rules. Here the main advantage is that it requires no database and is good for those languages with a phonetic orthography having a regular writing system. On the other hand, complexity of the rules grows substantially as the system takes into account irregular inputs.

8.9.1.3 Prosody analysis

Prosody or suprasegmental features determine the intonation, speaking rate, and loudness of the utterances which are ultimately represented at phoneme level as pitch (fundamental frequency), duration and stress (amplitude) over the time.

Intonation or pitch pattern is simply a version of speech while speaking. The pitch contour depends on the meaning of the sentence. For example, in normal speech the pitch slightly decreases towards the end of the sentence and when the sentence is in a question form, the pitch pattern will rise to the end of the sentence.

Also, the pitch contour is also affected by gender, physical and emotional state (happiness, angry, etc.) and attitude of the speaker. Hence, modelling of the intonation is an important task that affects intelligibility and naturalness of the synthesized speech.

To achieve high-quality text-to-speech conversion a good model of intonation is needed.

Generally, intonations are distinguished as

(i) Rising intonation: when pitch of the voice increases.
(ii) Falling intonation: when pitch of the voice decreases.
(iii) Dipping intonation: when pitch of the voice falls and then rises.
(iv) Peaking intonation: when pitch of the voice rises and then falls.

The duration or time characteristics can also be investigated at several levels from phoneme durations to sentence level timing. The segmental duration is determined by set of rules to determine correct timing.

Usually, some inherent duration for phoneme is modified by rules between maximum and minimum durations. For example, consonants in non-word initial positions are shortened, emphasized words are significantly lengthened. In general, the phoneme duration differs due to neighbouring phonemes.

The intensity pattern is perceived as loudness of speech over the time. At syllable level vowels are usually more intense than consonants and at a phrase level syllables at the end of an utterance can become weaker in intensity.

The speaker's feelings and emotional state affect speech in many ways and proper implementation of these features in synthesized speech may increase the quality considerably. For example, the deaf cannot express their feelings when communicating with speech synthesizer through telephone line. With TTS systems this is rather difficult because the written text usually contains no information of these features. However, this kind of information may be provided to a synthesizer with some specific control characters or character string. Hence, recent investigations suggest the identification of the vocal features which signals the emotional content may help to create a very natural synthesized speech.

8.9.2 TTS Backend

TTS backend is the second and final step involved in building a speech synthesis system. It accepts the control information from the frontend of the TTS system as input and provides the synthesized speech as output. Figure 8.7 shows the block diagram of the TTS backend.

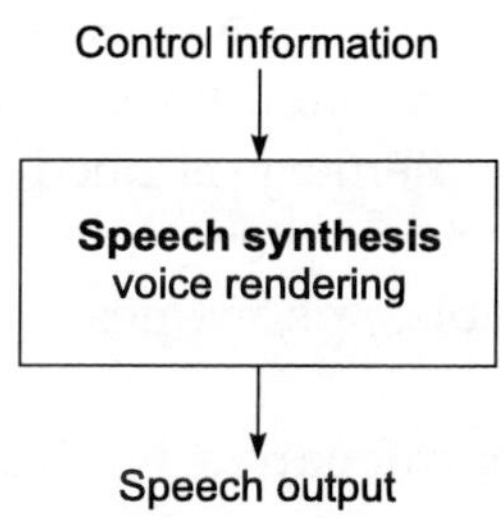

Figure 8.7 TTS backend.

The main task involved in TTS backend is speech synthesis (also called *acoustic processing*). The three primary technologies generating synthetic speech are articulation formant and concatenative speech synthesis. An important decision a designer of a TTS system needs to make is, which synthesis method to use. The most important quality of any speech synthesizer is to provide naturalness and intelligibility in the synthesized speech.

Benefits and Problems of Speech synthesis Methods

As discussed in section 8.3, there are many methods to produce speech sounds after text and prosodic analysis. All these methods have some benefits and problems of their own.

- In articulatory synthesis, the collection of data and implementation of rules to drive that data correctly is very complex.
- In formant synthesis, the set of rules controlling the formant frequencies and amplitudes and the characteristics of the excitation source is large. Also lack of naturalness is a major problem with formant synthesis.

- In concatenative synthesis, the collection of speech samples and labelling them is time consuming and may yield quite large waveform databases. Concatenative points between samples may cause distortion in the synthesized speech. Some problems with memory and system requirements may arise.

8.10 CONCATENATIVE SPEECH SYNTHESIS

Concatenative synthesizer strings together pieces of recorded speech extracted from the database of speech samples. Here the database consists of the actual short segments of recorded speech that were cut from the recordings. Hence, connecting, prerecorded natural utterances is probably the easiest way to produce intelligible and natural sounding synthetic speech. However, concatenative synthesizers are usually limited to one speaker and one voice and usually require more memory capacity than other methods. One of the most important aspects in concatenative synthesis is to find correct unit length. The selection is usually a trade-off between longer and shorter units.

With longer unit's high naturalness, less concatenation points and good control of coarticulation are achieved, but the amount of required units and memory is increased. With shorter units, less memory is needed but sample collection and labelling procedures are difficult and complex. In the present system units used are usually words, syllables, demi-syllables, phonemes, diaphones and sometimes even triphones.

There are several problems in concatenative synthesis:

- Distortion from discontinuities in concatenation points which can be reduced using some special methods for signal smoothing.
- Memory requirements are usually very high, especially when long concatenation units are used, such as syllables or words.
- Data collection and labelling of speech samples is usually time consuming.

8.11 CHALLENGES IN SPEECH SYNTHESIS

The challenges in speech synthesis are very wide. There are several challenges in text processing. Correct prosody and pronunciation analysis from written text is also a major challenge in text-to-speech synthesis.

8.11.1 Text Processing

- Text processing is usually a complex task and includes several language dependent problems.
- Digits and numerals must be expanded into full words. For example, in English, numeral 250 would be expanded as two hundred and fifty and 1750 as one thousand seven hundred and fifty (if measure) and as seventeen fifty (if year).

- Fractions and dates are also problematic. 5/16 can be expanded as five-sixteenths (if fraction) or may sixteenth (if date).
- Abbreviations may be expanded into full words, pronounced as written or pronounced letter by letter. These are some contextual problems. For example, kg can be either kilogram or kilograms depending on the preceding number, St. can be saint or street, ft. can be fort or feet.
- Special symbols and characters can also cause some special kind of problems. For example, $71.50 must be expanded as seventy one dollars and fifty cents and $100 million as one hundred million dollars, not as one hundred dollars million.

It can be inferred from these examples that for correct conversion in every possible situation, a very complicated set of rules is needed.

8.11.2 Pronunciation

The most challenging task is to find correct pronunciation for different contexts in the text. Some words called homographs may be the most difficult problems in TTS systems. Homograph is spelled the same way but they differ in meaning and usually in pronunciation (e.g., lies, fair). The word lives is, for example, pronounced differently in sentences, "three lives were lost" and "one lives to eat". With these kind of words semantic information is necessary to achieve correct pronunciation.

8.11.3 Prosody

Finding correct information, stress and duration from the written text is probably the most challenging problem for years to come. These features together are called prosodic or suprasegmental features and may be considered the melody, rhythm and emphasis of the speech at the perceptual level.

The prosody of continuous speech depends on many aspects such as the meaning of the sentence and the speaker characteristics and emotions. The prosodic dependencies are shown in Figure 8.8.

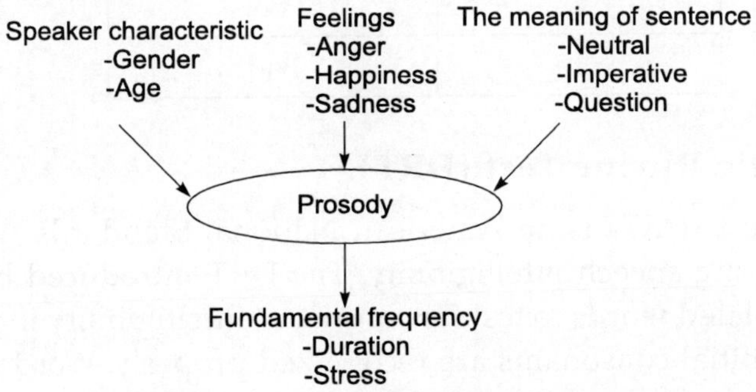

Figure 8.8 Prosodic dependencies.

Unfortunately, the written text usually contains limited information of these features and some of them change dynamically during speech. However, with some specific control characters this information may be given to a speech synthesizer.

8.12 SPEECH QUALITY AND EVALUATION

Synthetic speech can be compared and evaluated with respect to intelligibility and naturalness. Naturalness corresponds to the similarity to the human voice and the intelligibility corresponds to the ability to understand the synthesized speech. Several methods have been developed to evaluate speech quality in general and these methods are also suitable to measure overall quality or acceptability of synthetic speech.

8.12.1 Mean Opinion Score (MOS)

Mean opinion score (MOS) is probably the most widely used and simplest method to evaluate speech quality in general. MOS is the arithmetic mean of all the individual scores and it gives the numerical indication of the perceived audio quality. MOS tests for voice are specified by international telecommunication union-testing (ITU-T) recommendation. MOS is a five-level scale from bad (1) to excellent (5). To check the naturalness of the synthesized speech, listeners were asked to evaluate the synthesized speech with the scale described in Table 8.1. Hence, this subjective evaluation accounts for naturalness of the synthesized speech.

TABLE 8.1

MOS	Quality
5	Excellent
4	Good
3	Fair
2	Poor
1	Bad

8.12.2 Diagnostic Rhyme Test (DRT)

Diagnostic rhyme test (DRT) is an American National Standards Institute (ANSI) standard for measuring speech intelligibility. The DRT introduced by Fairbanks in 1958 uses a set of isolated words to test for consonant intelligibility in initial position. DRT tells how the initial consonants are recognized properly. Word pairs that have the same way out such as "dense-tense", "sheat-cheat", "read-deed", etc., which can be easily misunderstood are chosen to evaluate the intelligibility of the TTS system.

The listener hears one word at the time and marks to the answering sheet which one of the two words he thinks is correct. Finally, the results are summarized by averaging the error rates from answer sheets.

DRT is a quite widely used method and it provides lots of valuable diagnostic information on how properly the initial consonant is recognized and it is very useful as a developing tool. However, it does not test any vowels or prosodic features, so it is not suitable for any kind of overall quality evaluation.

8.12.3 Comprehension Tests (CT)

In comprehensive test a subject hears a few sentences or paragraphs and answers to the questions about the content of the test. It is not important to recognize one single phoneme, it is enough if the meaning of the sentence is understood as 100% segmental intelligibility is not crucial for text comprehension.

8.13 APPLICATION OF TTS SYSTEM

- Probably the most important and useful application field in speech synthesis is the reading and communication aids for the blind. It is easier to get information from computer with speech instead of using special symbol keyboard which is an interface for reading the braille characters.
- For individuals with speech impairment, synthesized speech gives an opportunity to communicate with people who do not understand the sign language.
- Synthesized speech can be used in many educational situations. It is almost impossible to learn, write and read without spoken help. With proper computer software, unsupervised training for these problems is easy and inexpensive to arrange. Also a speech synthesizer connected with word processor is also a helpful aid to proofreading.
- In telecommunication, TTS systems made it possible to listen to the text read by machines, for instance, from a database and queries to such databases can be transmitted via a telephone keypad (DTMF systems).
- In multimedia, TTS synthesis has made possible the existence of talking books, toys, and interactive games.

8.14 SUMMARY

This chapter discusses the reconstruction of speech signals using various synthesis techniques. Supplementing to this, challenges, architecture, speech quality and evaluation methods are discussed with a TTS system design.

EXERCISES

Review Questions

1. Explain different methods for speech synthesis.
2. Explain synthesis based on waveform coding. What are the advantages of this method?
3. With necessary block diagram, explain formant synthesizer.
4. With block schematic, explain cascade and parallel formant synthesizers.
5. What is articulatory synthesis?
6. With block schematic, explain LPC based synthesizer.
7. With block schematic explain text-to-speech synthesis system and explain the function of each block.
8. Explain two important phases in TTS synthesis.
9. Explain phonetic and prosody analysis in text-to-speech synthesis system.
10. List and explain challenges of text processing.
11. List and explain different methods to evaluate speech quality.
12. Explain important areas of application of TTS system.

9 CHAPTER

Speech Recognition

This chapter provides insights on

- Challenges in speech recognition
- Pattern recognition and knowledge based methods
- Basic model of pattern recognition
- Pattern recognition techniques
- Automatic speaker recognition

9.1 INTRODUCTION

Speech recognition is a process of automatically extracting and determining linguistic information conveyed by a speech wave using computers or electronic circuits. In speech wave the most important information is linguistic information and is also called *phonetic information*. In 1952, at Bell laboratories, Davis, Biddulph and Balshek built a system for isolated digit recognition for a single speaker. In 1956, at RCA laboratories, Olson and Belar tried to recognize 10 distinct syllables of a single talker. In the 1960s several fundamental ideas in speech recognition surfaced and were published.

The ultimate goal of most research in speech recognition is to develop a machine that has the ability to understand fluent, conversational speech, with unrestricted vocabulary from any speaker. The difficulties in speech recognition are as follows:

- The spectrum of a phoneme in a word or sentence is influenced by neighbouring phonemes as a consequence of coarticulation.
- Spectra continuously change from phoneme to phoneme to their mutual interaction and it is very difficult to precisely determine the phoneme boundaries.

- Acoustic features vary from speaker to speaker, even when the same words are uttered.
- The physical features of speech do not always convey enough information in and of themselves. The listener can usually predict the next word according to several linguistic constraints, and incomplete phonetic information is compensated with linguistic knowledge.

9.2 CLASSIFICATION OF SPEECH RECOGNITION SYSTEMS

Speech recognition systems can be classified into:

- Isolated word recognition
- Continuous speech recognition

Isolated word recognizers accept one word at a time. Hence, in isolated word recognition, words uttered in isolation are recognized. Continuous speech recognizers allow users to speak naturally, in the speech recognition phase, it is assumed that the speaker deliberately utters sentences with sufficiently long pauses between words (a minimum of 200 msec) such that the sentences are not confused with weak fricatives. Continuous speech recognition can be further classified into:

- Transcription
- Understanding

Transcription aims at recognizing each word correctly. Understanding is also called *conversational speech recognition*, focuses on understanding the meaning of sentences rather than recognizing each word.

9.3 STRUCTURE OF WORD RECOGNITION SYSTEMS (MATCHING TECHNIQUES)

Structure of isolated word recognition system is classified into two types, they are:

- Systems using words as a unit
- Systems using subwords (such as phonemes or syllables) as a unit

Systems using word unit structure is shown in Figure 9.1(a). In this system input speech is compared with pre-recorded template of the word and the word unit with the smallest distance from the input speech is selected. This technique takes much less processing than subword matching. But it requires the users' pre-recorded words (sometimes several hundred to thousand words).

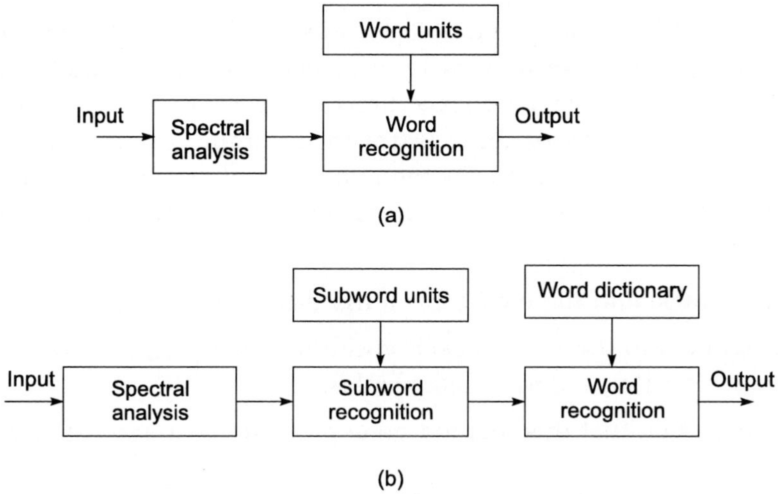

Figure 9.1 Structure of word recognition systems: (a) word-based recognition; (b) subword-based recognition

Systems using subword unit structure is as shown in Figure 9.1(b). Short periods of speech are compared with the subword usually phonemes to calculate the distances. This technique takes more processing than whole word matching. But it requires much less storage. Word dictionary represents each word by a concatenation of the subword units. Distances and the word dictionary are then combined to make the decision. Systems using words as unit is suited to smaller vocabulary. Systems using subwords is suited to a larger vocabulary.

9.4 TECHNIQUES FOR SPEECH RECOGNITION

Human beings are the best "machine" to recognize and understand speech. We are able to combine a wide variety of linguistic knowledge concerned with syntax and semantics and adaptively use this knowledge according to the difficulties and characteristics of the sentences. The speech recognition system is built with this aim to match or exceed human performance. Techniques for speech recognition can be broadly divided into two classes, namely:

- Pattern recognition technique
- Knowledge based technique

9.4.1 Pattern Recognition Technique

This approach involves manipulating the speech signals directly without explicit features of speech. There are two stages in this approach, mainly the training of speech patterns and recognition of patterns via the training procedure.

The training procedure should be able to adequately characterize the acoustic properties of the speech pattern. This type of classification is known as *pattern classification*. The recognition stage does a direct comparison between the unknown speech signal and the speech patterns learnt in the training phase. It generates a "accept" or "reject" decision based on the similarity of the two patterns. Pattern recognition has been the method of choice for many researchers around the world due to the fact that

- It is simple to use and the method is fairly easy to understand.
- It has robustness to different speech vocabularies, users, feature sets, pattern comparison algorithms, and decision rules.
- It has been proven that this method generates the most accurate results.

9.4.2 Knowledge-based Technique

Knowledge-based approach uses the information regarding linguistic, phonetic and spectrogram. Some speech researchers developed recognition systems that used acoustic phonetic knowledge to develop classification rules for speech sounds. The acoustic phonetic approach has been studied in depth for more than 40 years. It is based on the theory that suggests that there exist finite, distinctive phonetic units of spoken language and these units are widely characterized by a set of properties that are manifest in the speech signal, or its spectrum, over time. The first step in this approach is to segment the speech signal into discrete time regions where the acoustic properties of the speech signal are represented by one phonetic unit. The next step is to attach one or more phonetic labels to each segmented region according to the acoustic properties. Finally, the last step attempts to determine a valid word from phonetic labels generated from the first step.

Till date, the greatest degree of success in speech recognition has been obtained using pattern recognition approach. Hence in this book, we will be concerned primarily how pattern recognition techniques have been applied to speech recognition.

9.5 SPEECH RECOGNITION BASED ON PATTERN RECOGNITION APPROACH

Pattern recognition methods have two important steps; namely, training of data and recognition of pattern via pattern comparison. Figure 9.2 shows a block diagram of the pattern recognition model used for speech recognition. The pattern recognition approach has four steps, namely:

- Feature extraction techniques
- Reference patterns
- Pattern classifier
- Decision logic

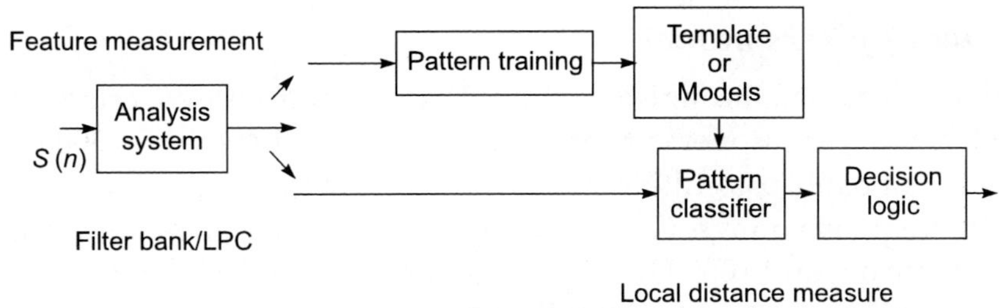

Figure 9.2 Pattern recognition model for speech recognition

Feature extraction techniques: The input speech signals are analyzed to identify the test patterns. For speech signals, the feature measurements are usually the output of some type of spectral analysis techniques such as filter bank analyzer, linear predictive coding analysis, Mel-frequency cepstrum, sub-band coding.

Reference patterns: Test patterns corresponding to speech sounds are used to create pattern representatives. The resulting pattern is called reference pattern and it is a template or it can be a model.

Pattern classifier: In pattern classifier, unknown test patterns are compared with reference patterns and a measure of similarity between the test pattern and reference pattern is computed.

Decision logic: In decision logic similarity scores are used to decide which reference pattern best matches the unknown test pattern.

9.5.1 Feature Extraction Techniques

Feature extraction is the most important part of speech recognition since it plays an important role to separate one speech from other. The purpose of this module is to convert the speech waveform to some type of parametric representation (at a considerably lower information rate) for further analysis and processing. This is often referred to as frontend in the signal processing. Parametric representation is basically a data reduction technique whereby a large number of data points are transformed into a smaller set of features which are equivalent in the sense that they faithfully describe the salient properties of the acoustic waveform.

The most popular parametric representations for speech recognition are:

- Bank of filters model
- Linear prediction coding (LPC) model
- Mel-frequency cepstrum coefficients (MFCC) model

9.5.1.1 *Bank of filters model*

The speech signal is passed through a bank of Q band-pass filters covering the speech band from 100 Hz to some upper cut-off frequency (typically between 3000 and 8000 Hz). The number of band-pass filters varies from as few as 5 to as many as 32 and it depends on the frequency range. Generally, filter spacings are linear until about 1000 Hz and logarithmic beyond 1000 Hz. The output of each band-pass filter is generally passed through nonlinearity, such as a full-wave rectifier. The nonlinearity shifts the band-pass signal spectrum to the low frequency band as well as creates some high frequency signals. Then the signal is low-pass filtered to eliminate high frequency signals. A logarithmic compressor is generally used to reduce the dynamic range of the intensity and the compressed signal is decimated for efficiency of storage. This analysis is a good model for voiced, quasiperiodic speech sounds.

9.5.1.2 *Linear prediction coding (LPC) model*

In this section, we describe how LPC model is used for speech analysis. The mathematical details and derivations are omitted here, since it is already discussed in chapter 7. LPC provides a good model of the speech signal. The method of LPC is mathematically precise and is simple to implement in either software or hardware. The steps involved in obtaining the vector of LPC coefficients are as follows:

- Frame blocking: In this step the continuous speech signal is blocked into frames of N samples, with adjacent frames being separated by $M (M < N)$. The first frame consists of the first N samples. The second frame begins M samples after the first frame, and overlaps it by N-M samples. Similarly, the third frame begins $2M$ samples after the first frame (or M samples after the second frame) and overlaps it by N-$2M$ samples. This process continues until all the speech is accounted for all the frames.
- Windowing: The next step in the processing is to window each individual frame so as to minimize the signal discontinuities at the beginning and end of each frame. The concept here is to minimize the spectral distortion by using the window to taper the signal to zero, at the beginning and end of each frame. Typically, the Hamming window is used, which has the form:

$$w(n) = 0.54 - 0.46\cos\left(\frac{2\pi n}{N-1}\right), \quad 0 \le n \le N-1$$

- Autocorrelation analysis: The windowed set of speech samples is auto-correlated to give set of $(P + 1)$ coefficients, where P is the order of the desired LPC analysis.
- LPC analysis: The vectors of LPC coefficients are computed from the autocorrelation vector using the Durbin recursive method.

9.5.1.3 *Mel-frequency cepstrum coefficient method*

The block diagram of the structure of the Mel-frequency cepstrum coefficient (MFCC) method is given in Figure 9.3, the speech input is typically recorded at sampling rate above 10000 Hz. These sampled signals can capture all frequencies up to 5 kHz. Which cover most energy of sounds that are generated by humans.

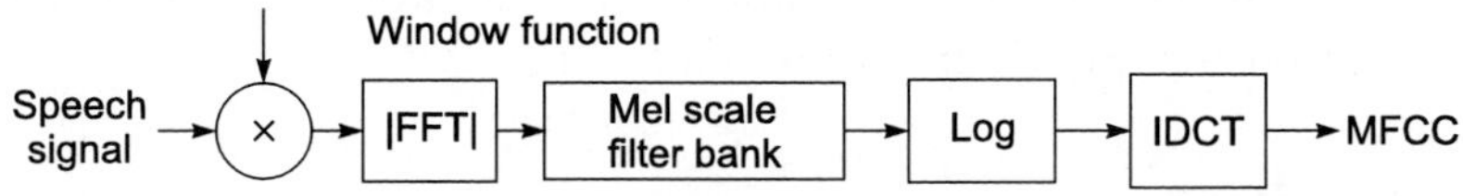

Figure 9.3 Processing steps of MFCC

The main purpose of MFCC is to mimic the behaviour of the human ears. The steps involved in obtaining the vector of MFCC coefficients are as follows:

Frame blocking: In this step, the continuous speech signal is blocked into frames of *N* samples, with adjacent frames being separated by $M (M < N)$. The first frame consists of the first *N* samples. The second frame begins *M* samples after the first frame, and overlaps it by *N-M* samples. Similarly, the third frame begins 2*M* samples after the first frame (or *M* samples after the second frame) and overlaps it by *N*-2*M* samples. This process continues until all the speech is accounted for all the frames.

Windowing: The next step is to window each individual frame so as to minimize the signal discontinuities at the beginning and end of each frame. The concept here is to minimize the spectral distortion by using the window to taper the signal to zero at the beginning and end of each frame. Typically, the Hamming window is used, which has the form:

$$w(n) = 0.54 - 0.46\cos\left(\frac{2\pi n}{N-1}\right), \quad 0 \le n \le N-1$$

Fast Fourier transform (FFT): The next step is the fast Fourier transform, which converts each frame of *N* samples from the time domain to frequency domain. The result after this step is often referred to as spectrum or periodogram.

Mel-frequency wrapping: Psychophysical studies have shown that human perception of the frequency contents of sounds for speech signal does not follow a linear scale. Thus for each tone with an actual frequency, *F*, measured in Hz, a subjective pitch is measured on a scale called the 'Mel scale'. The Mel-frequency scale is linear frequency spacings below 1000 Hz and a logarithmic spacing above 1000 Hz. Therefore, we can use the following approximate formula to compute the Mel (s) for a given frequency *F* in Hz.

$$\text{Mel}(F) = 2595\log_{10}\left(1 + \frac{F}{700}\right)$$

One approach to simulating the subjective spectrum is to use a filter bank, spaced uniformly on the Mel scale.

Cepstrum: In this final step, the log Mel spectrum is converted back to time. The result is called the Mel-frequency cepstrum coefficients (MFCC). The cepstral representation of the speech provides a good representation of the local spectral properties of the signal for the given frame analysis.

9.5.2 Reference Patterns

Pattern training is the method by which representative sound patterns are converted into reference patterns for use by the pattern classifier. There are several ways in which pattern training can be performed, including:

- *Causal training*: In this each individual training pattern is used directly to create either a non-parametric reference pattern or a statistical model. Causal training is the simplest, most direct method of creating reference patterns.
- *Robust training:* In this several (i.e., two or more) versions of each vocabulary entry are used to create a single reference pattern or statistical model.
- *Clustering training:* In this large number of versions of each vocabulary entry is used to create one or more reference pattern or statistical model. Clustering training is generally used for creating speaker independent reference patterns, in which case the multiple training patterns of each vocabulary entry are derived from a large number of different speakers.

The final result of the pattern training algorithm is the set of reference patterns used in the recognition phase of the model of Figure 9.1.

9.5.3 Pattern Classifier

A key step in pattern classifier of Figure 9.1, is the determination of similarity between the unknown test patterns and each of the stored reference pattern. There are three types of pattern classifier that are used for speech recognition, namely:

- Template matching (spectral distance measure, template matching add. Dynamic time warping (DTW)
- Clustering based approach (vector quantization)
- Statistical modeling (hidden Markov modelling)

9.5.3.1 Template matching

In template matching, the template generally consists of set features which could be represented by an *n*-dimensional vector. An important method used to classify speech patterns represented by a sequence of vectors is known as nearest neighbour classifier. This calculates the distance between corresponding pair of vectors in the input pattern and a template as follows:

$$D = \sum_{j=1}^{m}\sum_{i=1}^{n}\left(X_{ij} - Y_{ij}\right)^2$$

Where the testing patterns X and the template Y consist of a sequence of m vectors each contain n features. The class to which a particular pattern belongs is determined by choosing the template having the minimum distance from the pattern. The speech with the lowest distortion distance is chosen to be identified as the unknown speech.

Spectral distance measure: In almost all speech recognition systems, short-time spectral distances or similarities between input speech and stored units (models or templates) are calculated based on feature extraction techniques. Following methods have been used for obtaining spectral distances based on general spectral analysis techniques which do not incorporate modelling speech production mechanism.

Band-pass filter bank method: Centre frequencies of the band-pass filters are usually set with equal spaces along the logarithmic frequency scale. Differences of logarithmic output for each band-pass filter between the reference and input speech are averaged over all frequency range to produce the overall distance.

Linear prediction coding (LPC) method: In speech and speaker recognition applications widely LPC is used as common feature extraction method. In this following different distance measure are used:

- Itakura-Saito distance
- Cosh distance
- LPC cepsral distance

Itakura-Saito distance or maximum likelihood spectral distance: This method was proposed by Fumitada Itakura and Shuzo Saito in 1970. If $A(\lambda)$ is the spectral envelope of reference template LPC model and $B(\lambda)$ is the spectral envelope of input speech LPC model, then the distance is calculated by the equation

$$D = \frac{1}{2\pi}\int_{-\pi}^{\pi}\left(\log\frac{A(\lambda)}{B(\lambda)} + \frac{B(\lambda)}{A(\lambda)} - 1\right)d\lambda$$

Cosh distance: Cosh measure was developed in order to remove the asymmetry associated with the weighting for the spectral difference in the Itakura-Saito spectral distance. If $A(\lambda)$ is the spectral envelope of reference template LPC model and $B(\lambda)$ is the spectral envelope of input speech LPC model, then the distance is calculated by the equation

$$D = \frac{1}{2\pi} \int_{-\pi}^{\pi} 2\left\{\cosh\left(\log A(\lambda) - \log B(\lambda)\right) - 1\right\} d\lambda$$

LPC cepstral distance: The LPC cepstral distance is the distance between spectral envelopes represented by the LPC cepstral coefficients. If $C_1(n)$ is the spectral envelope of reference template LPC cepstral coefficients and $C_2(n)$ is the spectral envelope of input speech LPC cepstral coefficients, then the distance is calculated by the equation

$$D = \sum_{n=0}^{N-1} \left[C_1(n) - C_2(n)\right]^2$$

Mel-frequency cepstrum coefficient method: In this method Euclidean distance between lower order Mel-frequency cepstrum coefficients of reference template and input speech vectors corresponds to distance between logarithmic spectra.

9.5.3.2 *Dynamic time warping (DTW)*

If reference template and input template are of same length, then finding distance between two templates is simple. This leads to linear relation between two templates and is called linear time warping. In case of speech signals linear time warping is not useful, since length of the speech signals and the number of segments for the reference and the input template will be different. In this section, we explain the dynamic time warping (DTW) algorithm, one of the oldest and most important algorithms in speech recognition. The simplest way to recognize an isolated word sample is to compare it against a number of stored word templates and determine which the "best match" is. Hence, sequences of different length are compared using dynamic time warping (DTW) algorithm. In this the matching is done by stretching and/or compressing the sequences so that they are of same length. Features of the template and input are MFCC, PLP, etc.

The first step in DTW algorithm is to compute the distance between each vector of the template and each vector of the test pattern. This can be done by Euclidean distance, which is given as

$$d(i, j) = \sum_{q=1}^{p} \left[X_i(q) - Y_j(q)\right]^2$$

where, $X_i(q)$ are the features of the unknown speech patterns consisting of n p-dimensional vectors and $Y_j(q)$ are the elements of the reference template which consists of m p-dimensional vectors. The next step is to compute global distance, which is given by

$$D(i, j) = \min\left[D(i+1, j+1) + D(i+1, j) + D(i, j+1)\right] + d(i, j)$$

where, $d(i, j)$ is the local distance at the point (i, j) from its corresponding point in the input template.

The process of DTW is illustrated in Figure 9.4 with initial condition $D(1,1)=d(1,1)$. Here the input 'SIIGNALL' is a noisy version of the reference template 'SIGNAL'. The noisy input is matched against all the templates in the system, the best matching template which has a smaller distance from the input template is found. Figure 9.4 shows the path followed for dynamic matching (1, 1), (1, 2), (2, 3), etc., are matched, further, matching is linear for three points, then points (4, 4) and (4, 5) also get matched. Hence, the DTW algorithm works by keeping track of the best path to each point in the grid. Thus, the minimum distance is calculated for three points, namely, $(i+1, j+1)$, $(i+1, j)$, and $(i, j+1)$ as shown in Figure 9.4.

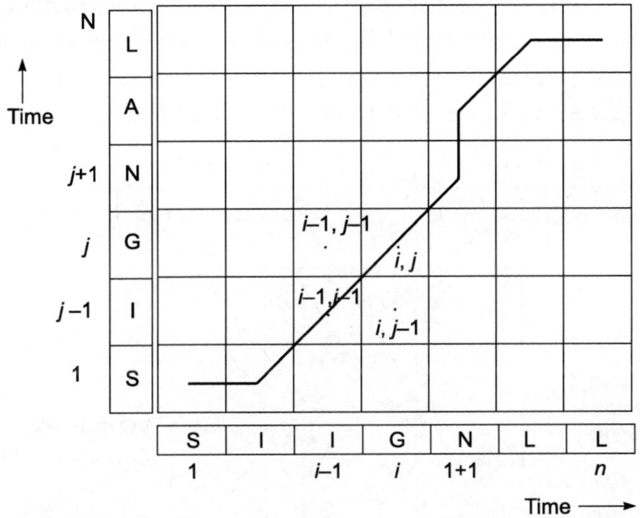

Figure 9.4 Process of DTW for matching two templates

Optimization of the DTW algorithm is based on different conditions, which can be summarized as follows:

- i and j indexes either stay the same or increase, they never decrease.
- i and j indexes advance one step at a point, it should not skip any points.
- Functions must match the end points of both input and reference templates.

9.5.3.3 Clustering based approach (vector quantization)

Clustering based approach uses the information regarding linguistic, phonetic and spectrogram. Vector quantization is often applied to speech recognition. It is useful in speech coders, i.e., for efficient data reduction. Vector quantization (VQ) is a process of mapping vectors from a large vector space to a finite number of regions in that space. Each region is called a cluster and can be represented by its centre called a codeword. The collection of all code words is called a codebook.

Figure 9.5 shows a conceptual diagram to illustrate this recognition process. In the figure, only two speech signals and two dimensions of the acoustic space are shown. The circles refer to the acoustic vectors from the speech signal 1 while the triangles belong the speech signal 2. In the training phase, using the clustering algorithm described in the next section, a speech-specific VQ codebook is generated for each known speech signal by clustering similar training acoustic vectors. The result codeword's (centroids) are shown by black circles and black triangles for speech signals 1 and 2, respectively. The distance from a vector to the closest codeword of a codebook is called a VQ-distortion. In the recognition phase, an input utterance of an unknown speech signal is "vector-quantized" using each trained codebook and the total VQ distortion is computed. The speech signal corresponding to the VQ codebook with smallest total distortion is identified as the speech signal of the input utterance. Speech signals can be discriminated based on the location of the centroid.

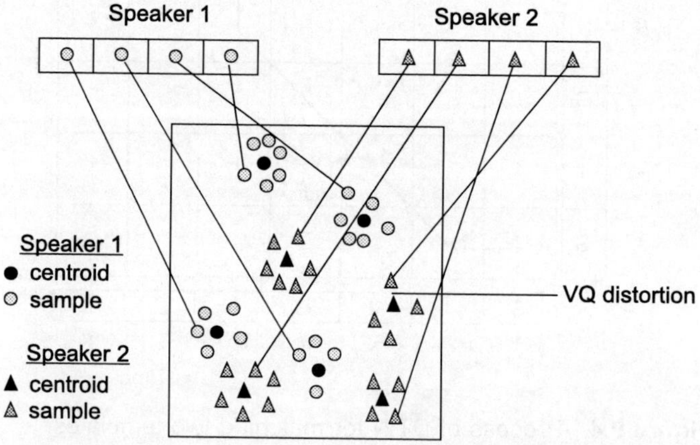

Figure 9.5 Vector quantization codebook formation.

Clustering of Training Vectors (LBG Algorithm)

The acoustic vectors extracted from input speech of each speech signal provide a set of training vectors for that specific speech signal. As described above, the next step is to build a speech-specific VQ codebook for each speech signal using those training vectors. For this, a well known algorithm, namely, LBG algorithm [Linde Buzo Gray] is used for clustering a set of L training vectors into a set of M codebook vectors. The LBG algorithm designs an M-vector codebook in stages. It starts first by designing a 1-vector codebook, then uses a splitting technique on the codewords to initialize the search for a 2-vector codebook, and continues the splitting process until the desired M-vector codebook is obtained. Flowchart of LBG algorithm is shown in Figure 9.6.

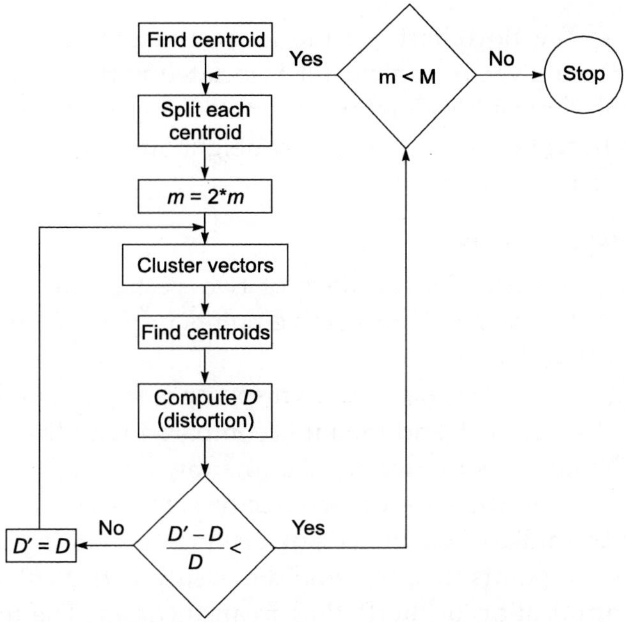

Figure 9.6 Flowchart of LBG algorithm

Implementation Of LBG Algorithm

The LBG algorithm is formally implemented by the recursive procedure which is explained below:

1. 1-vector codebook is designed. This is the centroid of the entire set of training vectors. No iteration is required here.
2. The size of the codebook is doubled by splitting each current codebook y_n according to the rule
   (a) $y_n^+ = y_n(1+\varepsilon)$
   (b) $y_n^- = y_n(1-\varepsilon)$
   where, n varies from 1 to the current size of the codebook, ε is the splitting parameter ($\varepsilon = 0.01$).
3. Nearest-neighbour search: For each training vector, the codeword is found in the current codebook that is closest (in terms of similarity measurement), and that vector is assigned to the corresponding cell (associated with the closest codeword).
4. Centroid update: The codeword is updated in each cell using the centroid of the training vectors assigned to that cell.
5. Iteration 1: Steps 3 and 4 are repeated until the average distance falls below a preset threshold.
6. Iteration 2: Steps 2, 3 and 4 are repeated until a codebook size of M is designed.

"Cluster vectors" in the flowchart are the nearest-neighbour search procedure which assigns each training vector to a cluster associated with the closest codeword. "Find centroids" is the centroid update procedure. "Compute D (distortion)" sums the distances of all training vectors in the nearest-neighbour search so as to determine whether the procedure has converged.

Euclidean Distance Measure

Euclidean distance is the distortion distance of two vector sets. It is nothing but measuring the distance between two respective points in two different vectors. The value with least distortion distance is chosen.

In the speech recognition phase, an unknown speech is represented by a sequence of feature vector $\{X_1, X_2 \ldots\ldots X_n\}$, and then it is compared with the codebooks from the database. Now the target is to identify the unknown speech. This can be done by measuring the distortion distance of two vector sets based on minimizing the Euclidean distance. As said earlier, Euclidean distance is nothing but "ordinary" distance between the two points that one would measure with a ruler, which can be proven by repeated application of the Pythagorean theorem. The formula used to calculate the Euclidean distance can be defined as below.

The Euclidean distance between two points $P = (P_1, P_2 \ldots\ldots P_n)$ and $Q = (q_1, q_2 \ldots\ldots q_n)$ is given by the equation

$$\sqrt{(P_1 - q_1)^2 + (P_2 - q_2)^2 + \ldots + (P_n - q_n)^2} = \sqrt{\sum_{i=1}^{n} (P_i - q_i)^2}$$

The speech with the lowest distortion distance is chosen to be identified as the unknown speech.

9.5.3.4 *Statistical modelling*

Real-world processes generally produce observable outputs which can be characterized as signals. The signals can be discrete in nature or continuous in nature. The signal source can be stationary (i.e., its statistical properties do not vary with time), or non-stationary (i.e., the signal properties vary over time). The signals can be pure or can be corrupted from other signal sources or by transmission distortions. A problem of fundamental interest is characterizing such signal in terms of signal model, signal model gives us:

- Theoretical description of a signal processing system which can be used to process the signal and so as to provide the desired output.
- It helps to understand great deal about signal source without having to have source available.

One can classify signal model into two types:

- **Deterministic model:** It exploits some known property of signal (like amplitude of wave).

- **Statistical model:** It takes statistical property of signal into account. Example of this type of model is Hidden Markov Model (HMM), Gaussian model, etc.

9.6 HIDDEN MARKOV MODEL (HMM) SYSTEM FOR WORD RECOGNITION

An alternative approach to template matching is to build a generative model of speech unit, whether it is a word or a sub-word. This can be done by using statistical models. This concept is based on the assumptions that human brain learns to identify sound pattern after exposure to many examples of that sound and other sounds. Therefore, the sequence of spectra representing that sound has statistical properties that distinguish it from other sounds. It is possible to get at these statistical properties by imagining that they were produced by a machine based on Markov model. The most popular and successful model is hidden Markov model (HMM).

Figure 9.7 shows the block schematic of an isolated word recognition using HMM, where each word is modelled by a distinct HMM, and V is the vocabulary size of the words. The essential blocks in the speech recognition systems are LPC feature analysis (or vector quantization) block, probability computation block for each word model and HMM for each word. To perform isolated word speech recognition, following procedure is used.

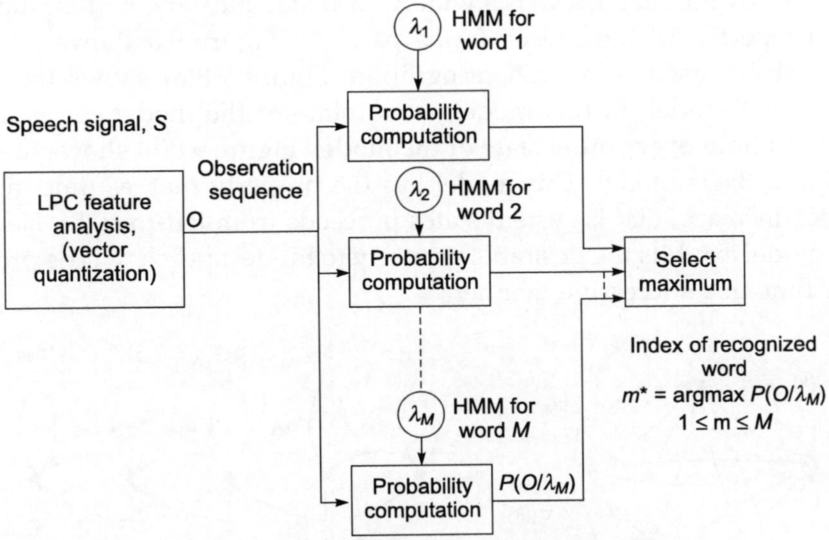

Figure 9.7 Isolated word HMM recognizer

1. Speech recognition will have two modes of operation, the training mode and recognition mode. In the training mode for each word (v) in the vocabulary, we must build an HMM, i.e., we must estimate the model parameters $\lambda = (A, B, \pi)$. Where A is the state transition probability distribution, B is observation probability distribution and Π is initial state distribution. It is

necessary to limit the parameter estimation to prevent them from becoming too smaller values.

2. In recognition mode, for each unknown word which is to be recognized, the processing must be carried out as shown in Figure 9.7, measurement of the observation sequence $O = (O_1, O_2 \ldots\ldots O_T)$, via a feature analysis of the speech corresponding to the word; followed by calculations of model likelihoods for all possible models, $P(O/\lambda_v)$, $1 \le v \le V$. Then selection of the word is based on whose model likelihood is highest.

$$v^* = \max\left[P\left(O/\lambda_v\right)\right]$$

The probability computation step is generally performed using the Viterbi algorithm (i.e., the maximum likelihood path is used). Viterbi procedure finds the best path for each model corresponding to every word to be recognized. It requires VN^2T computations. If V = 10 words, with an N = 5 state model and T = 40 observations for the unknown word, a total 10^4 computations are required for recognition.

9.6.1 Theory and Implementation of HMM

The hidden Markov model (HMM) is a widely used statistical method for modelling the spectral properties of the speech frame patterns. Figure 9.8 shows the typical structures of HMM used in speech recognition. Figure 9.8(a) shows the ergodic or fully connected model. In this model, every state of the model can be reached (in a single step) from every other state of the model. Figure 9.8(b) shows the left-to-right model or a Bakis model. This model has the property that, as time increases, the state index increase, that is system states proceeds from left-to-right. Hence, the left-to-right model exhibits the desirable property to model speech, whose properties change over time in a successive manner.

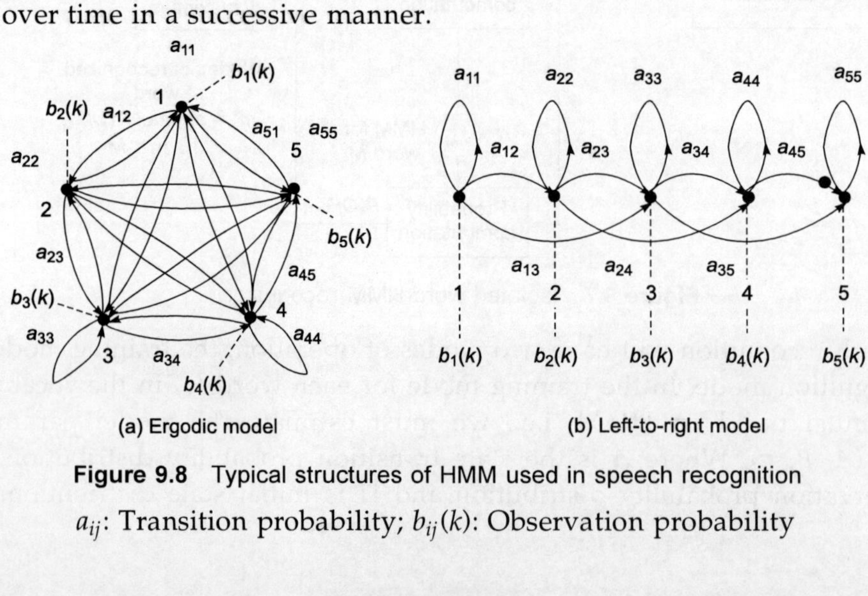

Figure 9.8 Typical structures of HMM used in speech recognition

a_{ij}: Transition probability; $b_{ij}(k)$: Observation probability

Elements of hidden Markov model

HMM is characterized by the following:

1. N = Number of states
2. $O = (O_1, O_2, \dots \dots O_T)$ = Observation sequence
3. T = Length of observation sequence
4. $Q = (q_1, q_2, \dots \dots q_T)$ = (hidden) states in the model
5. $V = (v_1, v_2, \dots \dots v_T)$ = discrete set of possible symbol observations (in VQ codebook)
6. M = Number of distinct observation symbol per state (in VQ codebook size)
7. State transition probability, $A = \{a_{ij}\} = \text{Prob}\left[q_j \text{ at } t+1 | q_i \; att\right]$,

   For the ergodic model, $a_{ij} > 0$ for all i, j. For the left-to-right model $a_{ij} > 0$ for $i < j$
8. Observation symbol probability distribution in state j, $B = \{b_j(k)\}$,

$$b_j(k) = \text{Prob}\left[v_k \text{ at } t \,|\, q_j att\right]$$

9. The initial state distribution $\pi = \{\pi_i\}$ where, π_i = $\text{Prob}\left[q_i \text{ at } t = 1\right] 1 \le i \le N$

Given appropriate value of N, M, A, B and π, HMM can be used as generator to give an observation sequence $O = (O_1, O_2, \dots \dots O_T)$. The observation sequence O is generated as follows:

Step 1: Set $t = 1$

Step 2: Choose the value of i

Step 3: Choose O_t according to $b_i(k)$, observation symbol probability distribution in state i

Step 4: Choose j according to a_{ij} $1 \le j \le N$, the state transition probability distribution for state j.

Step 5: Set $t = t + 1$, return to step 3 if $t < T$, otherwise terminate the procedure.

The above procedure can be used in the training phase for each word model and in the recognition phase for the unknown input.

9.6.2 Three Basic Problems for HMM

Problem 1: Evaluation Problem

Given the observation sequence $O = (O_1, O_2, \dots \dots O_T)$ and model $\lambda = (A, B, \pi)$, how to compute efficiently the observation sequence probability of the given model, $P(O/\lambda)$?

Problem 1 is evaluation problem, i.e., given observation sequence and model, how do we compute probability of the observed sequence produced by the model as shown in Figure 9.9. We can think it as a scoring problem. If you have to choose between many competing models, then one with maximum probability will give better result.

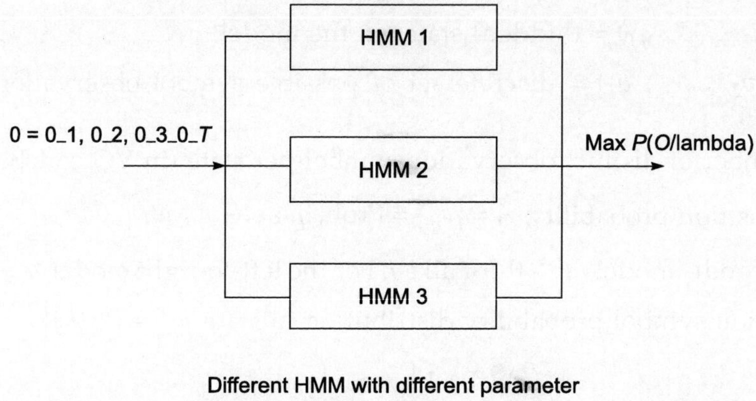

Figure 9.9 Evaluation problem

Problem 2: Hidden State Determination (Decoding)

Computing optimal state sequences $Q = (q_1, q_2, \ldots \ldots q_T)$ for the given observation sequences $O = (O_1, O_2, \ldots \ldots O_T)$ and model parameters $\lambda = (A, B, \pi)$.

Problem 2 is one which attempts to uncover the hidden part of the problem. There is no correct solution to this problem, in practice we usually use optimal criterion to find best possible solution. For coin tossing example with observation sequence O = H H T T H corresponding state sequence in each of the model can be

$$O = \mathrm{H\,H\,T\,T\,H}$$

Model 1: S = 1 1 2 2 1

Model 2: S = 2 1 1 1 2

Model 3: S = 3 1 2 3 2

Problem 3: Training Problem

Computing model parameter $\lambda = (A, B, \pi)$ to maximize $P(O/\lambda)$.

Problem 3 is one in which we try to optimize model parameter so as to best describe as to how given observation sequence comes out. The observation sequence used here is called *"**training**"* sequence since it is used for training HMM. It allows us to adjust model parameter as to create best model for the given training sequence.

Solution to three Problems: Given this problem, it's vital to solve them using efficient algorithm.

Forward Algorithm for Evaluation Problem [$P(O \mid \lambda)$]

We want to find $P(O/\lambda)$, given the observation sequence $O = (O_1, O_2, \ldots \ldots O_T)$, The most straightforward way to find the solution is enumerating every possible state sequence of length T. Consider one such state sequence $Q = (q_1, q_2, \ldots \ldots q_T)$ such that q_1 produces O_1 with some probability, q_2 produces O_2 with some probability and so on. So using chain rule

$$P\left(\frac{O}{\lambda}\right) = \sum_{q_1, q_2, \ldots\ldots q_T} \pi_{q_1} b_{q_1}(O_1) a_{q_1 q_2} b_{q_2}(O_2) \ldots\ldots a_{q_{T-1} q_T} b_{q_T}(O_T)$$

but the order of chain rule is N^T, since at every $t = 1, 2 \ldots\ldots T$, there are N possible states which can be reached. This is clearly an inefficient algorithm, to overcome this forward algorithm is used.

Forward algorithm is a dynamic algorithm which uses forward variable $\alpha_t(i)$ defined as

$$\alpha_t(i) = P(O_1, O_2 \ldots\ldots O_i, i_t = q_i \mid \lambda$$

i.e., the probability of partial observation sequence, $O_1, O_2, \ldots \ldots O_t$ and state q_i at time t given the model λ, as we are taking partial observation into account, we can solve $\alpha_i(i)$ inductively as

Initialization:

$$\alpha_1(i) = \pi_i b_i(O_1) \qquad 1 \le i \le N$$

Induction: For $t = 1, 2 \ldots\ldots, T - 1 \quad (1 \le i \le N)$

$$\alpha_{t+1}(j) = \left[\sum_{i=1}^{N} \alpha_t(i) a_{ij}\right] b_j(O_{t+1})$$

Termination:

$$P\left(\frac{O}{\lambda}\right) = \sum_{i=1}^{N} \alpha_t(i)$$

This algorithm can be derived by transforming the HMM into lattice (or trellis) diagram as shown in Figure 9.9.

Similarly, a backward variable $\beta_t(t)$ is defined as

$$\beta_t(i) = P(O_{t+1}, O_{t+2} \ldots O_T \mid i_t = q_i, \lambda)$$

This demonstrates the probability of the partial observation sequences, from $t + 1$ to the end for the given state q_i at time t and model λ. $\beta_t(i)$ can be solved recursively as follows:

Initialization:

$$\beta_T(i) = 1 \qquad 1 \le i \le N$$

Induction: For $t = T-1, T-2, \ldots \ldots 1 \quad (1 \le i \le N)$

$$\beta_t(i) = \left[\sum_{j=1}^{N} \beta_{t+1}(j) a_{ij}\right] b_j(O_{t+1})$$

Termination:

$$P\left(\frac{O}{\lambda}\right) = \sum_{i=1}^{N} \pi_i b_i(O_1) \beta_1(i)$$

Results show that this algorithm requires the order of N^2T calculation rather than N^T in case of chin rule.

Viterbi Algorithm for Decoding Hidden State Sequence $P(Q, O \mid \lambda)$

Problem 2 can be solved using the Viterbi algorithm. It says that to find the single best state sequence, $Q = (q_1, q_2 \ldots \ldots q_T)$ (which produces given observation sequence) for a given observation sequence $O = (O_1, O_2 \ldots \ldots O_t$, we define a quantity

$$\delta_t(i) = \max_{q_1, q_2 \cdots q_T} P[q_1, q_2 \ldots\ldots q_t = i, O_1, O_2 \ldots\ldots O_t \mid \lambda]$$

i.e., $\delta_t(i)$ is the best score along a single path, at time t, which account for the first t observations and ends in state q_i, by induction

$$\delta_{t+1}(j) = \left[\max_i \delta_t(i) a_{ij}\right] b_j(O_{t+1})$$

in order to find the state sequence, we need to keep track of state which maximizes the above equation. we do this via array $\psi_t(t)$ for each t and state j. Once the final state is reached, the corresponding state sequence can be found out using backtracking. Key point is, Viterbi algorithm is similar (except for the backtracking part) in implementation to the Forward algorithm. The major difference is maximization of the previous state in place of summing procedure in forward calculation. Veterbi algorithm is given as follows:

Step 1: Initialization

$$\delta_1(i) = \pi_i b_i(O_1) \qquad 1 \le i \le N$$

$$\psi_1(i) = 0$$

Step 2: Recursion

For $2 \le t \le T, 1 \le j \le N$,

$$\delta_t(j) = \max_{1 \le i \le N} \left[\delta_{t-1}(i) a_{ij}\right] b_j(O_t)$$

$$\psi_t(j) = \underset{1 \le j \le N}{\arg\max} \left[\delta_{t-1}(i) a_{ij} \right]$$

Step 3: Termination

$$P^* = \max_{1 \le i \le N} \left[\delta_T(i) \right]$$

$$q_t^* = \underset{1 \le i \le N}{\arg\max} \left[\delta_T(i) \right]$$

Step 4: State sequence backtracing

For $t = T-1, T-2, \ldots, 1$

$$q_t^* = \psi_{t+1}\left(q_{t+1}^*\right)$$

Here P^* is the maximum likelihood, and ψ indicates the maximum likelihood state sequence.

Figure 9.10 shows the calculation of word "ONE" whose phoneme sequence is *W AH* and *N*. In forward calculation we consider summing of all incoming node to find $P(O \mid \lambda)$. Since our phoneme model is bigram model, only its predecessor states input is considered to calculate the sum of probability. For better understanding refer to Figure 9.10, at time $t = 3$, input to state *N* is from state *AH* and *N* only and not from *W*. In this way, we can find the $P(O \mid \lambda)$, given an observation sequence and model, ψ.

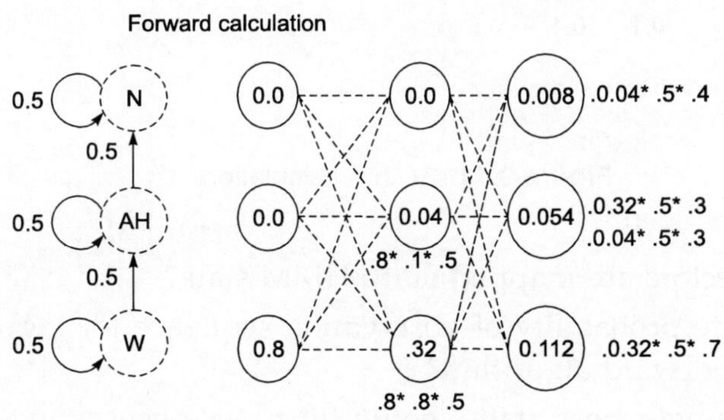

Output probability matrix

Time	1	2	3
W	0.8	0.8	0.7
AH	0.1	0.1	0.3
N	0.0	0.0	0.4

Figure 9.10 Forward calculation

Now we will look at Viterbi algorithm for finding the best state sequence. As explained above Viterbi algorithm is same (except backtracking) in implementation as of Forward algorithm. Major difference is in Viterbi we take MAX of all incoming input for the given state. As shown in Figure 9.11, line from state 1 at time $t = 1$ to state 2 at $t = 3$ is considered for probability calculation, while in Forward algorithm we sum up all probability.

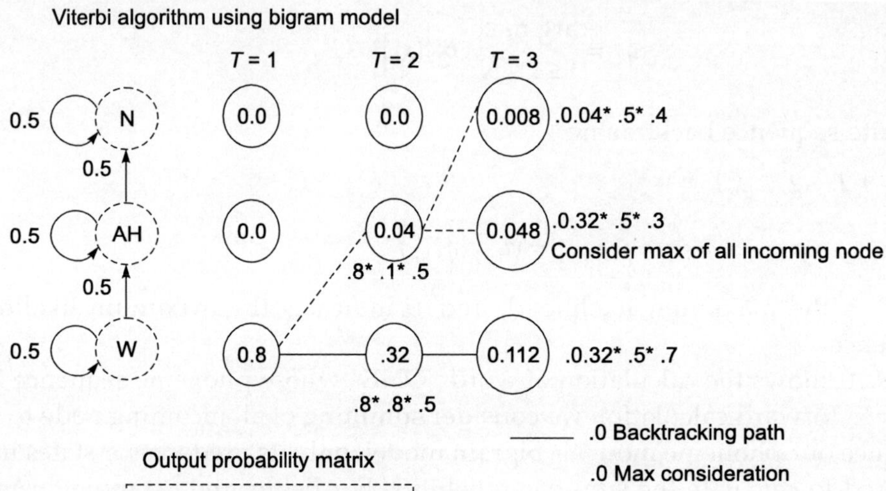

Output probability matrix

Time	1	2	3
W	0.8	0.8	0.7
AH	0.1	0.1	0.3
N	0.0	0.0	0.4

Figure 9.11 Viterbi calculation

So till now we have seen

- How feature vectors are mapped on to HMM state?
- How to calculate probability of observation sequence for a given model, i.e., $P(O \mid \lambda)$, using Forward algorithm?
- How to find single best state sequence given observation sequence and model, i.e., sequence of states which produces required output, using Viterbi algorithm?

Baum-Welch Algorithm for Training Problem

Most challenging of all is to adjust the model parameter (A, B, λ) to maximize the probability of the observation sequence given in the model. Given any finite observation sequence as training data, there is no optimal way of estimating the model parameters. But there are some heuristic procedures which try to find local optimization over global optimization. An iterative procedure, such as Baum-Welch method, or gradient technique for optimizing is used for solving this problem.

Re-estimation procedure: In order to define a procedure for re-estimation of HMM parameter, we first define

$$\xi_t(i, j) = P\left(i_t = q_i, i_{t+1} = q_j | O, \lambda\right)$$

i.e., the probability of being in state q_i, at time t, and state q_j at time $t + 1$ given the model and observation sequence, we can write

$$\xi_t(i, j) = \frac{\alpha_t(i) b_j(O_{t+1}) \beta_{t+1}(j)}{P\left(\frac{O}{\lambda}\right)}$$

where, $\beta_t(j)$ = backward variable, i.e., probability of the partial observation sequence $t+1$ to end (T), given state q_i at time t. Now we define $\gamma_t(i)$ as the probability of being in state q_i at time t, given the observation sequence and model; hence we can relate $\gamma_t(i)$ to $\xi_t(i, j)$ by summing over j, such that

$$\gamma_t(i) = \sum_{j=1}^{N} \xi_t(i, j)$$

if we sum $\gamma_t(i)$ over time index we get a quantity which can be interpreted as expected number of time state q_i is visited, or expected number of transitions made from q_i. Similarly, summation of $\xi_t(i, j)$ over time can be interpreted as the expected number of transitions made from state q_i to state q_j, that is,

$$\sum_{t=1}^{T-1} \gamma_t(i) = \text{Expected number of transitions from } q_i$$

$$\sum_{t=1}^{T-1} \xi_t(i, j) = \text{Expected number of transitions from } q_i \text{ to } q_j$$

Using the above formula we can give a method for re-estimation of the parameter of an HMM.

$$\overline{\pi_i} = \text{Expected number of times in state } q_i \text{ at time } (t = 1) = \gamma_1(i)$$

$$\overline{a}_{ij} = \frac{\text{Expected number of transition from state } q_i \text{ to } q_j}{\text{Expected number of transition from state } q_i} = \frac{\sum_{t=1}^{T-1} \xi_t(i, j)}{\sum_{t=1}^{T-1} \gamma_t(i)}$$

$$\overline{b_j}(k) = \frac{\text{Expected number of times in state } j \text{ and observing symbol } v_k}{\text{Expected number of times in state } j} = \frac{\sum_{t=1}^{T} \gamma_t(j) \ \text{S.t. } O_t = v_k}{\sum_{t=1}^{T-1} \gamma_t(j)}$$

Using the above three algorithms we can solve all three problems associated with HMM.

9.7 NEURAL NETWORK

Neural network is a mathematical model which is used to perform a particular function. Artificial neural networks consist of nodes (neurons) interconnected with each other. It works like the human brain. Neural networks are highly parallel in nature, and hence they permit very fast processing of speech data. Neural networks have many similarities with Markov model. Both neural network and Markov model are statistical models, Markov model uses probabilities for state transitions and neural network uses connection strengths and functions. Another important difference is neural networks are fundamentally parallel, while Markov chains are serial. In speech, formant frequencies occur in parallel, while syllable and words are in series. Hence, both techniques are powerful. In many speech recognition systems both techniques are implemented together.

There are different types of neural networks, which map the complex input into simple output. In this section, two types of neural networks, multilayer feed forward (back propagation) networks and radial basis functions.

9.7.1 Multilayer Feedforward (Back Propagation) Networks

Multilayer feedforward neural networks is one of the popular neural networks and is used in different types of applications. The network consists of three layers: input layer, hidden layer and output layer as shown in Figure 9.12. Neural networks have 10 to 20 inputs, data points from the spectrogram would be too much and hence Mel-frequency cepstrum coefficients are used as input for neural network. Input values are in the range –5 up to 1.5.

Hidden layer consists of nonlinear sigmoid activation function. In MATLAB NN toolbox function, this activation function is selected by using 'tansig'. Number of hidden layer neurons are decided by Oja rule of thumb and is given by

$$H = \frac{T}{5(N+M)}$$

where, H is the number of hidden layer neurons, N is the size of the input layer, M is the size of the output layer, and T is the training set size. For example, to recognize 10 words (with a training set of 75 samples per word). The neural network has 20 (each sample of word has 20 MFCCs). Neural network has 20 inputs and 5 outputs, then H is calculated by

$$H = \frac{10 \times 75}{5(20+5)} = \frac{750}{125} = 6$$

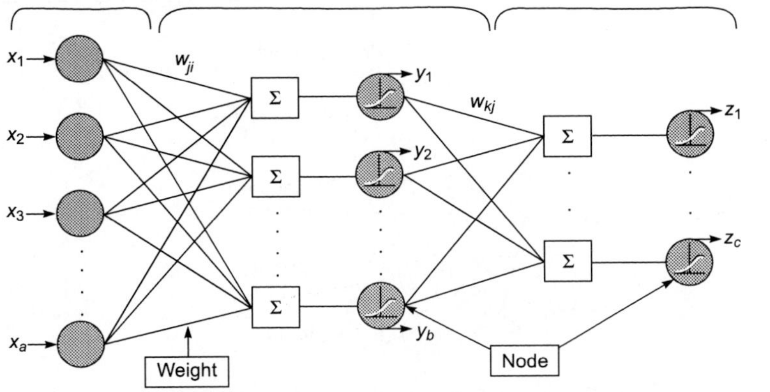

$$Y_j = f(\text{net}_j) = f\left(\sum_{i=1}^{a} W_{ji} x_i\right) \quad \text{where } i = 1, 2, \ldots a \qquad a = \text{number of input variables}$$

$$z_k = f(\text{net}_k) = f\left(\sum_{j=1}^{b} W_{kj} y_j\right) \quad \text{where } j = 1, 2, \ldots b \qquad b = \text{number of hidden nodes}$$

$$\text{where } k = 1, 2, \ldots c \qquad c = \text{number of output nodes}$$

Figure 9.12 Feed-forward, back-propagation neural network (multilayer perception)

Once the network is created, it can be trained by presenting training inputs and their corresponding targets. When the number of words to be recognized increases, the number of hidden layer neurons increases. Performance of the network mainly depends on the quality of the signal preprocessing. It works well with MFCCs as input having 90% successful classification rate.

9.7.2 Radial Basis Function Network

Another approach for speech classification is to use radial basis function network. Like feedforward, this network also consists of three layers: input layer, hidden layer, and output layer. Input layer is similar to the input layer of the multilayer feedforward network. Main difference to feedforward type is, RBF network consists of one hidden layer of neurons with radial basis function as shown in Figure 9.13. Radial means that they are symmetric around their centre, basis means that a linear combination of their function can generate an arbitrary function. Radial basis functions were designed with the MATLAB function 'newrbe'. Figure 9.12 shows a RBF capable of recognizing 9 words (with 9 outputs) with an input of MFCC (with 20 MFCC) is given. For this combination 450 hidden neurons are needed and in multilayer feedforward neural networks only 9 sigmoid hidden layer neurons are needed. This type of neural network is practical for large training sets and it performs very well for less training samples.

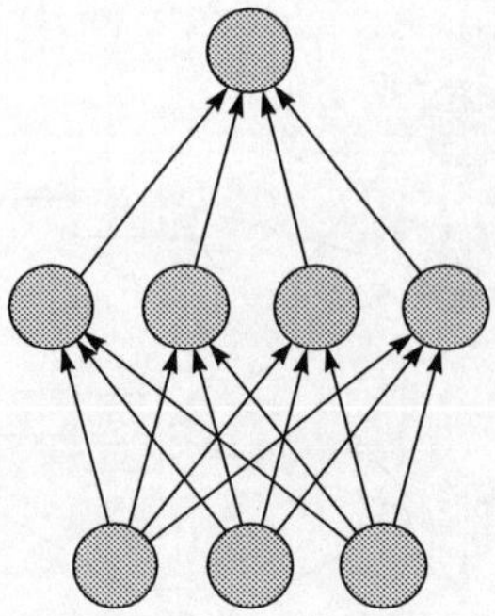

Figure 9.13 Radial basis function network

9.8 SPEAKER RECOGNITION

Speaker recognition is the automatic recognition of speaker (talker) with the help of characteristics of speaker voice signal. Speaker recognition is classified into speaker identification and speaker verification applications. Speaker identification is the process of determining which registered speaker provides a given utterance. Speaker verification, on the other hand, is the process of accepting or rejecting the identity claimed by a speaker. Speaker identification makes it possible to use the speaker's voice to verify their identity and control access to services such as voice dialing, banking by telephone, telephone shopping, database access services, information services, voice mail, security control for confidential information areas, and remote access to computers. AT&T and TI (with Sprint) have started field tests and actual application of speaker recognition technology; Sprint's voice phone card is used by many customers. In this way, speaker recognition technology is expected to create new services that will make our daily lives more convenient. Another important application of speaker recognition technology is for forensic purposes. Most applications in which a voice is used as the key to confirm the identity of the speaker are classified as speaker verification.

Figure 9.14 shows the basic structures of speaker identification and verification systems. In the speaker identification task, a speech utterance from an unknown speaker is analyzed and compared with speech models of known speakers. The unknown speaker is identified as the speaker whose model best matches the input utterance. In speaker verification, an identity is claimed by an unknown speaker, and an utterance of this unknown speaker is compared with a model for the speaker whose identity is being claimed. If the match is good enough, that is, above a threshold, the identity claim is accepted. A high threshold makes it difficult for impostors to be accepted by the system, but with the risk of falsely rejecting valid users. Conversely, a low threshold enables valid users to be accepted consistently, but with the risk of accepting impostors. To set the threshold at the desired level of customer rejection (false rejection) and impostor acceptance (false acceptance), data showing distributions of customer and impostor scores are necessary.

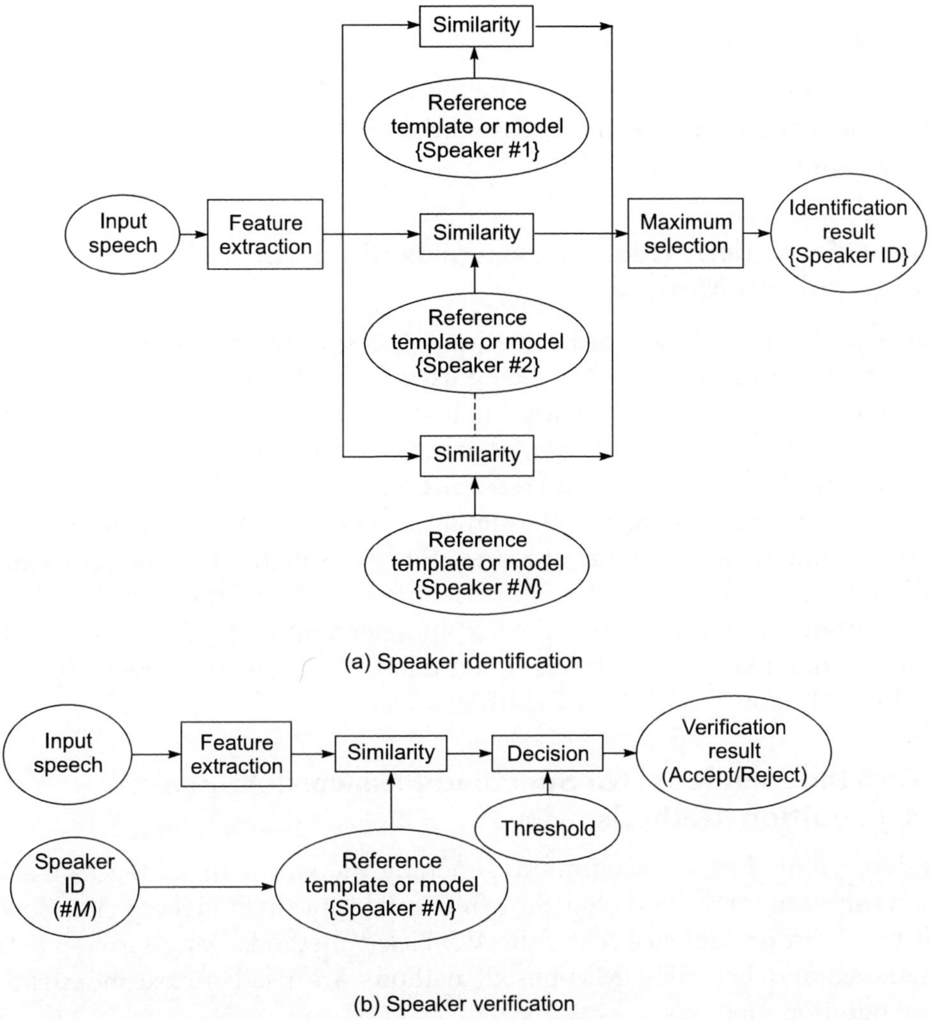

Figure 9.14 Basic structures of speaker identification and verification systems

The fundamental difference between identification and verification is the number of decision alternatives. In identification, the number of decision alternatives is equal to the size of the population, whereas in verification there are only two choices, acceptance or rejection, regardless of the population size. Therefore, speaker identification performance decreases as the size of the population increases, whereas speaker verification performance approaches a constant independent of the size of the population, unless the distribution of physical characteristics of speakers is extremely biased.

9.9 TEXT-DEPENDENT, TEXT-INDEPENDENT AND TEXT-PROMPTED METHODS

Speaker recognition methods can also be divided into the following methods:

- Text-dependent (fixed passwords) methods
- Text-independent (no specified passwords) methods
- Text-prompted methods

9.9.1 Text-dependent (Fixed Passwords) Speaker Recognition Methods

In text-dependent (fixed passwords) methods the speaker provide utterances of key words or sentences, the same text being used for both training and recognition, whereas the latter do not rely on a specific text being spoken. The text-dependent methods are usually based on template/sequence matching techniques in which the time axis of an input speech sample and reference templates or reference models of the registered speakers are aligned, and the similarities between them are accumulated from the beginning to the end of the utterance. Since this method can directly exploit voice individuality associated with each phoneme or syllable, it generally achieves higher recognition performance than the text-independent method. Text-dependent speaker recognition methods can be classified into DTW (dynamic time warping) or HMM (hidden Markov model) based methods.

9.9.2 Text-Independent (No Specified Passwords) Speaker Recognition Methods

In text-independent speaker recognition, generally the words or sentences used in recognition trials cannot be predicted. Since it is impossible to model or match speech events at the word or sentence level, the VQ-based methods, long-term-statistics-based methods, and Ergodic-HMM-based methods are used in text-independent speaker recognition methods.

9.9.3 Text-prompted Speaker Recognition

In this method, key sentences are completely changed every time. The system accepts the input utterance only when it determines that the registered speaker uttered the prompted sentence. Because the vocabulary is unlimited, prospective impostors cannot know in advance the sentence they will be prompted to say. This method not only accurately recognizes speakers, but can also reject an utterance whose text differs from the prompted text, even if it is uttered by a registered speaker. Thus, a recorded and played back voice can be correctly rejected. This method uses speaker-specific phoneme models as basic acoustic units. One of the major issues in this method is how to properly create these speaker-specific phoneme models when using training utterances of a limited size. The phoneme models are represented by Gaussian-mixture continuous HMMs or tied-mixture HMMs, and they are made by adapting speaker-independent phoneme models to each speaker's voice.

9.10 SUMMARY

In this chapter, we have understood challenges of speech recognition, speech recognition algorithms like pattern recognition and knowledge-based techniques with some of the applications of speech signal processing.

EXERCISES

Review Questions

1. What are the difficulties in speech recognition system?
2. What are the differences between isolated word and continuous speech recognition?
3. Explain the different techniques for speech recognition.
4. With necessary block schematic, explain pattern recognition model used for speech recognition. Explain the function of each block.
5. Explain the various parametric feature extraction methods of speech recognition.
6. With block schematic, explain parametric representations for speech recognition using Mel-frequency cepstrum coefficient (MFCC) method.
7. Explain different distance measure used in linear prediction coding (LPC) method.
8. With an example explain dynamic time warping (DTW) and list its limitations.
9. Explain vector quantization with LBG algorithm.
10. Discuss LBG algorithm.
11. With necessary block schematic explain isolated word recognition using HMM.
12. State three problems associated in case of HMM for speech recognition.
13. Explain the procedure to calculate the probability of any observation sequence using forward or backward algorithm.
14. Explain Viterbi algorithm to recognize a word.
15. Explain Baun-Welch algorithm of HMM used for training.

10

CHAPTER

Speech Enhancement

This chapter provides insights on

- Evaluation techniques for speech intelligibility
- Speech enhancement algorithms
- Applying Weiner filter in frequency domain
- Principles of adaptive filtering
- Adaptive filtering by LMS and RLS
- Single-channel adaptive filtering

10.1 INTRODUCTION

The term speech enhancement refers to the improvement of the quality or intelligibility of speech. Background interference regrades the quality or intelligibility of speech. The problem of enhancing speech degraded by additive background noise has received considerable attention in the last two decades. Speech enhancement aims to improve speech quality by using various algorithms. Enhancement of speech signals is of great interest in many applications ranging from speech recognition to hearing aids and hands-free mobile communication. The aims of speech intelligibility vary according to the application and may include:

- Improvement in the intelligibility of speech to human listeners.
- Improvement in the quality of speech that makes it more acceptable to human listeners.
- Modifications to the speech that lead to improved performance of automatic speech or speaker recognition systems.
- Modifications to the speech so that it may be encoded more for storage or transmission.

Speech enhancement means improving the clarity and intelligibility, pleasantness, or compatibility with some other method in speech processing. Intelligibility and pleasantness are difficult to measure by any mathematical algorithm. Usually, listening tests are employed. Although this problem has attracted significant research efforts for several decades, many aspects remain open and require further research. Following are the challenges in speech enhancement.

- Completely removing the background noise is practically impossible, as we do not have access to the noise signal (only the corrupted signal).
- The majority of speech enhancement algorithms introduce some type of speech distortion.
- There exists speech and noise distortion trade-off associated with most algorithms, that is, it is difficult to completely reduce the background noise without introducing speech distortion. Conversely, it is difficult to minimize speech distortion without introducing noise distortion.

The speech enhancement refers to the class of algorithms which aim to remove or reduce the background noise. The noisy signal can be acquired using a single sensor (microphone) or multiple sensors (microphone array). Removing or reducing the background noise present in single-microphone recordings is extremely challenging. The typical sources of degradations for speech enhancement are due to

(a) Additive acoustic noise
(b) Acoustic reverberation
(c) Convolutive channel resulting in an uneven or band limited response
(d) Nonlinear distortion such as arising from clipping
(e) Additive broadband electronic noise
(f) Electrical interference
(g) Codec distortion
(h) Distortion introduced by recording apparatus.

Additive noise that is uncorrelated with the wanted speech signal may arise in either the acoustic or electronic domain. Its perceived effect is to degrade listenability and intelligibility and in extreme cases, completely mask the wanted signal. For some types of additive noise, the spectral characteristics are stationary or change slowly with time. This is typically true of hum and amplifier noise as well as of some environmental acoustic noise sources. Spectral subtraction and single-channel adaptive filtering have been successful in reducing the perceived level of such stationary noise sources. Other forms of additive noise are intermittent or highly non-stationary and their identification and deletion is the subject of model-based and missing data methods. Such non-stationary noise sources include media interference, unwanted co-talkers and some forms of electrical interference.

Convolutive effects are perceived as reverberation and poor spectral balance; they differ from the previous group because the added noise is strongly correlated with the wanted signal. Reverberation and echo normally arise from acoustic reflections and can seriously degrade intelligibility. The increasing use of distant microphones in hands-free telephony has prompted extensive research into reducing the effects of reverberation. Bandwidth restrictions and uneven spectral response may arise from microphone placement, microphone characteristics and CODEC limitations. There has been some work on expanding the bandwidth narrowband telephone signals in order to improve listenability but there is little evidence of any intelligibility benefits.

Nonlinear distortion frequently arises from amplitude limiting or clipping in the microphone, amplifier or CODEC. This is perceived as harsh distortion that varies with the signal amplitude. A similar perceptual effect can result from high bit errors in the coded signal used by some CODECs. Clipped portions of a waveform are easy to identify provided that no subsequent phase distortion is present and techniques exist for reconstruction the corrupted portions of the waveform.

10.2 EVALUATION OF SPEECH INTELLIGIBILITY

The main evaluation methods for speech enhancement systems look at the effect of the system on the intelligibility of the speech signal and the improvement in the overall quality of the signal.

There are two categories of evaluation techniques for speech intelligibility

- Subjective measures
- Objective measures

10.2.1 Subjective Methods

In subjective measure there are three types of evaluation techniques for speech intelligibility, they are:

- Phonetic tests
- Word intelligibility
- Sentence intelligibility

10.2.1.1 Phonetic tests

In 1929, Fletcher and Steinberg proposed articulation test material consisting of a list of 66 consonant-vowel-consonant (CVC) nonsense syllables. Since the list was balanced for each consonant and vowel phoneme, the number of syllables heard correctly gave an indication of phoneme intelligibility. Miller and Nicely [1955] used a similar approach, but reduced the consonants to a size 16 subset: /p, t, k, b, d, g, f, T, s, S, v, D, z, Z, m, n/in a VCV frame. This subset was designed so that intelligibility could be broken down into five articulator dimensions: voicing, nasality, frication, sibilance and place. Since then VCV materials have been widely used within speech perception research, because intelligibility results can be analyzed in terms of the effectiveness of the primitive acoustic patterns necessary for consonant discrimination. However, for overall assessment of intelligibility, most effort has gone into word-level tests.

10.2.1.2 Word intelligibility

One problem with the use of nonsense syllables is that listeners require training to be able to identify component phonetic units, and may be confused by phonemes which don't readily accord with spelling. Limiting listener responses to real words allows them to respond in ordinary spelling, but introduces other difficulties: first, that different listeners may have different degrees of familiarity with the words being used and their potentially confusing competitors; secondly that words are memorable, and that, having heard a word once, listeners may be biased in their usage of the word another time. One solution to these problems includes creating multiple lists of balanced difficulty, so that a listener can be used more than once. Another is to create tests with closed response sets, so every listener needs to make the same choices about the word under test. One of the earliest word lists created for intelligibility testing was that by Egan in 1948, who constructed 20 lists of 50 monosyllabic words. His aim was to balance average difficulty and range of difficulty over the lists, while ensuring that the phonetic units present were equally represented. Egan introduced the idea of "*phonetic balance*" by which he meant that the relative frequency of phonemes in the word lists matched the relative frequency of phonemes in conversational speech. In 1959, Lehiste and Peterson built a similar set of lists of monosyllabic words, but which were balanced for the relative frequency of phonemes in monosyllabic words. Fairbanks, in 1958 modified Egan's idea so that instead of identifying the whole word, listeners merely had to identify the leading consonant (the rest of the word was written down on the response sheet), leading to the concept of a "rhyme" test. House et al. in 1965, modied Fairbanks' idea by introducing a closed response set, where listeners only had to choose amongst six alternatives: this was called the modified rhyme test (MRT). Kreul et al., in 1968, attempted to modify the MRT to make it a clinically useful tool, while a phonetically-balanced version of the MRT was developed by Clarke [1965]. In the diagnostic rhyme test (DRT), Voiers [1983] extended the MRT by requiring that listeners choose between a pair of words differing in only one phonetic feature. Voiers chose features that had appeared in the analysis by Miller and Nicely (1955) of consonant confusions. This allowed listener judgements to be analysed in terms of their performance on the different feature dimensions as well as on their overall score.

10.2.1.3 Sentence intelligibility

Words in isolation are not typical of the way in which speech is used for communication. In everyday listening to speech, knowledge of the topic, semantic, grammatical and prosodic constraints provide additional information which allows us to recognize words. However, the use of sentences in intelligibility testing only amplifies the problems discussed earlier in relation to the use of words: listeners will vary even more in their ability to extract higher level information from sentences to aid recognition, and sentences are even more memorable. It is also much harder to create balanced lists of sentences, since the amount of benefit to be gained from sentence context is hard to quantify.

Kalikow et al. (1977) developed the speech perception in noise (SPIN) test to try and assess the extent to which listeners were able to extract contextual information from sentences to aid word recognition. The SPIN test comprises 8 lists of 50 sentences, with one keyword per sentence. There are 25 key words in each list, occurring once in a high predictable context (e.g., "The boat sailed across the bay"), and once in a low predictable context (e.g., "John was talking about the bay"). The relative scores on the two types of sentence provide a measure of the listener's ability to use context.

In 1994, Nilsson et al., developed a revised version of the SPIN test known as the hearing in noise test (HINT). The HINT test comprises 25 balanced lists with 10 sentences per list. All words in the list are scored, but alternatives are allowed for grammatical words (e.g., a, the, is, was, . . .). A standard recording of the lists by a professional male actor has been made available and the HINT test remains popular for the assessment of hearing impairment.

Two testing methodologies are in common use for the measurement of intelligibility with speech materials such as HINT. In the first method, the speech signals are first processed into a number of fixed levels of presentation: where the level is either the speech amplitude or the signal to noise ratio. Then speech materials from all levels are presented to the listeners in random order, and the intelligibility at each level is recorded and plotted as a function of the presentation level. This graph usually takes the shape of an S-shaped psychometric function as intelligibility changes from 0 to 100% across a range of levels. From this graph we can estimate the presentation level which achieves the required level of intelligibility. The level at which 50% intelligibility is obtained is called the *speech reception threshold* (SRT) [Plomp and Mimpen, 1979].

This fixed presentation method has a number of disadvantages: first, the same amount of speech is presented at every level—even though we may be only interested in one point on the psychometric curve, this makes the test rather long; secondly, to ensure that the levels bracket 50% intelligibility say, we may need to use a wide range of levels, many of which will have 0% or 100% intelligibility for a particular listener; thirdly, we need to ensure that the speech material presented at each level is equally difficult.

An alternative to this method is to use an adaptive procedure where presentation levels change dynamically within the test to determine the level which delivers an intelligibility score of required size. In a typical procedure, the speech material is first played at a good level or SNR such that listeners can readily identify the words. Then the level is adjusted such that the presentation level or SNR falls by (say) 2 dB when the listener gets the words correct, and rises by 2 dB when the words are incorrect. Typically, the average of the levels at which the last eight reversals in direction were made is taken to be the overall SRT score. Studies into the effectiveness of this procedure have shown that it seems to work well [Levitt, 1971, Dirks et al., 1982].

A disadvantage of an adaptive procedure is that it relies on assumptions about the shape of the psychometric function. A curve that is not very steep at 50% intelligibility will produce more variable results, and a curve that is not monotonic may not produce sensible results at all. Where the systems under test are available as real-time implementations, it is possible to process stimuli while a test is taking place; this avoids the need to capture and process the stimuli off-line [Worrall et al., 2006]. In combination with an adaptive procedure, this makes intelligibility testing fast enough for rapid assessment of systems in the field.

10.2.2 Objective Measures

The SNR is the most widely used measure for analog and waveform coding systems. It is also been used for assessing enhancement algorithms for noise distortions. Let $s(n)$ denote a noise free signal at time n and $\hat{s}(n)$ denote the processed speech signal at time n. Then the SNR in dB is defined by

$$\text{SNR} = 10\log_{10}\frac{\sum_n s^2(n)}{\sum_n [s(n)-\hat{s}(n)]^2} \tag{10.1}$$

Segmental signal-to-noise ratio is a refinement with respect to conventional SNR measure and is created to handle the dynamic nature of non-stationary speech signals. The SSNR is defined by

$$\text{SSNR} = \frac{1}{M}\sum_{j=0}^{M-1} 10\log_{10}\frac{\sum_{n=m_j-N+1}^{m_j} s^2(n)}{\sum_{n=m_j-N+1}^{m_j} [s(n)-\hat{s}(n)]^2} \tag{10.2}$$

where $m_0, m_1, \ldots \ldots \ldots n_{M-1}$ are the end times for the M frame, each of which is of length, N. Each frame is typically 15-25 msec. Final SNR measure is obtained by averaging these measurements over all segments of the waveform.

10.3 SPEECH ENHANCEMENT ALGORITHMS

A number of books devoted to speech enhancement have been published including Davis [2002], Benesty et al. [2005] and Loizou [2007]. In addition, there are a number of survey papers on the subject including Lim [1986], Ephraim and Cohen [2006] and Ephraim et al. [2005]. The performance of several enhancement algorithms was tested by Song et al. [2006] by means of a standardised speech recogniser; they found that some enhancement techniques reduced the recognition performance. A subjective assessment of 13 enhancement algorithms is given in Hu and Loizou [2007] which, following ITU [2003], asked listeners for separate judgments on signal distortion, background noise, and overall quality. Following algorithms are discussed in this book.

- Spectral subtraction
- Weiner filter in frequency domain
- Adaptive filtering
- Wavelet transforms

10.3.1 Spectral Subtraction

The spectral subtraction algorithm is historically one of the first algorithms proposed for noise reduction. More papers have been written describing variations of this algorithm than any other algorithm. It is based on a simple principle. Assuming additive noise, one can obtain an estimate of the clean signal spectrum by subtracting an estimate of the noise spectrum from the noisy speech spectrum. The noise spectrum can be estimated, and updated, during the periods when the signal is absent or when only noise is present. The simple subtraction processing comes with a price. The subtraction process needs to be done carefully to avoid any speech distortion. If too much is subtracted, then some speech information might be removed, while if too little is subtracted then much of the interfering noise remains.

The spectral subtraction approach can be categorized as a nonparametric approach, which simply needs an estimate of the noise spectrum. It is assumed that there is an estimate of the noise spectrum which is obtained during periods of speaker silence. Let $x(n)$ be a noisy speech signal:

$$x(n) = s(n) + N(n) \tag{10.3}$$

where, $s(n)$ is the clean signal, and considered noise $N(n)$ is additive. In this case, the noise and the clean signal can be assumed uncorrelated. So, the spectral subtraction approach can be used to estimate the short-term magnitude spectrum of the clean signal $|S(e^{jw})|$ by the subtraction of the estimated noise magnitude spectrum $|N(e^{jw})|$ from the noisy signal magnitude spectrum $|X(e^{jw})|$ as shown in Figure 10.1.

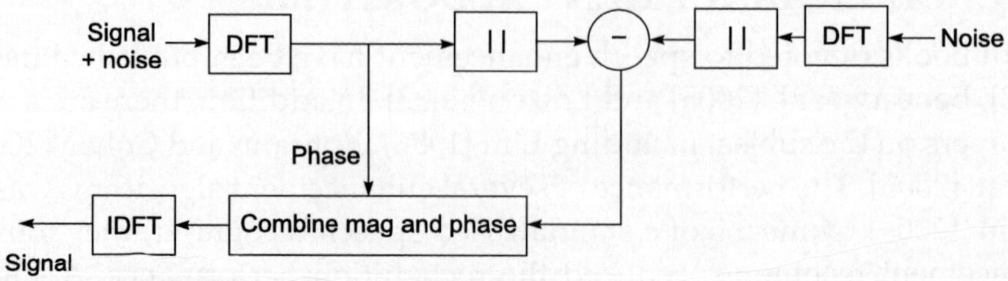

Figure 10.1 Block schematic of the spectral subtraction method

It is sufficient to use the noisy signal phase spectrum as an estimate of the clean speech phase spectrum as follows:

$$\hat{S}(e^{j\omega}) = (|X(e^{j\omega})| - |\hat{N}(e^{j\omega})|)\exp(j\angle X(e^{j\omega})) \tag{10.4}$$

The estimated time-domain speech signal is obtained as the inverse Fourier transform of $\hat{S}(e^{j\omega})$.

Another way to recover the clean signal, $s(n)$, from the noisy signal $x(n)$ using the spectral subtraction approach is performed by assuming that there is an estimate of the power spectrum of the noise $P_N(e^{j\omega})$, which is obtained by averaging over multiple frames of a known noise segment. An estimate of the short time squared magnitude spectrum of the clean signal using this method can be obtained as follows:

$$|S(e^{j\omega})|^2 = \begin{cases} |X(e^{j\omega})|^2 - P_N(e^{j\omega}), & \text{if } |X(e^{j\omega})|^2 - P_N(e^{j\omega}) \geq 0 \\ 0 & \text{otherwise} \end{cases} \tag{10.5}$$

It is possible to combine this magnitude spectrum estimate with the phase of the noisy signal and then get the short time fourier transform (STFT) estimate of the clean signal as follows:

$$\hat{S}(e^{j\omega}) = |\hat{S}(e^{j\omega})|\exp(j\angle X(e^{j\omega})) \tag{10.6}$$

Noise-free signal can then be obtained with the inverse Fourier transform, as shown in Figure 10.2.

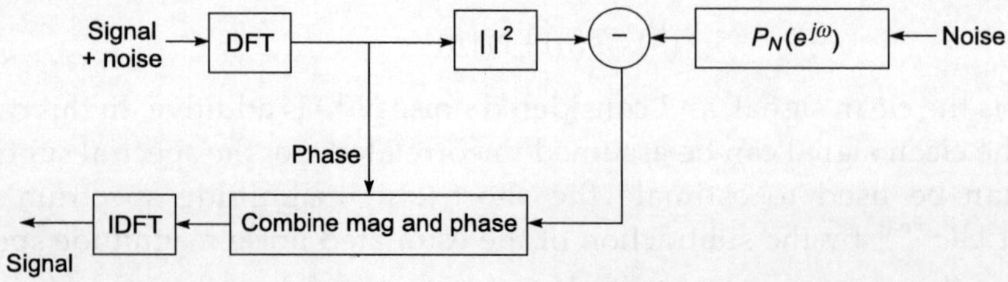

Figure 10.2 Block schematic of the modified spectral subtraction method

10.3.2 Wiener Filter in Frequency Domain

The Wiener filter is a popular technique that has been used in many signal enhancement methods. The basic principle of the Wiener filter is to obtain an estimate of the clean signal from that corrupted by additive noise. This estimate is obtained by minimizing the mean square error (MSE) between the desired signal, $s(n)$, and the estimated signal, $\hat{s}(n)$. The frequency domain solution to this optimization problem gives the following filter transfer function:

$$H(e^{j\omega}) = \frac{P_S(e^{j\omega})}{P_S(e^{j\omega}) + P_N(e^{j\omega})} \tag{10.7}$$

where, $P_S(e^{j\omega})$ and $P_N(e^{j\omega})$ are the power spectral densities of the clean and the noise signals, respectively. This formula can be derived considering the signal, $s(n)$ and the noise, $N(n)$, as uncorrelated and stationary signals. The SNR is defined by

$$\text{SNR} = \frac{P_S(e^{j\omega})}{P_N(e^{j\omega})} \tag{10.8}$$

This definition can be incorporated to the Wiener filter equation as follows:

$$H(e^{j\omega}) = \frac{P_S(e^{j\omega})}{P_S(e^{j\omega}) + P_N(e^{j\omega})} = \frac{1}{1 + \dfrac{1}{P_S(e^{j\omega}) \big/ P_N(e^{j\omega})}} \tag{10.9}$$

$$H(e^{j\omega}) = \left[1 + \frac{1}{\text{SNR}}\right]^{-1} \tag{10.10}$$

The drawback of the Wiener filter is the fixed frequency response at all frequencies and the requirement to estimate the power spectral density of the clean signal and noise prior to filtering.

10.3.3 Adaptive Filters

Adaptive filters are used whenever there is a requirement to process the signal whose statistics are not known. Fixed-frequency-response digital FIR filters were discussed in chapter 3. This chapter looks at filters with a frequency response or transfer function that can change over time to match desired system characteristics. In many applications, time variable filters whose characteristics can be varied with time are required. Such time-variable filters that incorporate adaptation mechanism by which filter coefficients can be adjusted are called *adaptive filters*.

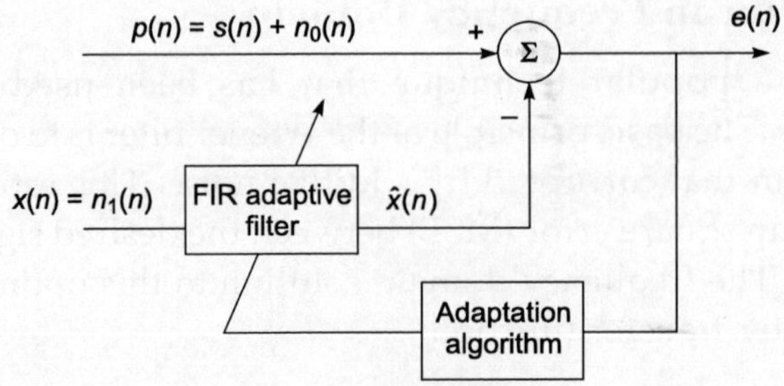

Figure 10.3 Structure of adaptive filtering system

Based on the principle proposed by Widrow and Hoff, in adaptive filtering system, primary and at least one reference input are required. Adaptive filter configuration is illustrated in Figure 10.3. There are two input signals to the adaptive filters: $p(n)$ and $x(n)$ which are called the primary input and the reference input respectively. $p(n)$ includes desired signal $s(n)$ and noise $n_0(n)$ which is uncorrelated with $s(n)$. The reference input $x(n)$ is another noise $n_1(n)$ which is correlated with $n_0(n)$ and uncorrelated with $s(n)$. Let $\hat{x}(n)$ be the output of variable filter. The error signal, $e(n)$, is used to update filter coefficients, $h_n(n)$ through an adaptation algorithm that minimizes the error $e(n)$. The error signal is the difference between $p(n)$ and $\hat{x}(n)$. Output of adaptive filter $e(n)$ is given by

$$e(n) = p(n) - \hat{x}(n) = s(n) + n_0(n) - \hat{x}(n) \tag{10.11}$$

But
$$\hat{x}(n) = h_n(n) * x(n) = \sum_{k=0}^{N-1} h_n(k)x(n-k) \tag{10.12}$$

Squaring both the sides of Eq. 10.11 and taking expectation, we get expectation error $\varepsilon(h)$ and given by

$$\varepsilon(h) = E[e^2(n)] = E[p(n) - \hat{x}(n)]^2 \tag{10.13}$$

$$\varepsilon(h) = E[e^2(n)] = E[s(n) + n_0(n) - \hat{x}(n)]^2$$

$$\varepsilon(h) = E[e^2(n)] = E[s(n)]^2 + E[n_0(n) - \hat{x}(n)]^2 + 2E[s(n)(n_0(n) - \hat{x}(n))]$$

As the filter parameters are adjusted to minimize $E[e^2(n)]$, $E[s(n)]^2$ will be unaffected and since $s(n)$ is uncorrelated with $n_0(n)$ and $n_1(n)$, last term in the above equation is zero. Hence, minimum output power is given by

$$\min E[e^2(n)] = E[s(n)]^2 + \min E[n_0(n) - \hat{x}(n)]^2 \tag{10.14}$$

Hence, from Eq. 10.14, minimizing the total output power $E[e^2(n)]$, minimizes the output noise power. Since signal in the output remains constant, minimizing the total output power maximizes the output signal to noise ratio. The minimization condition for Eq. 10.14 is given by

$$\frac{d\varepsilon(h)}{dh_n(k)} = \frac{dE[e^2(n)]}{dh_n(k)} = 2E\left[e(n)\frac{de(n)}{dh_n(k)}\right] \tag{10.15}$$

Since $e(n) = p(n) - \sum_{k=0}^{N-1} h_n(k)x(n-k)$, hence Eq. 10.15 is given by

$$\frac{d\varepsilon(h)}{dh_n(k)} = 2E[e(n)(-x(n-k))] = -2E[e(n)x(n-k)] \tag{10.16}$$

In adaptive filters, filter parameter, $h_n(k)$, is time dependent. Hence, $h_n(k)$ value is updated from one time instant to the next and new value, $h_{n+1}(k)$, is given by the equation

$$h_{n+1}(k) = hn(k) + \Delta hn(k)$$

The dependence of the error (ε) on the filter parameter (h) is parabolic; simplest way to chose $\Delta h_n(k)$ is by gradient descent method. If $\Delta h_n(k)$ is selected as negative gradient and if μ is the step size parameter then

$$\Delta h_n(k) = -\mu\frac{d\varepsilon(h)}{dh_n(k)} \tag{10.17}$$

Then

$$h_{n+1}(k) = h_n(k) + \Delta h(k) = h_n(k) - \mu\frac{d\varepsilon(h)}{dh_n(k)}$$

$$h_{n+1}(k) = h_n(k) + \Delta h(k) = h_n(k) + 2\mu E[e(n)x(n-k)] \tag{10.18}$$

10.3.3.1 LMS adaptive algorithm

A very useful LMS adaptive algorithm has evolved from simply approximating the expectation of Eq. 10.16 with the instantaneous value and is given by

$$E[e(n)x(n-k)] = e(n)x(n-k)$$

Substituting the above in Eq. 10.19, weight adjustment algorithm becomes

$$h_{n+1}(k) = h_n(k) + 2\mu e(n)x(n-k) \tag{10.19}$$

Hence, in LMS algorithm, updating equation has simple form and μ governs both the rate of convergence and stability of the algorithm. Following steps are used to implement LMS algorithm:

1. Assume initial filter weights $h_0(k) = 0$, for $k = 0 \ldots N-1$ and $\mu = 0.1$
2. Calculate the filter output
   $\hat{x}(n) = h_n(n) * x(n)$
3. Calculate the error signal $e(n) = p(n) - \hat{x}(n)$

4. Update filter coefficients by using the following equation

$$h_{n+1}(k) = h_n(k) + 2\mu e(n)x(n-k)$$

5. Repeat from step 2, for $n = 0, 1, 2\ldots$, until error is minimized.

10.3.3.2 Normalized LMS adaptive algorithm

Normalized least mean square (NLMS) algorithm is the first derivative or variant of LMS algorithm. LMS algorithm is still widely used adaptive algorithm for many applications. But major drawback of the LMS algorithm is selection of step size or converging factor (μ). When step size (μ) is too large, stability of the algorithm decreases and also introduces mean square error. On the other hand, the step size is small then adaption process becomes slower. Fixed step size algorithm also introduces gradient amplification noise. Along with these disadvantages when LMS algorithm is used in adaptive noise canceller for speech signals the steady state excess mean square error introduces and it increased linearly as power of the primary signal increases. Fixed step size will be best suitable when statistical characteristics of the inputs are well known. To overcome this problem a variable step size was first introduced in 1967 independently by Nagumo et al. and Albert et al. In NLMS algorithm the step size will keep on adjusting by iteration to iteration. The main advantage of the NLMS algorithm is that, the algorithm does not require step size readjustment when power of the input signal changes, because NLMS algorithm normalizes the step size with input signal power. Compare to LMS algorithm, this algorithm reduces the gradient amplification noise.

The output, error and weight equation of conventional NLMS algorithms are shown below.

Output: $\hat{x}(n) = h_n(n) * x(n)$

Error: $e(n) = p(n) - \hat{x}(n)$

$$\text{Weight equation: } h_{n+1}(k) = h_n(k) + \frac{\mu}{\varepsilon + \|x(n)\|^2} e(n)x(n-k) \tag{10.20}$$

where, w is the adaptive filter vector weight and $\|x(n)\|^2$ is equal to $x(n)*x^T(n)$, const (ε) is used in the denominator to prevent the division by a very small number.

Example 10.1 Develop an adaptive filtering system for the noise cancellation application using LMS algorithm with two coefficients (for $N = 2$). Perform the adaptive filtering to obtain outputs $e(n)$ and updated filter coefficients for $n = 0, 1, 2$ for the following primary input, $p(n) = \left\{ \underset{\uparrow}{4}, 2.2, -1 \right\}$, secondary input, $x(n) = \left\{ \underset{\uparrow}{1}, 1, -1 \right\}$, and convergence parameter, $\mu = 0.1$.

Solution Adaptive filtering system for noise cancellation application using LMS algorithm is shown in Figure 10.4.

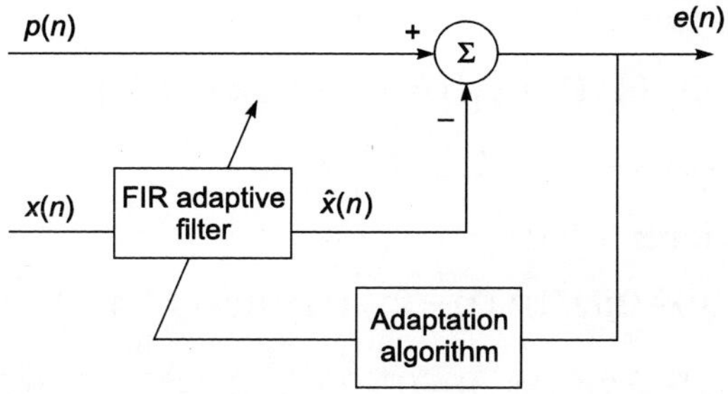

Figure 10.4 Structure of adaptive filtering system

The adaptive LMS algorithm is set up for $N = 2$ as:

1. Initialization: $h_0(0) = 0$ and $h_0(1) = 0$, $\mu = 0.1$
2. Compute FIR adaptive filter output:

$$\hat{x}(n) = h_n(n) * x(n)$$

$$\hat{x}(n) = \sum_{k=0}^{N-1} h_n(k)x(n-k) = \sum_{k=0}^{1} h_n(k)x(n-k) = h_n(0)x(n) + h_n(1)x(n-1)$$

3. Compute the output error: $e(n) = p(n) - \hat{x}(n)$
4. Update the filter weight for the next coming sample:

$$h_{n+1}(k) = h_n(k) + 2\mu e(n)x(n-k) \quad \text{for } k = 0, 1$$

$$\therefore \begin{bmatrix} h_{n+1}(0) \\ h_{n+1}(1) \end{bmatrix} = \begin{bmatrix} h_n(0) \\ h_n(1) \end{bmatrix} + 2\mu e(n) \begin{bmatrix} x(n) \\ x(n-1) \end{bmatrix}$$

5. Repeat steps 2 to 4 for $n = 0, 1, 2$.

Given $p(n) = \left\{ \begin{matrix} 4, & 2.2, & -1 \\ \uparrow & & \end{matrix} \right\}, x(n) = \left\{ \begin{matrix} 1, & 1, & -1 \\ \uparrow & & \end{matrix} \right\},$

for $N = 2$, $\hat{x}(n) = h_n(0)x(n) + h_n(1)\,x(n-1)$

Step 1: For $n = 0$ $\quad \hat{x}(0) = h_0(0)x(0) + h_0(1)x(-1) = 0 \times 1 + 0 \times 0 = 0$

$$e(0) = p(0) - \hat{x}(0) = 4 - 0 = 4$$

Updating filter coefficients

$$h_1(0) = h_0(0) + 2\mu e(0)x(0) = 0 + 2 \times (0.1) \times 4 \times 1 = 0.8$$

$$h_1(1) = h_0(1) + 2\mu e(0)x(-1) = 0 + 2 \times (0.1) \times 4 \times 0 = 0$$

or $$\begin{bmatrix} h_1(0) \\ h_1(1) \end{bmatrix} = \begin{bmatrix} 0 \\ 0 \end{bmatrix} + 2\times(0.1)\times 4\times \begin{bmatrix} 1 \\ 0 \end{bmatrix} = \begin{bmatrix} 0.8 \\ 0 \end{bmatrix}$$

Step 2: For $n = 1$ $\quad \hat{x}(1) = h_1(0)x(1) + h_1(1)x(0) = 0.8\times 1 + 0\times 1 = 0.8$

$$e(1) = p(1) - \hat{x}(1) = 2.2 - 0.8 = 1.4$$

Updating filter coefficients

$$h_2(0) = h_1(0) + 2\mu e(1)x(1) = 0.6 + 2\times(0.1)\times 1.4\times 1 = 0.88$$

$$h_2(1) = h_1(1) + 2\mu e(1)x(0) = 0 + 2\times(0.1)\times 1.4\times 1 = 0.28$$

or $$\begin{bmatrix} h_2(0) \\ h_2(1) \end{bmatrix} = \begin{bmatrix} 0.6 \\ 0 \end{bmatrix} + 2\times(0.1)\times 1.4\times \begin{bmatrix} 1 \\ 1 \end{bmatrix} = \begin{bmatrix} 0.88 \\ 0.28 \end{bmatrix}$$

Step 3: For $\quad \hat{x}(2) = h_2(0)x(2) + h_2(1)x(1) = 0.88\times(-1) + 0.28\times 1 = -0.6$

$$e(2) = p(2) - \hat{x}(2) = -1 + 0.6 = -0.4$$

Updating filter coefficients

$$h_3(0) = h_2(0) + 2\mu e(2)x(2) = 0.88 + 2\times(0.1)\times(-0.4)\times(-1) = 0.96$$

$$h_3(1) = h_2(1) + 2\mu e(2)x(1) = 0.28 + 2\times(0.1)\times(-0.4)\times 1 = 0.2$$

$$\begin{bmatrix} h_3(0) \\ h_3(1) \end{bmatrix} = \begin{bmatrix} 0.88 \\ 0.28 \end{bmatrix} + 2\times(0.1)\times(-0.4)\times \begin{bmatrix} -1 \\ 1 \end{bmatrix} = \begin{bmatrix} 0.96 \\ 0.2 \end{bmatrix}$$

Hence, the adaptive filter outputs for the first three samples are

$$e(0) = 4, \quad e(1) = 1.4, \quad e(2) = -0.4$$

And updated filter coefficients are

$$h_3(0) = 0.96 \text{ and } h_3(1) = 0.2$$

10.3.3.3 Recursive least square (RLS) algorithm

Compared to least mean squares (LMS) algorithms, recursive least squares (RLS) algorithms have a faster convergence speed and do not exhibit the eigenvalue spread problem. However, RLS algorithms involve more complicated mathematical operations than LMS algorithms. As discussed in section 10.3.3, the output of adaptive filter, $e(n)$, for Nth FIR filter is given by

$$e(n) = p(n) - \hat{x}(n)$$

$$\hat{x}(n) = \sum_{k=0}^{N-1} h_n(k)x(n-k)$$

In RLS algorithm error function is modified by introducing exponential weighing

$$\varepsilon(h) = \sum_{i=0}^{n} \lambda^{n-i} e(i)^2 = \lambda^n e(0)^2 + \ldots + \lambda e(n-1)^2 + e(n)^2 \tag{10.21}$$

were, $$e(i) = p(i) - \hat{x}(i) = d(i) - \sum_{k=0}^{N-1} h_n(k) x(i-k) = p(i) - X^T(i) h_n$$

here, $h_n = [h_n(0), h_n(1) \ldots h_n(N-1)]^T$ and $X(i) = [x(i), x(i-1), \ldots\ldots. x(i-(N-1))]^T$

where forgetting factor, λ, is positive and its value is $(0 < \lambda \leq 1)$, which gives exponentially less weight to older error samples. Substituting $\varepsilon(h)$ value and it is minimized by taking the partial derivatives for all entries of k and setting the result to zero, we get

$$\frac{d\varepsilon(h)}{dh_n(k)} = \frac{d\left[\sum_{i=0}^{n} \lambda^{n-i} e(i)^2\right]}{dh_n(k)} = \sum_{i=0}^{n} \lambda^{n-i} 2e(i) \frac{de(i)}{dh_n(k)} = 0$$

Substituting $e(i)$ and $\frac{de(i)}{dh_n(k)}$ from the above equations, we get

$$\frac{d\varepsilon(h)}{dh_n(k)} = \sum_{i=0}^{n} \lambda^{n-i} [p(i) - X^T(i) h_n] X(i) = 0 \tag{10.22}$$

Rearranging the Eq. 10.22 yields

$$\sum_{i=0}^{n} \lambda^{n-i} p(i) X(i) = h_n \sum_{i=0}^{n} \lambda^{n-i} X(i) X^T(i) \tag{10.23}$$

$$R_X(n) = \sum_{i=0}^{n} \lambda^{n-i} X(i) X^T(i) \tag{10.24}$$

$$r_{dX}(n) = \sum_{i=0}^{n} \lambda^{n-i} p(i) X(i) \tag{10.25}$$

Hence, Eq. 10.25 can be written as

$$r_{dX}(n) = h_n R_X(n)$$

here, $R_X(n)$ is covariance matrix, $r_{dX}(n)$ is cross-correlation vector and h_n depends on $R_X(n)$ and $r_{dX}(n)$, which makes h_n dependent on n and is given by

$$h_n = \frac{r_{dX}(n)}{R_X(n)} = R_X^{-1}(n) r_{dX}(n) \tag{10.26}$$

By expanding Eq. 10.24, we get

$$R_X(n) = \sum_{i=0}^{n-1} \lambda^{n-i} X(i)X^T(i) + X(n)X^T(n)$$

Hence,

$$R_X(n) = \lambda R_X(n-1) + X(n)X^T(n)$$

$$R_X^{-1}(n) = [\lambda R_X(n-1) + X(n)X^T(n)]^{-1} \tag{10.27}$$

Right side of the Eq. 10.27 is a $n \times n$ matrix, hence by using matrix inversion lemma and by substituting $C(n) = R_X^{-1}(n)$ and $C(n-1) = R_X^{-1}(n-1)$, Eq. 10.27 can be written as

$$C(n) = \lambda^{-1}C(n-1) - \lambda^{-1}C(n-1)X(n)\{\lambda + X^T(n)C(n-1)X(n)\}^{-1}X^T(n)C(n-1) \tag{10.28}$$

Let $g = C(n-1)X(n)\{\lambda + X^T(n)C(n-1)X(n)\}^{-1} = \dfrac{C(n-1)X(n)}{\{\lambda + X^T(n)C(n-1)X(n)\}}$

Hence, Eq. 10.28 can be written as

$$C(n) = \lambda^{-1}C(n-1) - \lambda^{-1}gX^T(n)C(n-1) \tag{10.29}$$

By expanding $r_{dX}(n)$, Eq. 10.25, we get

$$r_{dX}(n) = \sum_{i=0}^{n-1} \lambda^{n-i} p(i)X(i) + p(n)X(n) = \lambda r_{dX}(n-1) + p(n)X(n)$$

By substituting $r_{dX}(n)$ in Eq. 10.29, we get

$$h_n = R_X^{-1}(n) r_{dX}(n) = C(n) r_{dX}(n) = C(n)[\lambda r_{dX}(n-1) + p(n)X(n)]$$

$$h_n = \lambda C(n) r_{dX}(n-1) + p(n)C(n)X(n)$$

Substituting $P(n)$ values from Eq. 10.29 and simplifying, we obtain a recursive equation for updating the least-square estimate leads to the important relation

$$h_n = h_{n-1} + g[p(n) - X^T(n)h_{n-1}] \tag{10.30}$$

Let $$a(n) = [p(n) - X^T(n)h_{n-1}] \tag{10.31}$$

$\therefore$ $$h_n = h_{n-1} + ga(n) \tag{10.32}$$

This is the recursive definition. The above equation shows how to update the filter coefficients with each new data value. Once the error is minimized, $h_n = h_{n-1}$, and hence $e(n) = a(n)$. In RLS algorithm, updating equation has complex form. The standard RLS algorithm performs the following operations to update the coefficients of an adaptive filter.

1. Assume $h_0 = [h_n(0), h_n(1), \ldots \ldots .. h_n(N-1)]^T = [0, 0, \ldots \ldots .. 0]^T, \lambda$, and δ.

   Initial value of $C(0) = \delta^{-1}I$, I is the identity matrix of rank N

   For $N = 2$, the initial value of $C(0) = \begin{bmatrix} \delta^{-1} & 0 \\ 0 & \delta^{-1} \end{bmatrix}$

2. $X(n) = [x(n), x(n-1), \ldots \ldots .. x(n-(N-1))]^T$

3. Calculate $a(n) = [p(n) - X^T(n)h_{n-1}]$

4. Calculate

$$g = \frac{C(n-1)X(n)}{\{\lambda + X^T(n)C(n-1)X(n)\}}$$

$$C(n) = \lambda^{-1}[C(n-1) - gX^T(n)C(n-1)]$$

   In practice, λ is usually chosen between 0.98 and 1.

5. Updates the filter coefficients by using the equation $h_n = h_{n-1} + ga(n)$

6. Repeat steps 2 to 4 for $n = 1, 2\ldots$, until error becomes minimized.

Comparison between LMS and RLS algorithms: LMS algorithm is very simple and easy to implement whereas RLS algorithm is complex for implementation. RLS algorithm has faster convergence speed than LMS algorithm. Both LMS and RLS algorithm are stable.

Example 10.2 Develop an adaptive filtering system for the noise cancellation application using RLS algorithm for $N = 2$. Perform the adaptive filtering to obtain filter weights for $N = 1, 2$ for the following primary input $p(n) = \left\{\underset{\uparrow}{1}, 2, 1\right\}$, secondary input $x(n) = \left\{\underset{\uparrow}{1}, 1, -2\right\}$ and convergence parameter $\lambda = 0.98$ and $\delta = 0.5$.

Solution Given

$$p(n) = \left\{\underset{\uparrow}{1}, 2, 1\right\}, x(n) = \left\{\underset{\uparrow}{1}, 1, -2\right\} \quad N = 2 \quad \lambda = 0.98 \text{ and } \delta = 0.5$$

$$\therefore \text{ Initial value of } C(0) = \begin{bmatrix} \delta^{-1} & 0 \\ 0 & \delta^{-1} \end{bmatrix} = \begin{bmatrix} 2 & 0 \\ 0 & 2 \end{bmatrix}$$

$$\therefore \text{ For } N = 2, X(n) = [x(n), x(n-1)]^T \text{ and } h_0 = \begin{bmatrix} 0 \\ 0 \end{bmatrix}$$

Step 1: For $n = 1$

$$\therefore \quad X(1) = [x(1), x(0)]^T = [1, 1]^T = \begin{bmatrix} 1 \\ 1 \end{bmatrix}$$

$$a(n) = [p(n) - X^T(n)h_{n-1}] \quad \therefore a(1) = [p(1) - X^T(1)h_0] = \left[2 - [1, 1]\begin{bmatrix} 0 \\ 0 \end{bmatrix}\right] = 2$$

$$g = \frac{C(n-1)X(n)}{\{\lambda + X^T(n)C(n-1)X(n)\}} = \frac{C(0)X(1)}{\{\lambda + X^T(1)C(0)X(1)\}}$$

$$\therefore \quad g = \frac{\begin{bmatrix} 2 & 0 \\ 0 & 2 \end{bmatrix}\begin{bmatrix} 1 \\ 1 \end{bmatrix}}{\left\{0.98 + [1, 1]\begin{bmatrix} 2 & 0 \\ 0 & 2 \end{bmatrix}\begin{bmatrix} 1 \\ 1 \end{bmatrix}\right\}} = \begin{bmatrix} 0.4016 \\ 0.4016 \end{bmatrix}$$

$$C(n) = \lambda^{-1}[C(n-1) - gX^T(n)C(n-1)] \quad \therefore C(1) = \lambda^{-1}[C(0) - gX^T(1)C(0)]$$

$$\therefore \quad C(1) = \frac{1}{0.98}\left[\begin{bmatrix} 2 & 0 \\ 0 & 2 \end{bmatrix} - \begin{bmatrix} 0.4016 \\ 0.4016 \end{bmatrix}[1, 1]\begin{bmatrix} 2 & 0 \\ 0 & 2 \end{bmatrix}\right] = \begin{bmatrix} 1.2212 & -0.8196 \\ -0.8196 & 1.2212 \end{bmatrix}$$

$$h_n = h_{n-1} + ga(n) \quad \therefore h_1 = h_0 + ga(1) = \begin{bmatrix} 0 \\ 0 \end{bmatrix} + \begin{bmatrix} 0.4016 \\ 0.4016 \end{bmatrix} 2 = \begin{bmatrix} 0.8032 \\ 0.8032 \end{bmatrix}$$

Step 2: For $n = 2$

$$\therefore \quad X(2) = [x(2), x(1)]^T = [-2, 1]^T = \begin{bmatrix} -2 \\ 1 \end{bmatrix}$$

$$a(n) = [p(n) - X^T(n)h_{n-1}] \quad \therefore a(2) = [p(2) - X^T(2)h_1] = \left[-1 - [-2, 1]\begin{bmatrix} 0.8032 \\ 0.8032 \end{bmatrix}\right] = 1.8032$$

$$g = \frac{C(n-1)X(n)}{\{\lambda + X^T(n)C(n-1)X(n)\}} = \frac{C(1)X(2)}{\{\lambda + X^T(2)C(1)X(2)\}}$$

$$\therefore \quad g = \frac{\begin{bmatrix} 1.2212 & -0.8196 \\ -0.8196 & 1.2212 \end{bmatrix}\begin{bmatrix} -2 \\ 1 \end{bmatrix}}{\left\{0.98 + [-2, 1]\begin{bmatrix} 1.2212 & -0.8196 \\ -0.8196 & 1.2212 \end{bmatrix}\begin{bmatrix} -2 \\ 1 \end{bmatrix}\right\}} = \begin{bmatrix} -0.3147 \\ 0.2760 \end{bmatrix}$$

$$C(n) = \lambda^{-1}[C(n-1) - gX^T(n)C(n-1)] \quad \therefore C(2) = \lambda^{-1}[C(1) - gX^T(2)C(1)]$$

$$\therefore\ C(2)=\frac{1}{0.98}\left[\begin{bmatrix}1.2212 & -0.8196\\ -0.8196 & 1.2212\end{bmatrix}-\begin{bmatrix}-0.3147\\ 0.2760\end{bmatrix}\begin{bmatrix}-2, 1\end{bmatrix}\begin{bmatrix}1.2212 & -0.8196\\ -0.8196 & 1.2212\end{bmatrix}\right]=\begin{bmatrix}0.2195 & 0.0639\\ 0.0639 & 0.4562\end{bmatrix}$$

$$h_n=h_{n-1}+ga(n)\quad \therefore\ h_2=h_1+ga(2)=\begin{bmatrix}0.8032\\ 0.8032\end{bmatrix}+\begin{bmatrix}-0.3147\\ 0.2897\end{bmatrix}(1.8032)=\begin{bmatrix}0.2464\\ 1.3256\end{bmatrix}$$

Hence, updated filter coefficients are: $h_2(0)=0.2464$ and $h_2(1)=1.3256$ and errors are $a(1)=2$ and $a(2)=1.8032$

10.4 SINGLE-CHANNEL ADAPTIVE FILTERING

A single channel adaptive filter can be used to identify components of a signal that are correlated with the previous samples; such correlations arise in particular from periodicity in the speech or noise. The general structure of a single-channel adaptive filter is shown in Fig. 10.5.

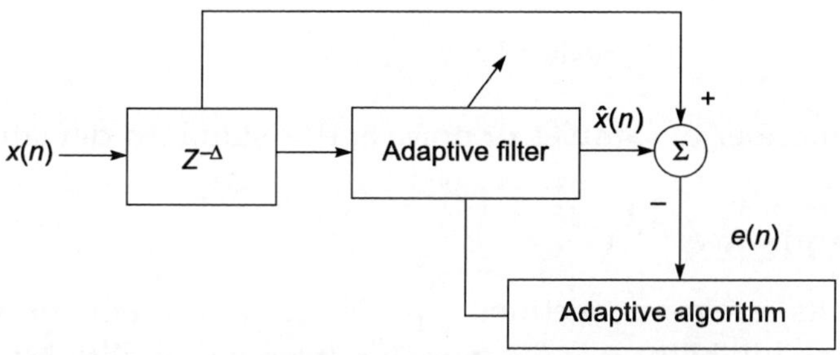

Figure 10.5 Single-channel adaptive filter

The noisy speech signal, $x(n)$, is delayed by D samples and passed through the filter to give $y(n)$ which is subtracted from $x(n)$ to generate the error signal, $e(n)$. The filter response is adjusted via the feedback path to reduce the power of $e(n)$ and the output of the filter is formed as a mixture of $e(n)$ and $y(n)$ according to whether the periodic signal components should be enhanced or suppressed. The filter adaptation is most often performed using either the LMS or more commonly, the NLMS gradient descent algorithm [Haykin, 2002]. Alternatively, it is also possible to use the RLS or sliding-window RLS algorithms [Haykin, 2002] which, at the cost of additional computation. The computation can be reduced by processing the signal in blocks and evaluating equation only once per block.

10.5 NOISE REDUCTION USING WAVELET TRANSFORMS

Removing noise from audio signal is a challenging task. Although different techniques are discussed in the previous section, these algorithms have limitations. Fourier transform assumes the signal is stationary, but audio signal is always non-stationary. To overcome this a wavelet is used. In this section, an audio de-noising technique based on wavelet transformation is proposed. Wavelet analysis involves filtering and down-sampling. As discussed in section 3.16.3.4, output of the wavelets is the new coefficient named as $a_k[n]$ & $d_k[n]$. $a_k[n]$ is approximation coefficient and $d_k[n]$ is detail coefficient. Wavelet thresholding is applied to the approximation and detail coefficients and these coefficients are again combined to reconstruct the noise-free signal. Steps used for audio denoising is shown in Figure 10.6.

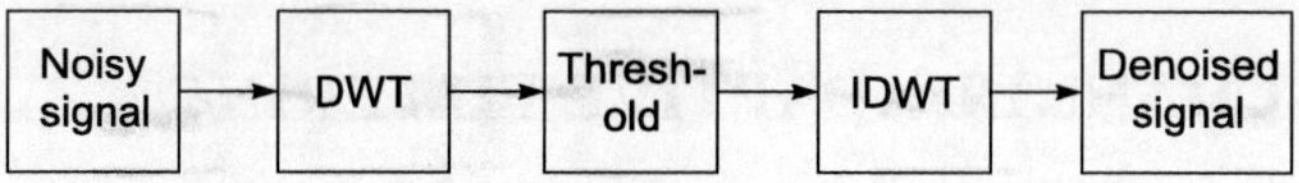

Figure 10.6 Steps used for audio denoising

One of the methods for selection of threshold was developed by Donoho and Jonstone and it is called universal threshold. It is given by

$$\text{Threshold} = \sigma_n \sqrt{2\log_2 N}$$

where, N denotes number of samples of noise and is standard deviation of noise.

10.6 APPLICATIONS

As discussed in the previous sections, speech enhancement aims to improve speech quality in intelligibility and/or overall perceptual quality by using various algorithms. Enhancing of speech degraded by noise, or noise reduction, is the most important field and used for many applications such as mobile phones, VoIP, teleconferencing systems, speech recognition and hearing aids.

10.6.1 Noise Reduction for the Hearing Impaired

Hearing impairment is a major chronic disability affecting people all over the world. If the hearing impairment is sensorineural loss, then they fail to understand speech in everyday communication in a noisy environment. LMS noise canceller algorithm was applied to this problem of improving speech communication for the hearing impaired. Mark Weiss used two channel adaptive noise cancellation methods by placing the primary and reference microphones in opposite directions just above one ear on the head of kemar mannikan. The primary signal is the contaminated speech signal obtained by omnidirectional microphone and reference signal (noise) is obtained by hypercardioid microphone. After LMS processing, output power of interference was reduced by 20 to 25 dB and improved the intelligibility of speech about 30 to 50%.

10.6.2 Echo Cancellation

One of the main problems associated with long distance telephone communication is the generation of echoes. Main source of echo is in the junction between four-wire long distance link and two-wire local subscriber link. Two-wire and four-wire links are joined by a device called hybrid. Adaptive filters are used as shown in Figure 10.7 to cancel the unwanted echoes. Echo generation can be accurately modelled by echo generation process. Hence, by minimizing the error, $e(n)$, echo is subtracted from the far end signal and echo is minimized.

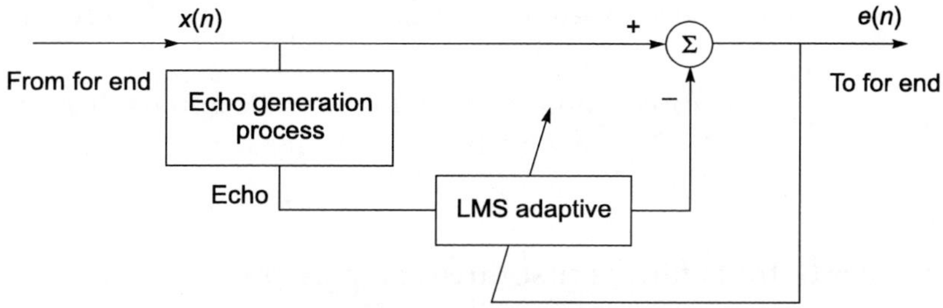

Figure 10.7 Structure of adaptive filters to cancel echo

10.7 SUMMARY

In this chapter, we have studied different evaluation techniques for speech intelligibility. Different speech enhancement algorithms like spectral subtraction method, Wiener filter theory and adaptive filtering techniques are also discussed.

EXERCISES

Review Questions

1. List and explain challenges in speech enhancement.
2. With block schematic, explain spectral subtraction method for noise cancellation.
3. Explain different evaluation techniques for speech intelligibility.
4. What is intelligibility test? Explain word and sentence intelligibility test.
5. Explain the basic principle of the Wiener filter to obtain an estimate of the clean signal from signal corrupted by additive noise.
6. With block schematic explain adaptive noise canceller for speech enhancement.
7. Develop a adaptive filtering system for the noise cancellation application using LMS algorithm with two coefficients (for $N = 2$). Perform the adaptive filtering

to obtain outputs $e(n)$ and updated filter coefficients for n = 0, 1, 2 for the following primary input, $p(n)=\begin{Bmatrix}3, & 2, & -1\\ \uparrow & & \end{Bmatrix}$, secondary input, $x(n)=\begin{Bmatrix}1, & 1, & -1\\ \uparrow & & \end{Bmatrix}$ and convergence parameter, $\mu = 0.1$.

8. Develop an adaptive filtering system for the noise cancellation application using RLS algorithm for N = 2. Perform the adaptive filtering to obtain filter weights for n = 1, 2 for the following primary input, $p(n)=\begin{Bmatrix}3, & 2, & -1\\ \uparrow & & \end{Bmatrix}$, secondary input, $x(n)=\begin{Bmatrix}1, & 1, & -1\\ \uparrow & & \end{Bmatrix}$, and convergence parameter, $\lambda = 0.98$ and $\delta = 0.5$.

ANSWERS

7. Adaptive filter outputs for the first three samples are:

$$e(0) = 3, e(1) = 1.4, e(2) = -0.4$$

Filter coefficients are: $h_3(0) = 0.96$ and $h_3(1) = 0.2$

8. Adaptive filter outputs for the first two samples are: $a(1) = 2,\ a(1) = -1$

Filter coefficients are: $h_2(0) = 1.2064$ and $h_2(1) = 0.4000$

11 CHAPTER

Auditory Signal Processing

This chapter provides insights on

- Anatomy of the ear
- Types of hearing impairments
- Analog hearing aids and its limitations.
- Design of digital & DSP based hearing aid.
- ASIC based programmable hearing aid
- Cochlear implants—advantages and disadvantages.

11.1 INTRODUCTION

Hearing is one of the major senses, like vision. **Hearing** is the ability to perceive sound by detecting vibrations through an organ such as the ear. It is one of the traditional five senses. The inability to hear is called hearing impairment. The function of the hearing mechanism is to convert physical vibration into an encoded nervous impulse. It can be thought of as a biological microphone. Like a microphone the ear is stimulated by vibration, in the ear nervous impulse is processed by the central auditory pathways of the brain. The mechanism to achieve this is complex. This chapter will deal mainly with the ear, first its structure and then its functions, types of hearing impairment, digital hearing aids and concludes with cochlear implants.

11.2 ANATOMY OF THE EAR

The ears are paired organs, one on each side of the head with the sense organ itself. The ears are paired sensory organs comprising the **auditory system**, involved in the detection of sound, and the **vestibular system**, for maintaining body balance/ equilibrium. The ear is divided anatomically and functionally into three regions: the **external ear**, the **middle ear**, and the **inner ear** as shown in Figure 11.1.

All three regions are involved in hearing. Only the inner ear functions in the vestibular system.

Hence, the auditory system is comprised of three components: the outer, middle, and inner ear, all work together to transfer sounds from the environment to the brain.

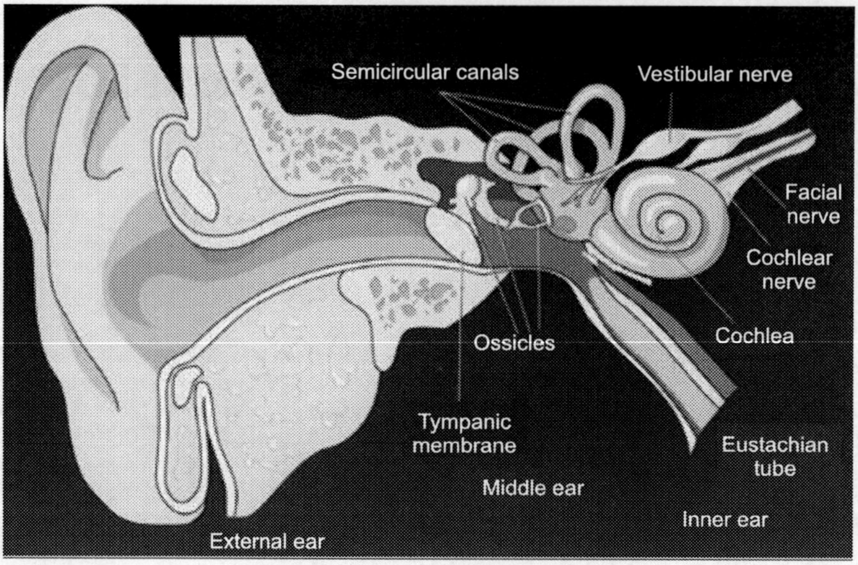

Figure 11.1 Anatomy of the ear

11.2.1 The Outer Ear

The external ear (or pinna) serves to protect the **tympanic membrane** (eardrum), as well to collect and direct sound waves through the **ear canal** to the eardrum. About 1¼ inch long, the canal contains modified sweat glands that secrete cerumen, or earwax. Too much cerumen can block sound transmission. The outer ear transmits sound to the tympanic membrane. Hence, the outer ear consists of:

- Pinna
- Ear canal

The pinna protrudes from the side of the skull, made of cartilage covered by skin, collects sound and channels it into the ear canal. The pinna is angled so that it catches sounds that come from in front more than those from behind and so is helpful in localizing sound.

The ear canal is about 1.25 inches long and 0.25 inch in diameter and consists of outer and inner parts. The outer portion is lined with hairy skin containing sweat glands and oily sebaceous glands which together form earwax. The inner two-thirds of the ear canal is imbedded in the temporal bone. The outer one-third is cartilage. The ear canal directs airborne sound waves towards the tympanic membrane (eardrum). The ear canal resonates sound waves and increases the loudness of the tones in the 3000-4000 Hz range. The ear canal also maintains the proper conditions of temperature and humidity necessary to preserve the elasticity of the tympanic membrane. Glands, which produce cerumen (earwax) and tiny hairs in the ear canal, provide added protection against insects and foreign particles from damaging the tympanic membrane.

11.2.2 The Middle Ear

The middle ear consists of:

- Tympanic membrane
- Eustachian tube
- Ear ossicles

The tympanic membrane or eardrum serves as a divider between the outer ear and the middle ear structures. The eardrum is very sensitive to sound waves and vibrates back and forth as the sound waves strike it. The eardrum transmits the airborne vibrations from the outer to the middle ear and also assists in the protection of the delicate structures of the middle ear cavity and inner ear.

The eustachian tube acts as an air pressure equalizer and ventilates the middle ear. Normally, the tube is closed but opens while chewing or swallowing. When the eustachian tube opens, the air pressure between the outer and middle ear is equalized. The transmission of sound through the eardrum is optimal when the air pressure is equalized between the outer and middle ear. When the air pressure between the outer and middle ear is unequal, the eardrum is forced outward or inward causing discomfort and the ability of the eardrum to respond to sound is reduced.

The ear ossicles amplifies a signal approximately 25 decibels as it transfers signals from the tympanic membrane to the inner ear. The middle ear is connected and transmits sound to the inner ear. The ossicles consists of the three smallest bones in the body: malleus, incus and stapes. The malleus is attached to the tympanic membrane. The footplate of the stapes inserts into the oval window of the inner ear. The incus is between the malleus and the stapes. Attached to the incus are two tiny muscles, the stapedius and tensor tympani muscles. These muscles contract to protect the inner ear from loud sounds by reducing the intensity of sound transmission to the inner ear.

11.2.3 The Inner Ear

The inner ear has two functions; the first is hearing and the second is balance. Their primary function is mechano-transduction, or conversion between mechanical and neural signals. This function is done by the **inner ear**. The inner ear is composed of sensory organ of hearing. The inner ear consists of:

- Vestibular apparatus
- Cochlea

Vestibular apparatus is the balance part of the ear (referred to as the vestibular apparatus). It is composed, in part of three semicircular canals located within the inner ear. The vestibular system helps to maintain balance, regardless of head position or gravity, in conjunction with eye movement and somatosensory input. The semicircular canals are innervated by the VIII cranial nerve or auditory nerve.

Cochlea is the part of the inner ear involved in hearing mechanism. The cochlea is the auditory portion of the inner ear as shown in Figure 11.2. It is spiral-shaped, similar to the shape of a snail, making 2.5 turns around its axis. The cochlea is composed of three fluid-filled chambers that extend the length of the structure. The two outer chambers are filled with a fluid called perilymph. Perilymph acts as a cushioning agent for the delicate structures that occupy the central chamber. Perilymph is connected to the cerebrospinal fluid that surrounds the brain and the spinal column. The third fluid filled chamber is the central chamber, called the cochlear duct. It secretes a fluid called endolymph, which fills this chamber.

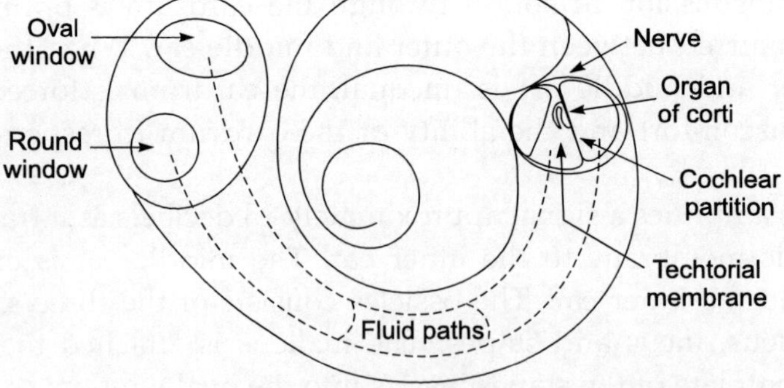

Figure 11.2 Structural diagram of the cochlea showing how fluid pushed in at the oval window moves, deflects the cochlear partition, and bulges back at the round window.

The cochlear duct contains the basilar membrane upon which lies the organ of corti. The organ of corti is a sensory organ essential to hearing. It consists of approximately 30,000 finger-like projections of cilia that are arranged in rows. These cilia are referred to as hair cells. Each hair cell is connected to a nerve fibre that relays various impulses to the cochlear branch of the VIII cranial nerve or auditory nerve. The "pitch" of the impulse relayed depends on the areas of the basilar membrane, and hence, portions of the organ of corti are stimulated. The apical portion of the basilar membrane (the most curled area of the cochlea) transfers lower frequency impulses. The basal end relays higher frequency impulses. The auditory nerve carries the impulses generated from the organ of corti to the brain stem. From the brain stem, nerve pathways extend through numerous nuclei to the cerebral cortex in the temporal lobes of the brain. It is in the temporal lobes of the brain that meaning is associated with the various patterns of nerve impulses.

11.3 FUNCTIONING OF THE EAR

Hearing begins when sound waves that travel through the air reach the outer ear or pinna. The sound waves then travel from the pinna through the ear canal to the middle ear, which includes the eardrum or tympanic membrane (a thin layer of tissue) and three tiny bones called ear ossicles. When the tympanic membrane vibrates, the ossicles amplify these vibrations and carry them to the inner ear. The inner ear is made up of a snail-shaped chamber called the cochlea which is filled with fluid and lined with thousands of tiny hair cells (outer hair cells and inner hair cells). When the vibrations move through the fluid, the tiny outer hair cells amplify the vibrations. The amplification is important because it allows hearing soft sounds, like whispering and birds.

Then, the inner hair cells translate the vibrations into electrical nerve impulses and send them to the auditory nerve, which connects the inner ear to the brain. When these nerve impulses reach the brain, they are interpreted as sound. The cochlea is like a piano so that specific areas along the length of the cochlea pick up gradually higher pitches.

This may seem like a long process, but it happens almost instantly. The school bell rings and we know we need to get to your next class. The phone rings and we automatically pick it up. We hear a question and immediately respond to it. But in reality, every time we hear a sound, the various structures of the ear have to work together to make sure the information gets to our brain.

11.4 HEARING IMPAIRMENT

Hearing impairment or hearing loss is the inability of an individual to hear sounds adequately. This may be due to improper development, damage or disease to any part of the hearing mechanism. Deafness is an invisible impairment. There are 4 major types of hearing losses. They are:

- Coductive loss

- Sensori neural loss
- Mixed loss
- Central loss

11.4.1 Conductive Loss

Conductive loss is the hearing loss due to problems with the outer ear (ear canal and eardrum) or middle ear and its little bones (the malleus, incus, and stapes). Conductive hearing loss occurs when sound is not conducted efficiently through the outer ear canal to the eardrum and through the tiny bones, or ossicles, of the middle ear. Conductive hearing loss usually involves a reduction in sound level. All sounds heard will appear for the individual to be weak and/or muffled. This type of hearing loss can often be medically or surgically corrected. The conditions that cause conductive loss are as follows:

Causes:

- Malformation of outer ear, ear canal or middle ear.
- Fluid in the middle ear from colds.
- Ear infection.
- Allergies.
- Poor eustachian tube function.
- Perforated eardrum.
- Tumours.
- Impacted earwax.
- Foreign body in the ear.

Treatment:

This can be treated by medications, antibiotics, surgery and/or hearing aids.

11.4.2 Sensorineural Loss

It is caused due to damage or disease of the inner ear (cochlea) or auditory nerve. The most common type is caused by malfunctioning of the outer hair cells. In severe cases, both outer and inner hair cells will not function properly. This is also a type of permanent hearing loss.

Causes:

- Exposure to loud noise.
- Head trauma.
- Virus or disease.

- Hearing loss that runs in the family (heredity).
- Aging.
- Malformation of the inner ear.
- Tumors

Treatment:

In mild conditions it is treated with hearing aids and during severe conditions, cochlear implant is preferred.

11.4.3 Mixed Loss

Mixed hearing loss is the combination of conductive and sensorineural losses.

Causes:

- Long lasting ear infection.
- If conductive loss is not treated properly, immediately and regularly, it may yield to sensorineural loss and eventually central loss.

Treatment:

Depending on the severity of the loss, hearing aids are used. Surgery is often attempted, but not always successful.

11.4.4 Central Loss

It is caused when the auditory pathways or auditory areas of brain are damaged. This is less often seen but more difficult to treat.

Causes:

- Head trauma.
- Infections.
- Malfunctioning of auditory nerves.
- Tumors.

11.5 CATEGORIES IN HEARING LOSS

Hearing impairment may be of various degrees. The degree of hearing impairment (in dBHL) is calculated by averaging the hearing thresholds at 500 Hz, 1 kHz & 2 kHz obtained from pure tone audiometry. Speech recognition score is obtained by administering speech audiometry. Hearing disability is calculated in percentage and is given in Table 11.1. The nature and quantum of benefits/facilities provided by the government is based on the degree and percentage of disability.

TABLE 11.1 Severity categories in hearing impairments

Category	Degree of Impairment	dB Level (HL)	Speech Recognition	Hearing in %
I	Mild hearing impairment	26 to 40 dB in better ear	80 to 100% in better ear	Less
II (a)	Moderate hearing impairment	41 to 60 dB in better ear	50 to 0% in better ear	40%
II (b)	Severe hearing impairment	60 to 70 dB in better ear	40 to 50% in better ear	51%
III (a)	Profound hearing impairment	71 to 90 dB in better ear	Less than 40% in better ear	71%
III (b)	Total deafness	91 dB and above in better ear	Very poor discrimination	100%

11.6 HEARING AIDS

Hearing aids amplify the signals of specific range of frequency. They come in various forms that fit inside or behind the ear and make sound louder. They are adjusted by the audiologist so that the sound coming in is amplified enough to allow the person with a hearing impairment to hear it clearly. Sometimes, the hearing loss is so severe that the most powerful hearing aids cannot amplify the sound enough. In those cases, a cochlear implant may be recommended.

Advantages:

- Available at different price ranges
- Easy maintenance.
- Can have a backup.
- Can be selected by the client on qualitative preference.
- Can take advantage of new technology as it becomes available.
- Battery gives a few hours warning that it is "dying".

Disadvantages:

- Loud noises bother for those using linear amplification.
- Hearing aids for those with severe loss need to be fitted carefully.
- Limited hearing assistance in high frequency range.
- Poor clarity while hearing sounds from a difference.
- Telephone conversation is not convenient.

Types of present-day hearing aids

Different types of hearing aids are available in the market like the body-worn aids, the spectacle aids, behind-the-ear aids, in-the-ear aids, in the canal aids, completely in the canal aids and receiver in the canal aids.

The body-worn aids, although require a connecting cord from the hearing aid to the earphone and the ear mould, however have certain distinct advantages. They permit high gains and high output, provides a stable platform large enough for the user (especially the aged and the bed ridden) to operate the controls.

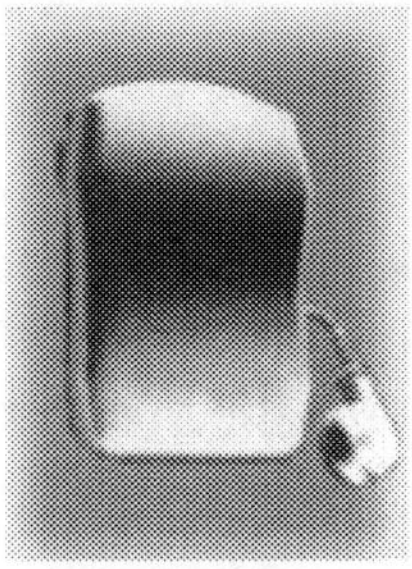

The behind the ear aids use expensive button power cells. The body of the instrument is worn behind the ear. It 'hooks' over the pinna. It is attached via plastic tubing to an earmould, which holds it in place in the ear.

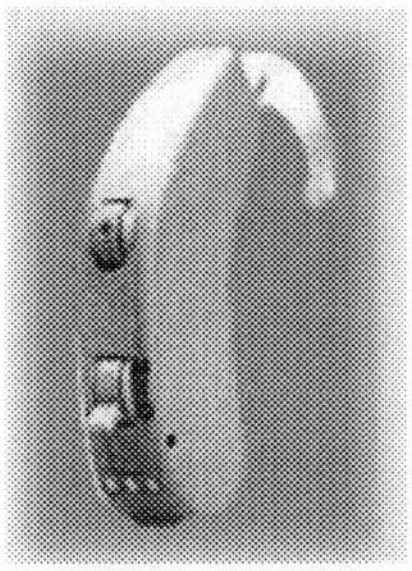

In-the-ear aids are very costly and there is little opportunity for post-fitting modifications. The complete hearing aid is in the ear. The hearing aid is housed in a hard plastic shell which is often custom made by taking an ear impression. The closeness of the microphone and the earphone means acoustic feedback is likely and this limits the gain and the maximum output available.

11.7 ANALOG VERSUS DIGITAL HEARING AIDS

The primary difference between the conventional **analog hearing aid** and the more modern **digital hearing aid** is the method of sound conversion that is used. An analog hearing aid converts sound waves into electrical waves that are mostly accurate. However, digital hearing aids use exact mathematical calculations to convert sound waves for an exact, precise duplication of sound.

11.7.1 Analog Hearing Aids

The incoming acoustic signal is changed into an electrical signal. Analog hearing aids do not differentiate different sounds, so they amplify every sound at an equal volume. Then electric signal is converted back into audible acoustic signal. This causes some sounds to be uncomfortably loud and others to be annoyingly undistinguishable. This causes a constant need for volume adjustment. Block diagram of analog hearing instrument is shown in Figure 11.3.

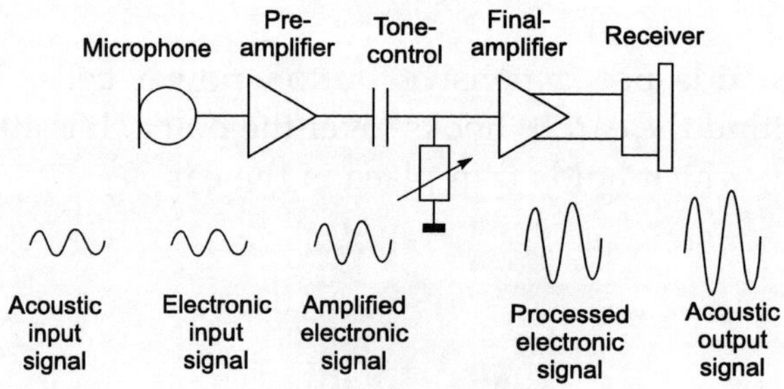

Figure 11.3 Block diagram of analog hearing instrument

They do make analog hearing aids that are programmable. These devices can have programmable environments for different listening situations. For instance, one can be used for conversation, while another can be used for the movies.

Limitations of Conventional Analog Hearing Aids

In the past, conventional analog hearing aids were the only option available to the hearing impaired. They have the following drawbacks:

- Conventional analog hearing aids provide the same amplification to all frequencies. But the hearing loss may be different for different frequencies. So an aid that can provide a variable gain with respect to frequency is to be provided.
- The element of programmability is absent in analog circuits, which puts serious limitation on the design in that it is virtually impossible to set the hearing aid characteristics to suit a particular user's hearing profile. Customization for individual profiles would entail a unique circuit design for each individual. This would be costly and impracticable for high-volume manufacture.

- Indiscriminate amplification of all audio signals regardless of frequency and signal intensity raises the sound levels of high intensity signals to exorbitantly high values resulting in uneasiness for the user. These necessitate a hearing aid that can adjust its gain in accordance with the intensity of sound signals.
- The user is unable to understand the speech signals in the presence of background noise.
- A purely analog approach lacks the flexibility and is difficult to interface with data analysers like computers.
- Present-day analog hearing aids are priced between Rs. 1300/- to Rs. 25000/-. High technology hearing aids incorporating the features mentioned above are prohibitively costly.

Advantages of Analog Hearing Aids

- Long time hearing aid users sometimes find the sound more acceptable because they are used to it.
- Cost is less than digital hearing aid.

11.7.2 Digital Hearing Aids

The first commercial digital hearing aid was created in 1987 by the Nicolet Corporation. The aid contained a body-worn processor that had a hardwire connection with an ear mounted transducer. While the Nicolet Corporation's hearing aid was not successful and the company shortly folded, it was able to start a competition to create more effective hearing aids. Two years later, in 1989, the behind-the-ear (BTE) digital hearing aid was launched.

Digital hearing aids promise many advantages over conventional hearing aids. These include:

- programmability;
- much greater precision in adjusting electro-acoustic parameters;
- self-monitoring capabilities;
- logical operations for self-testing and self-calibration;
- control of acoustic feedback (a serious practical problem with high-gain hearing aids);
- the use of advanced signal processing techniques for noise reduction; and
- automatic control of signal levels; and self-adaptive adjustment to changing acoustic environments.

Digital hearing aids convert sound exactly as it is using digital code and retransmit it using precise mathematical calculations. This means that they produce an exact replication of sound and they are accurate.

It is of common thought that digital technology revolutionized the hearing aid. Plus, digital hearing aids have a variety of features that make them more attractive to users. However, many people prefer analog hearing aids simply because of the lower price tag.

Digital hearing aids represent the latest technology in hearing devices. They provide a more accurate, higher quality sound than any other type of hearing device. Hearing aids that use analog technology convert sound waves into electrical waves for a pretty accurate sound quality. That said, digital hearing aids use exact mathematical calculations to produce an exact duplication for the most accurate sound quality.

Although many people realize that digital hearing aids provide a better sound quality many people stick with analog. This is because digital hearing aids are considerably more expensive than analog hearing aids. In many cases a hearing aid model that is digital can cost twice the cost of a comparable analog hearing aid.

Most people that experience hearing loss feel that the cost of the digital hearing aid is worth it. This is not just because of the better sound quality; digital hearing aids also have many features that analog hearing aids do not have. Let's look at some of the reasons why people choose a digital hearing aid versus an analog hearing aid.

- Better sound quality because of the exact mathematical calculations used for sound conversion.
- More features that make the digital hearing aid more user friendly than the analog hearing aid.
- Digital hearing aids can often be customized for each individual user.
- They are more cosmetically appealing because they are smaller and less noticeable.

Individuals experiencing hearing loss will find that overall the quality of their hearing is more improved with a digital hearing aid. They will get to experience an exact duplication of the sounds the device is converting, which makes the devices more user friendly and comfortable. The bottom line is that users must weigh how much the quality means to them because for most quality outweighs cost for most.

11.8 DSP BASED HEARING AIDS

A digital signal processor (DSP) based hearing aid takes the incoming signal from the microphone, converts it into a digital format, and then processes the signal using digital signal processor and then converts it back into an analog sound to be delivered to the ear (Figure 11.4). Digital hearing aids are normally more flexible and can be used to process sound more selectively than an analog.

Advantages of DSP

- Ability to adjust gain and output to subsequent changes in hearing loss.
- Certain "noise reduction algorithms" can be programmed into the circuit to help reduce background noise.
- Digital feedback reduction can make more gain available without whistling.
- Some instruments are able to have multiple memories for different listening situations.
- Some instruments select and follow the sound of voices with a self-focusing directional microphone.
- The greatest degree of flexibility in adjustments is available with a good instrument.

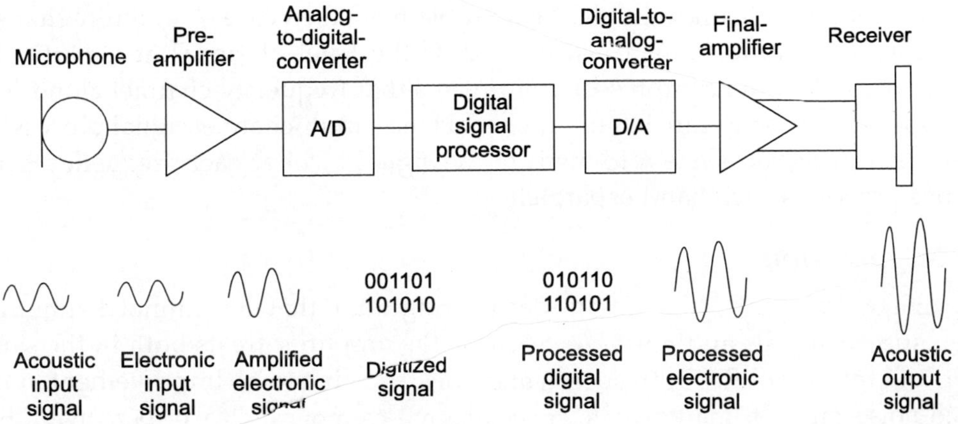

Figure 11.4 Block diagram of DSP based digital hearing instrument

Hearing Profile Compensation

As already explained, for a hearing impaired person the hearing threshold rises and it varies with respect to frequency. To overcome the difficulties, a hearing aid has to provide more amplification for frequencies components where hearing loss is more and thus becomes profile dependent. In conventional analog hearing aids, the aid has a fixed frequency response. However, in digital hearing aids this frequency dependent gain variation can be easily achieved by implementing a digital filter. The frequency response can also be easily customized to a particular user by changing some memory values.

Multi-channel Signal Processing

In multi-channel hearing aids the incoming sound is split into different frequency channels and is processed separately. There are mainly two reasons for processing different frequency bands separately. They are:

- Hearing thresholds of a person with hearing impairment usually varies with frequency.
- The energy distribution of signals and noises in the environment depends on frequency.

In a single compression channel hearing aid, when the compressor turns the gain down, signal components at all the frequencies are decreased in level. As a result a higher level of signal at a particular frequency can attenuate signals at other frequencies also. This is undesirable. This problem is taken care of in multi-channel hearing aids. In a multichannel hearing aid, if the level of signal at a particular frequency rises, then the compressor compresses that frequency channel alone. The rest of the frequency bands are left untouched. Hence, multichannel signal processing enables the hearing aid system to analyze the signal level at each frequency band and hence processes each band separately.

Noise Suppression

Another major drawback of the present-day hearing aids is that they cannot distinguish between speech signals and noise. As a result, the amplifier treats both in the same manner, and thus amplifies both speech and noise, leading to no improvement in the signal-to-noise ratio. In many cases, speech signal cannot be clearly perceived. For speech sensitive processing, after separating the input signal into different frequency channels, each channel is analysed individually. Speech has a temporal structure different from most noises. Digital signal processing can identify this different temporal structure to a degree not possible with analog technology. In each of the frequency ranges these results are used to decide whether the signal is more likely speech or noise. Each frequency range is then amplified or attenuated accordingly. This sophisticated processing technique makes it possible to suppress background noise without influencing speech signals. It is basically an adaptive filter that adapts its gain in accordance with the noise content present in the incoming signal.

Feedback Reduction

Another common problem with the present-day hearing aids, is acoustic feedback. A portion of the output of hearing aid may get fed back to the input of the aid through the microphone. The signal fed back will be processed along with the incoming sound. As a result the hearing aid output will keep on increasing at any frequency at which the forward gain is greater than leakage attenuation, and at which the phase shift around the entire loop is an integer multiple of 360. The hearing aid output at that frequency will go on increasing and will be limited only by the saturation level of the hearing aid. This acoustic feedback produces audible whistling which will be annoying to the hearing aid user and others in the room. There are several digital techniques employed to solve the problem of feedback oscillation. All these methods can improve the situation but none can banish feedback oscillations entirely. One common method to avoid feedback oscillation is to adjust the gain frequency response. In single channel hearing aids this is achieved by reducing the overall gain. In multichannel hearing aids it is possible to reduce the gain only at those frequencies where the feedback occurs. It is also possible to limit the maximum gain in each channel, the limiting value being selected suitably such that it does not cause oscillations.

Digital filters can provide finer control of gain frequency response shape. Once the frequencies that cause feedback oscillations are identified, narrow notches are placed in the gain frequency curve around these frequencies. However, the frequencies, which cause oscillations, depend on the sound path and the environment. Hence, the feedback causing frequencies varies with time. The best implementation of this method is when the hearing aid continually monitors its own output to determine feedback oscillation and automatically adjusts the gain frequency curve to prevent oscillation from growing. This technique is called *adaptive feedback cancellation*.

Hearing Aid with Multiple Programs

Digital hearing aids can have more than one programs. This provides the user access to different environments. Properties of sounds entering a hearing aid vary from environment to environment. For optimum listening, the hearing aid should have different amplification characteristics in different environments. In digital programmable hearing aids, all the parameters that can be adjusted in one program can be independently adjusted in other programs. This enables the user to switch between multiple listening programs and select the most suited one.

Dynamic Range Compression

Major role of compressor in a hearing aid is to decrease the range of sound levels in the environment to match the decreased dynamic range of a hearing impaired person. Hence, it is called dynamic range compression. The important parameters of a compression system are compression threshold, compression ratio, attack time, and release time.

Compression threshold is the input level above which compression starts.

Compression ratio can be defined as the ratio between variations in output level to variations in input level.

The time taken for the compression hearing aid to change from linear gain to within 3 dB of the compressed steady state is called **attack time**.

The time taken for the compression hearing aid to change from compressed state to within 4 dB of linear steady state is called **release time**.

Compression system can be input controlled, which means that the compressor is controlled by a signal prior to volume control or output controlled which means that the compressor is controlled by a signal subsequent to the volume control. By altering compression parameters, different compression systems can be designed.

Microphone Noise Reduction

Microphone noise reduction can be achieved by including an expansion region in the I/O curve. Expansion minimizes the amount of amplification for sounds below the compression threshold.

In most hearing aids the inaudible noise originates from the microphone. It flows through the signal path and gets amplified to a level audible to the user. If expansion region is included, then the sound levels falling below the threshold will be provided lesser gain. Depending on the users hearing characteristics, the noise level may not be audible.

11.9 ASIC BASED HEARING AIDS

A digital programmable hearing aid (DPHA) employing ASIC technology will offer numerous benefits such as programmability and complex frequency selectivity and is shown in Figure 11.4. The DPHA based on CMOS ASIC technology consume less power compared to designs using discrete electronic circuits (i.e., of the order of 100 mW). In addition, the ASIC will also incorporate power saving features for idle conditions. These enable a portable battery-powered DPHA with sophisticated technological features. ASIC-based products enable lower material, labour and testing costs for high volume manufacture. It is expected that the ASIC based DPHA reduces the production cost to a value lower than the currently available analog hearing aid, incorporating enhanced features. The cost of the fully integrated ASIC will be two to three times less than the total cost of the individual parts needed in analog or discrete digital-based approach, with the attendant advantages of higher reliability and ease of manufacture and testing. Since the entire product technology is embedded within the ASIC, the DPHA can be manufactured even by small-scale industry with low investments in capital and technical manpower. The initial cost for ASIC development may be high but once developed the cost for each piece will be considerably less. This results in low cost for high volume production.

The block diagram of the ASIC based programmable hearing aid is shown in Fig. 11.5. The microphone output is amplified, converted to digital and is fed to the FPGA/ASIC block. The FPGA/ASIC block manipulate the incoming digital data and generate the digital output data, after digital signal processing.

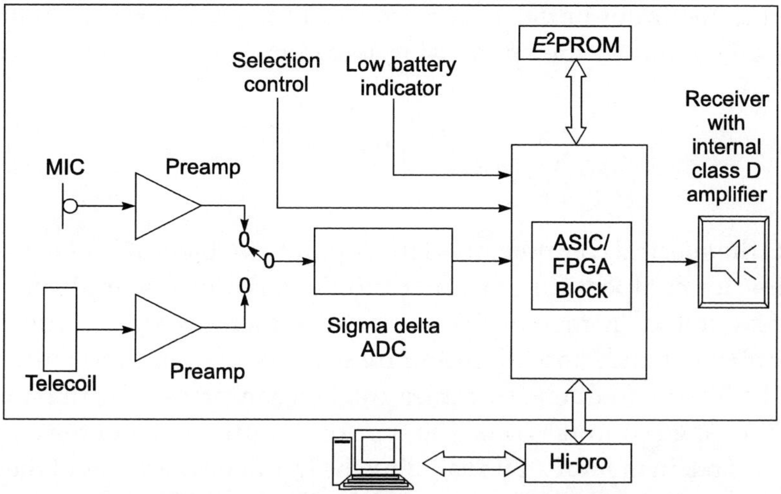

Figure 11.5 Block diagram of the ASIC based programmable hearing aid

The sound is captured by microphone and is converted to electrical signal. Tele-coil senses electromagnetic field produced by hearing aid compatible telephones and assistive listening devices (ALD) and converts it into electrical signal. It is used as auxiliary input and is selected when there is lot of background noise. The signal is amplified by the pre-amplifier. An analog-to-digital converter is required to convert the analog signals to digital signals. Further processing is done in the digital domain. By using over-sampling technique along with noise-shaping circuit, sigma delta ADC increases the effective resolution. The sigma-delta ADC converts the signal into digital signals. The FPGA block reads the data and stores it. It processes the digital signal and sends it to the Class D amplified receiver, which converts the electrical signal to sound. HI-PRO (hearing instrument programmer) contains level converter and the necessary logic for enabling serial communication between PC and hearing aid. Through this link, the ASIC/FPGA can be programmed to achieve a target response from the hearing aid.

ASIC/FPGA Block: The FPGA/ASIC block contains a processor core and a DSP. The various functions carried out by the FPGA/ASIC block are discussed below. The DSP engine uses the data stored in EEPROM to customize the DSP algorithm and process the digital signal. There are basically two approaches in DSP. One is the time domain approach and the other frequency domain approach. We follow the frequency domain approach in the development of DSP algorithm for DPHA. The output of ADC is read and stored in memory. Once the input buffer is full, it is read, processed and the output samples are generated. The processing should be completed before the input buffer is again filled. Discrete Fourier transform of the input block is taken and further processing techniques are performed on the DFT values.

$$X(k) = \sum_{n=0}^{N-1} x(n) e^{-j\left(\frac{2\pi}{N}\right)kn} \; ; \; k = 0, 1, \ldots N-1$$

$X(k)$ represents the spectral components of the N point $x(n)$ block under consideration. The digital filter for compensating hearing profile variations is implemented using filter parameters stored in memory. The signal is then separated into frequency bands. The number of bands and the cut-off frequencies of each band are provided in the memory. The filtered frequency samples are then compressed by the compression system (dynamic range compression). The type of compression and the compression parameters are stored in memory. By analyzing the frequency samples of the incoming signal, the system can detect the presence of feedback oscillation in a channel and can reduce the gain. It can also analyze the noise energy present in the incoming signal in each channel and can reduce the gain of the channel that has less speech content. After carrying out the entire signal processing functions, the modified DFT values are converted back into time domain before transmitting it to the digital output amplifier.

$$y(n) = \frac{1}{N} \sum_{k=0}^{N-1} Y(k) e^{j\left(\frac{2\pi}{N}\right)kn} \; ; \; n = 0, 1, \ldots N-1$$

where, $Y(k)$ represents the modified spectral samples obtained by performing various signal processing operations on $X(k)$. All the parameters required for implementing the different DSP functional blocks are fetched from the memory. Digital signal processing blocks (Figure 11.6). Facility is provided to change the memory contents through PC and an interface (HI-PRO). Using Hi-PRO interface the audiologist can program the hearing characteristics of the patient into the hearing aid. HI-PRO contains level converter and the necessary logic for enabling serial communication between PC and hearing aid. More than one memory areas can be assigned for storing different sets of parameters. The different memory areas are selectable through a push button switch.

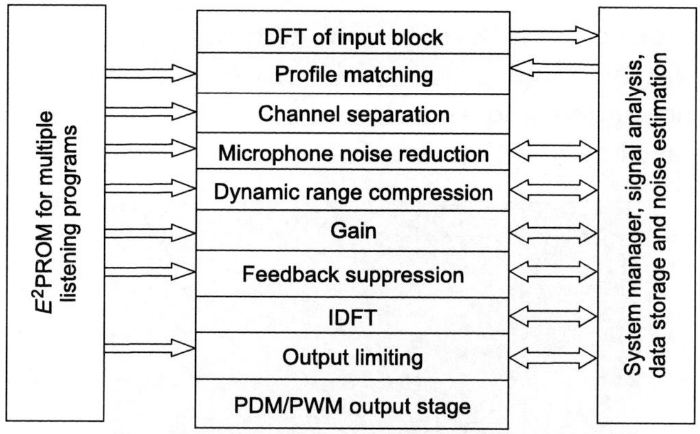

Figure 11.6 Digital signal processing blocks

11.10 COCHLEAR IMPLANT

Cochlear implants are surgically implanted devices that bypass the damaged inner ear and send signals directly to the auditory nerve to a person who is profoundly deaf or severely hard of hearing. Cochlear implants are often referred to as a bionic ear and are shown in Figure 11.7. A small microphone behind the ear picks up sound waves in a speech processor processes the signal and sends them to the receiver that has been placed under the scalp. This receiver then transmits impulses directly to the auditory nerve. These signals are perceived as sound and allow the person to hear. Cochlear implants may help provide hearing in patients that are deaf due to damage to sensory hair cells in their cochlea. In those patients, they can often enable sufficient hearing to allow better understanding of speech.

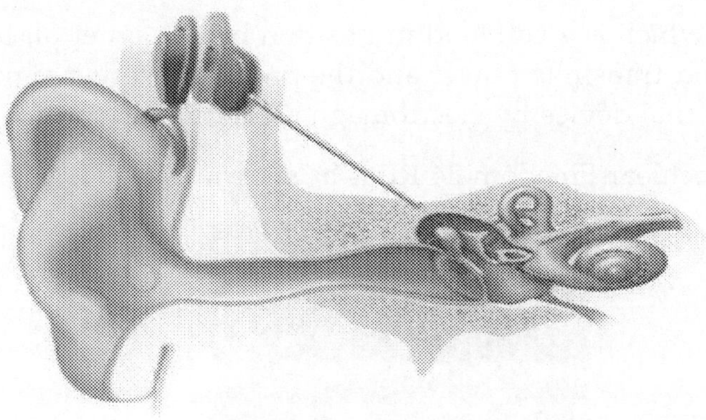

Figure 11.7 Cochlear implants

The quality of sound is different from natural hearing, with less sound information being received and processed by the brain. However, many patients are able to hear and understand speech and environmental sounds. Infant with cochlear implant is as shown in Figure 11.8.

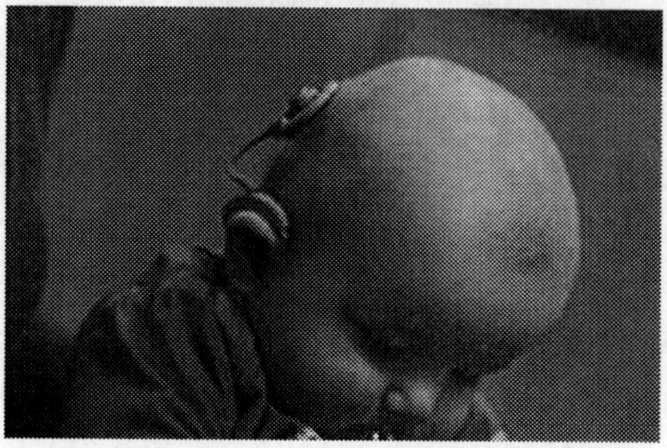

Figure 11.8 Infant with cochlear implant

11.10.1 Parts of the Cochlear Implant

The implant is surgically placed under the skin behind the ear. The basic parts of the device include:

External:

- one or more **microphones** which pick up sound from the environment;
- a **speech processor** which processes the speech using appropriate speech processing strategy and sends the electrical signals through a thin cable to the transmitter; and
- a **transmitter**, which is a coil held in position by a magnet placed behind the external ear, and transmits power and the processed sound signals across the skin to the internal device by electromagnetic induction.

Internal: Model Cochlear Freedom 24 RE is as shown in Fig. 11.9.

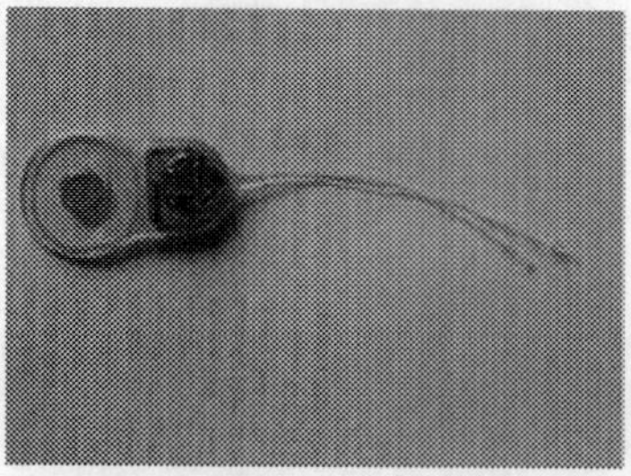

Figure 11.9 Internal part of a cochlear implant

- a **receiver and stimulator** secured in the bone beneath the skin, which converts the signals into electric impulses and sends them through an internal cable to electrodes; and
- an array of up to 22 electrodes wound through the cochlea, which send the impulses to the nerves in the scala tympani and then directly to the brain through the auditory nerve system. There are many manufacturers for cochlear implants, and each one produces a different implant with a different number of electrodes. The number of channels is not a primary factor upon which a manufacturer is chosen; the signal processing algorithm is also another important block.

11.10.2 Functionality

In individuals with sensorineural hearing loss, hair cells are often fewer in number and/or damaged. Hair cell loss or absence may be caused by a genetic mutation or an illness such as meningitis. Hair cells may also be destroyed chemically by an ototoxic medication, or simply damaged over time by excessively loud noises. The cochlear implant bypasses the hair cells and stimulates the cochlear nerves directly using electrical impulses. This allows the brain to interpret the frequency of sound as it would if the hair cells of the basilar membrane were functioning properly.

Sound received by the microphone must next be processed to determine how the electrodes should be activated. Filter bank strategies use fast Fourier transforms to divide the signal into different frequency bands. The algorithm chooses a number of the strongest outputs from the filters, the exact number depending on the number of implanted electrodes and other factors. These strategies emphasize transmission of the spectral aspects of speech.

Speech processors are the components of the cochlear implant that transforms the sounds picked up by the microphone into electronic signals capable of being transmitted to the internal receiver. The coding strategies programmed by the user's audiologist are stored in the processor, where it codes the sound accordingly. The signal produced by the speech processor is sent through the coil to the internal receiver, where it is picked up by radio signal and sent along the electrode array in the cochlea.

Transmitter transmits the processed sound information over a radio frequency link to the internal portion of the device. Radio frequency is used so that no physical connection is needed, which reduces the chance of infection and pain. The transmitter is held in place to the line of receiver using a magnet.

Receiver receives directions from the speech processor by way of RF link from the transmitter. (It also receives its power through the transmission.) The receiver is also a sophisticated computer that translates the processed sound information and controls the electrical current sent to the electrodes in the cochlea. It is embedded in the skull behind the ear.

Electrode array is made from a type of silicone rubber, while the electrodes are of platinum or a similar highly conductive material. It is connected to the internal receiver on one end and inserted into the cochlea deeper in the skull. (The cochlea winds its way around the auditory nerve, which is tonotopically organized as is the basilar membrane.) When an electrical current is routed to an intracochlear electrode, an electrical field is generated and auditory nerve fibres are stimulated.

Speech processors are the components of the cochlear implant that transforms the sound picked up by the microphone into electronic signals capable of being transmitted to the internal receiver. The coding strategies programmed by the user's audiologist are stored in the processor, where it codes the sound accordingly. The signal produced by the speech processor is sent through the coil to the internal receiver, where it is picked up by the radio signal and sent along the electrode array in the cochlea.

11.10.3 Advantages and Disadvantages of Cochlear Implants

Advantages

- Bypasses the cochlea which is the likely source of deafness.
- It eases the individual for telephone conversation.
- It is best preferred for person suffering from severe/profound hearing loss.
- Distance hearing is likely better compared to hearing aids.

Disadvantages

- Cochlear implantation requires surgery which is expensive.
- The person will be under risk of partial facial nerve paralysis.
- Batteries have a dramatically shorter lifespan than do hearing aid batteries.
- If the stimulator (the internal part of device) fails or is damaged, then another surgery may become mandatory.
- Person is not preferred for scuba diving, MRI treatment, should stay away from radar detectors, playing football, boxing, etc.
- Loud buzzing sound when the implant user is within close proximity to neon lights; interferes with speech signal.
- Cochlear implant is expensive (costs 6.5 to 11 lakh rupees) so that it is hard for middle-class people and almost impossible for lower class people to afford.
- Cochlear implant is quiet risky and does not assure 100% success.
- It is very difficult to treat people suffering from central loss.
- Hearing aids cannot be used for all kinds of disorders.

- A low cost alternative for cochlear implant would definitely help middle and lower class people of India.

11.11 SUMMARY

This chapter discussed the design considerations for designing hearing impairment aids by considering the anatomy of the ear with the DSP based hearing model.

EXERCISES

Review Questions

1. With the help of a diagram, explain the functions of the external, middle, and the inner ear.
2. Explain the classification of hearing impairment or hearing loss.
3. List the causes for conductive, sensorineural, mixed and central loss.
4. List the treatment for conductive, sensorineural, mixed and central loss.
5. Distinguish between analog and digital hearing aids.
6. List the limitations of conventional analog hearing aids.
7. With necessary diagram, explain analog hearing instrument and its limitations.
8. With necessary diagram, explain DSP based hearing instrument and its advantages.
9. With necessary diagram, explain ASIC based hearing instrument.
10. What is cochlear implant? List its advantages and disadvantages.

12

CHAPTER

Speech Processing Chips

This chapter provides insights on

- Pin configuration and operations of SP0264 speech processor
- Features and applications of RC8660
- Floating point DSP 320C6713
- Architecture and applications of 320C6713

12.1 INTRODUCTION

Speech processing chips are powerful, specialized integrated circuits used to provide synthesized voice output and/or voice recognition input for different types of computerized systems. This chapter covers the most popular speech synthesis, speech recognition, and digital signal processing (DSP) chips on the market and summarizes the important characteristics of each. SP0264 speech processor (SP) is a single-chip N-channel silicon gate MOSLSI device that is able, to synthesize speech. The RC8660 is a versatile voice and sound synthesizer, integrating a text-to-speech (TTS) processor, audio recording and playback. TMS 320C6X, announced in 1997 and TMS 320C62X is the first member of C6X family of fixed point digital signal processor. Subsequently, TMS 320C6713 floating point DSP was introduced as another member of the C6X family of processor. In this chapter we will discuss features, architecture, of SP0264, RC8660 and TMS 320C6713 processor.

12.2 SP0264

The SP0264 speech processor (SP) is a single-chip N-channel silicon gate MOSLSI device that is able to synthesize speech or complex sounds, using its stored program. The SP0264 speech processors use the method of synthesis by rule. They also support LPC synthesis, formant synthesis and allophone synthesis. Phoneme synthesis works by combining basic sound elements to form complete words and sentences. Following are the features of SP0264.

- Natural speech
- On-chip 64-kbit ROM
- Stand-alone operations with inexpensive support components.
- Wide operating voltage
- Directly expandable to a total 480 K ROM
- Simple interface to most microcomputers or microprocessors
- Supports LPC synthesis, formant synthesis, and allophone synthesis
- Pin compatible with the SP0256B speech synthesizer
- Available in –40°C to 80°C version

The speech process is initiated by addressing the ROM address that contains the phoneme desired. This device contains a microcontroller and a vocal tract model. The vocal-tract model is a digital filter. The SP0264 has a 64-kbit memory. The device supports a controller for external ROM SPR128A, which is a 128-kbit ROM. The processor incorporates four basic functions

- A software programmable digital filter that can be made to model a vocal tract.
- A 64-kbit ROM which stores both data and instructions.
- A microcontroller which controls the data flow from the ROM to the digital filter.
- A pulse width modulator that creates a digital output which is converted to an analog signal when filtered by an external low-pass filter.

When amplified, this analog output signal will drive a loud speaker. The digital output is equivalent to a flat frequency response varying from 0 to 5 kHz, a dynamic range of 42 dB, and a signal-to-noise ratio of approximately 35 dB. A nice feature of this synthesizer is the natural speech, and the external ROM directly expandable to a total of 480 k.

12.2.1 Pin Configuration of SP0264

Table 12.1 gives the pin configuration of the SP0264, which is identical to the SP0256B. Functions of the pins are as follows:

TABLE 12.1 Pin configuration of the SP0264

Pin Number	Name	Function
1	Vss	Ground
2	RESET	A logic 0 resets that portion of the SP powered by VDD. Must be returned to a logic 1 for normal operation.
3	ROM DISABLE	For use with an external serial speech ROM, a logic 1 disables the external ROM.
4, 5,6	Cl, C2, C3	Output control lines for use with an external serial speech ROM. Refer to the SPR016 Data Sheet for details.
7	VDD	Power supply for all portions of the SP except the microprocessor interface logic.
8	SBY	STANDBY. A logic 1 output indicates that the SP is inactive and VDD can be powered down externally to conserve power. When the SP is reactivated by an address being loaded, SBY will go to a logic 0.
9	LRQ LOAD REQUEST	LRQ is a logic 1 output whenever the input buffer is full. When LRQ goes to a logic 0, the input port may be loaded by placing the 8-address bits on A1-A8 and pulsing the ALD output.
10,11,13,14,15, 16,17,18	A8,A7,A6,A5,A4, A3,A2,A1	8-bit address which defines any one of 256 speech entry points.
12	SER OUT SERIAL ADDRESS OUT	This output transfers a 16-bit address serially to an external speech ROM.
19	SE	STROBE ENABLE. Normally held in a logic 1 state. When tied to ground, ALD is disabled and the SP will automatically latch in the address on the input bus approximately just after detecting a logic 1 on any address line.
20	ALD	ADDRESS LOAD. A negative pulse on this input loads the 8-address bits into the input port. The negative edge of this pulse causes LRQ to go high.
21	SER IN	SERIAL IN. This is an E-bit serial data input from an external speech ROM.
22	TEST	This pin should be grounded for normal operation.
23	VD1	Power supply for the microprocessor interface logic and controller.
24	DIGITAL OUT	Pulse width modulated digital speech output which, when filtered by a 5 kHz low-pass filter and amplified, will drive a loudspeaker.

Contd.

25	SBY RESET	STANDBY RESET. A logic 0 resets the microprocessor interface logic and the address latches. Must be returned to a logic 1 for normal operation.
26	ROM CLOCK	This is a 1.56 MHz clock output used to drive an external serial speech ROM
27	OSC1 XTAL IN.	Input connection for a 3.12 MHz crystal
28	OSC2 XTAL OUT.	Output connection for a 3.12 MHz crystal.

12.2.2 Operation of SP0264

Addressing of SP0264 is controlled by the address pins (A1-A8), Address Load ($\overline{ALD}$), Strobe Enable (SE). Speech data SP0264 may be stored within internal 64K ROM and/or an external serial speech ROM. There are two modes available for loading an address into the chip. Strobe Enable (SE) controls which mode will be used.

In Mode 0 (SE=0), speech processor will latch an address, when any one (or more) address pin makes a low to high transition. All address lines must be returned to the low state prior to entering a new address. Address zero (0000 0000) is not a valid address in this mode operation. In order to best utilize this mode of operation, a vocabulary should consist of no more than eight individual words or phrases such that single address line transitions can be made. These words or phrases must store using the following entry point address: 1,2,4,8,16,32,64 and 128. NOTE: There is a 2-byte overhead penalty paid for each dummy entry point between the entries point actually used.

In Mode 1 (SE-1), SP0264 will latch address using $\overline{ALD}$ pin. The desired address is set up on the address bus (A1-A8) and then $\overline{ALD}$ is pulsed low. Any address can be loaded using this mode, but certain set-up and hold times are required. Two microprocessor interface pins are available for loading of addresses using mode 1. They are $\overline{LRQ}$ and SBY. Load Request $\overline{LRQ}$ tells the processor when address input buffer is full. Standby (SBY) tells processor that the chip has stopped talking and no new address has been loaded. When $\overline{LRQ}$ is low, a new address loaded may be loaded onto the address bus. Strobing $\overline{ALD}$ will cause the new address to be loaded and $\overline{LRQ}$ to go high. $\overline{LRQ}$ will return low when the address buffer is available to accept a new address. The SP0264 is capable of latching one new address while speaking last utterance (word or phrase). The time between address load requests is variable, depending on the length of the utterance currently spoken. Standby (SBY) goes low when an address is loaded and will stay low until all internal instructions have been completed (i.e., speech chip stops talking). If a second address has been loaded before the chip stops speaking first the first utterance, SBY will stay low through the completion of the second utterance. The SP0264 may be partially powered down when SBY is high to conserve battery life, provided V_{D1} remains powered. During power down and power up, reset should be held low to ensure the proper reset condition.

Internal ROM disable mode is entered by wiring TEST to a permanent high. This causes SP0264 to ignore its internal ROM to fetch data. From external speech ROMs, encoded device select mask.

12.3 RC8660

RC8660 is a versatile voice and sound synthesizer, integrating a text-to-speech (TTS) processor, audio recording and playback, musical and sinusoidal tone generators, telephone dialer and A/D converter, into an easy to use chipset. Using a standard serial or 8-bit bus interface, virtually any ASCII text can be streamed to the RC8660 for automatic conversion into speech by the TTS processor. The audio record and playback modes augment the TTS processor for applications requiring high voice quality and a relatively small, fixed vocabulary, applications requiring special sounds or sound effects, and/or the recording of voice memos. The audio output is delivered in both analog and digital PCM audio formats, which can be used to drive a speaker or digital audio stream. RC8660 is used in the following:

- Robotics
- Talking OCR systems
- ATM machines
- Talking pagers and PDAs
- GPS navigation systems
- Vending and ticketing machines
- Remote diagnostic reporting
- Dial-up information systems
- Hand-held barcode readers
- Electronic test and measurement
- Security systems
- Aids for the orally or visually disabled
- Meeting federal ADA requirements

12.3.1 Features of RC8660

RC8660 comprises two surface-mounted devices. Both operate from either a + 3.3 V or + 5 V supply. Most applications require only the addition of a low-pass filter/audio power amplifier to implement a fully functional system. Features of RC8660 are:

- Integrated text-to-speech processor:
  - High voice quality, unlimited vocabulary
  - Converts any ASCII text into speech automatically
  - Capable of very high reading rates
  - Add/modify messages by simply editing a text file
  - On-the-fly control of speed, pitch, volume, etc.
- On-chip recording, storage and playback of sound files:
  - Record to chip via microphone
  - Upload, download, and erase recordings and sound files, even in the field
  - Data logging mode allows analog quantities to be sampled and stored for later retrieval
  - Recording times from 2 min to 33 min available
- Tone generation:
  - Three voice musical
  - Dual sinusoidal
  - DTMF (Touch-Tone) dialer
- On-chip A/D converter:
  - Four channels, 8-bit resolution
  - One-shot, continuous, single sweep, and continuous sweep modes of operation
  - Software and hardware triggering
  - Support for external op amp
- Analog and digital audio outputs
- Stop, pause, and resume controls
- Standard serial (UART) and 8-bit bus interfaces
- User programmable greeting and default settings
- Flexible user exception dictionary:
  - Change the pronunciation of any input string based on spelling and context
  - Convert encrypted data into meaningful messages

  - Trigger tone generation, recorded message playback, voice parameter changes

- In-circuit, field programmable
- 8 kB input buffer for virtually no-overhead operation
- Available in 3.3 V and 5 V versions
- Low power (typ @ 3.3 V):
  - 3.8 mA active
  - 700 μA idle
  - 0.7 μA standby

12.3.2 Applications of RC8660

The RC8660 chipset includes a number of features that make it ideally suited for any design requiring voice output. The RC8660's applications are described below.

Text-to-Speech Synthesizer

The RC8660 provides text-to-speech conversion with its integrated doubletalk text-to-speech synthesizer. Any English text written to the RC8660 is automatically converted into speech. Commands can be embedded in the input stream to dynamically control the voice; even at the phoneme level (phonemes are the basic sound units of speech). A greeting message can be stored in the RC8660 that is automatically spoken immediately after the RC8660 is reset. Any of the commands recognized by the RC8660 may be included as part of the greeting message, which can be used to set up custom default settings and/or play a prerecorded message or tone sequence. An integrated nonvolatile memory area is also provided for storing a custom pronunciation dictionary, allowing the pronunciation of any character string to be redefined.

Audio Recording and Playback

Up to 33 minutes of recorded messages and sound effects can be stored in the RC8660 for on-demand playback. Recordings are stored in on-chip flash memory, providing zero-power message storage. Additionally, the RC8660 can play 8-bit PCM and ADPCM audio in real time, such as speech and/or sound effects stored in an external memory or file system.

12.3.3 Musical Tone Generator

An integrated, three-voice musical tone generator is capable of generating up to three tones simultaneously over a four-octave range. Simple tones to attention-getting sounds can be easily created.

12.3.4 Touch-Tone Generator

The RC8660 includes an integrated DTMF (Touch-Tone) generator. This is useful in telephony applications where standard DTMF tones are used to signal a remote receiver, modem, or access the public switched telephone network.

Sinusoidal Tone Generator

A precision, dual sinusoidal tone generator can synthesize the tones often used in signalling applications. The tone frequencies can be independently set, allowing signals such as call-progress tones to be generated.

Analog-to-Digital Converter

The four channel, 8-bit A/D converter can be used to monitor battery cell voltages, temperature, and other analog quantities. The ADC can be programmed on the fly to convert any single channel, or scan up to four channels, repetitively. ADC data can be read via the TXD pin, or stored in the RC8660's recording memory. With the appropriate signal conditioning, the ADC can be used for audio recording to the recording memory.

Versatile I/O

All data is sent to the RC8660 through its built-in serial and/or parallel ports. For maximum flexibility, including infield update capability, use of the serial port is recommended whenever possible. The RC8660's audio output is available in both analog and digital formats. The analog output should be used in applications where no further processing of the audio signal is required, such as driving a speaker or headphones (the output still needs to be filtered and amplified, however). The digital output is for applications that require further processing of the audio signal, such as digital mixing or creating sound files for later playback.

12.4 TMS320C6713

The TMS320C6713's high performance CPU and rich peripheral set are tailored for multichannel audio applications such as broadcast and recording mixing, home and large venue audio decoders, and multi-zone audio distribution. TMS 320C6713 Floating point DSP developed by Texas Instruments is based on very long instruction-word (VLIW) architecture and capable of both fixed and floating point processing. It operates with 225 MHz clock signal, and requires 3.3 V for I/O and 1.2 V for the core internal. Operating at 225 MHz, TMS320C6713 delivers up to 1350 million floating-point operations per second (MFLOPS), 1800 million instructions per second (MIPS), and with dual fixed-/floating-point multipliers up to 450 million multiply-accumulate operations per second (MMACS). Internal memory of C6713 includes two-level cache architecture and has rich peripheral set. Important features of TMS 320C6713 are internal memory of C6713 includes two-level cache architecture and has a rich peripheral set.

12.4.1 Features of TMS 320C6713

- It is capable of both fixed and floating point processing.
- Very long instruction word (VLIW) architecture CPU
  - 8 instructions can be fetched per clock cycle (e.g., clock rate 225 MHz, clock period = 1/225 MHz = 4.44 ns, hence 8 instructions can be fetched every 4.44 ns.)
  - Composed of eight independent functional units, they are:
    - 6 arithmetic and logic units (ALUs) (Four ALU's (fixed and floating point) and two ALU's (fixed point))
    - Two multipliers (fixed and floating point)
- Hardware support for IEEE 754 single and double precision floating point operations
- Load and store architecture with two sets of 32-bit registers
- 264 Kbyte of internal memory with two-level L1/L2 program/data memory architecture
    - 4 Kbyte of level 1 (L1) program cache (L1P) memory
    - 4 Kbyte of level 1 (L1) data cache (L1D) memory
    - 256 Kbyte of RAM of level 2 (L2) is shared between program/data space.
- 32-bit address bus to address 4 G byte memory (4G byte = 4×2^{30})
- General-purpose input/output (GPIO) module with 16 pins
- 16 independent channels, enhanced direct-memory-access (EDMA) controller
- 16-bit Host-Port Interface (HPI)
- To support multi-master and slave interfaces, two inter-integrated circuit buses (I2C bus)
- Two multichannel buffered serial ports and two multichannel audio serial ports

- Two 32-bit general-purpose timers
- Clock generator module with flexible phase-locked loop (PLL)
- Device configuration
    - Boot mode: HP1, 8,16, 32-bit ROM boot
    - Endiánness: Little Endian, Big Endian

12.4.2 General-Purpose Registers

It consists of two sets of register file: register file A (A0 – A15) and register file B (B0 – B15) as shown in Figure 23.1. These 32 registers are considered general-purpose registers. The instruction set architecture (ISA) is load-store (register-register) architecture, i.e., all the operands are in registers, not in the memory. Data transfer is faster and it reduces the data traffic. The function of the registers is as follows:

- Registers A0, A1, B0, B1, and B2 are used as conditional registers.
- Registers A4-A7 and B4-B7 are used for circular addressing.
- Registers A0-A9 and B0-B9 (except B3) are used as temporary registers.
- Registers A10-A15 and B10-B15 are used for saving main program address before calling a subroutine.

All data transfers make use of the .D units. The data-addressing unit's .D1 and .D2 are responsible for data transfer between registers and the memory.

These registers provide 32-bit and 40-bit fixed point data. The 32-bit data can be stored in any register. For 40-bit data, processor stores least significant 32 bits in an even register and remaining 8-bits in upper (odd) register. A similar scheme is used to hold a 64-bit double precision value within a pair of registers (even and odd).

12.4.3 Central Processing Unit (CPU)

CPU consists of eight independent functional units: four ALU's (fixed and floating point), two ALU's (fixed point), two multipliers (fixed and floating point), and also includes program fetch unit, instruction dispatch unit, instruction decode unit, test, emulation and interrupt logic. It is divided into two paths A and B as shown in Figure 12.1. Data path A includes a set of sixteen 32-bit general-purpose registers and functional units: Multiply operations (.M1), Logical and arithmetic operations (.L1), branch, bit manipulation, and arithmetic operations (.S1) and loading/storing and arithmetic operations (.D1). Data path B includes a set of sixteen 32-bit general-purpose registers and functional units: Multiply operations (.M2), logical and arithmetic operations (.L2), branch, bit manipulation, and arithmetic operations (.S2) and loading/storing and arithmetic operations (.D2). All data (not program) transfers go through .D1 and .D2. Each path can read directly from or write directly to the register file within its own path. Each functional unit can access data from the registers on the opposite side using a cross path. There can be a maximum of two cross path source read per cycle. Functions and operations performed by .L, .S, .M and .D are listed in Table 12.2.

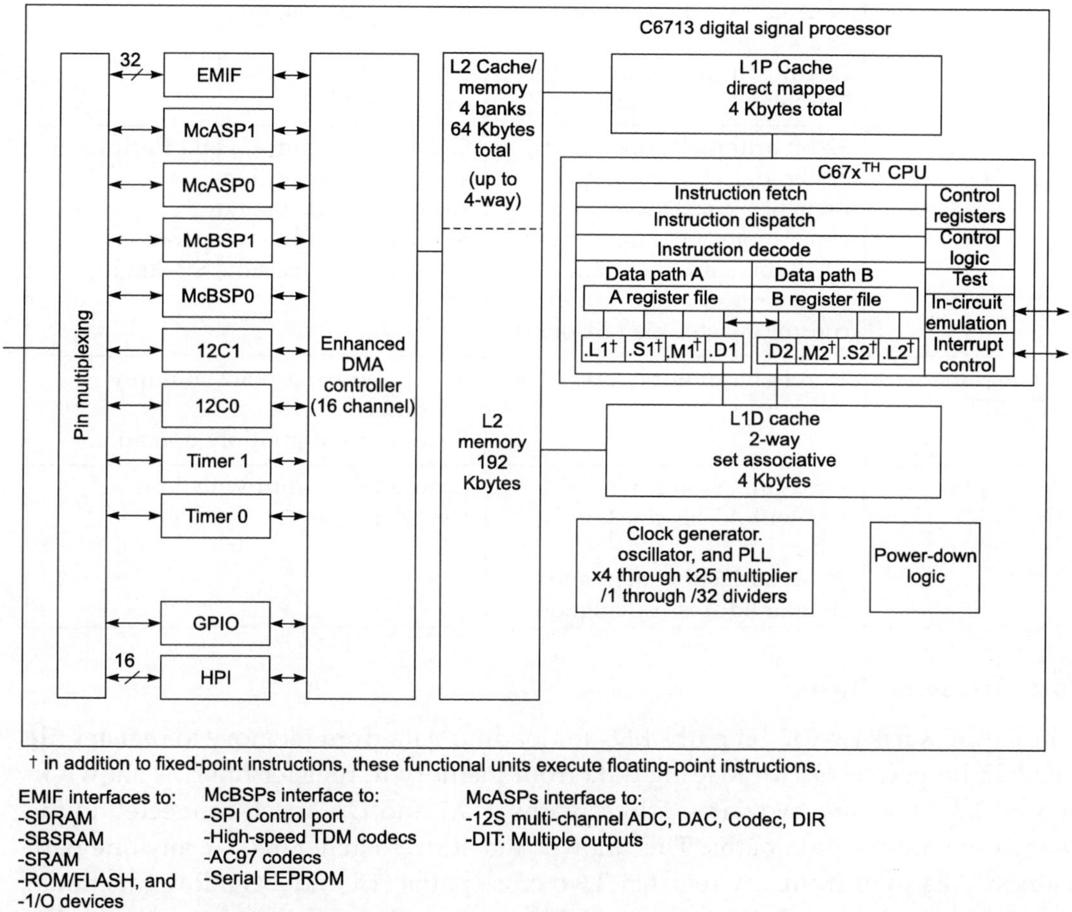

Figure 12.1 Function block diagram of TMS 3206713 digital signal processor (Courtesy: Texas Instruments)

TABLE 12.2 Functional units and operations performed (Courtesy of Texas Instruments)

Functional Unit	Fixed point operations	Floating point operations
.L Fun. unit (.L1, .L2)	32/40-bit arithmetic operations; 32/40-bit logical and compare operations; Normalization count for 32 and 40 bits; Leftmost 1 or 0 counting for 32 bits	Arithmetic operations; Conversion operations DP → SP, INT → DP, INT → SP (INT-integer, DP-Double precision, SP-Single precision)
.S Fun. unit (.S1, .S2)	32-bit arithmetic operations; 32-bit logical operations, branches, constant generations, field operations; 32/40-bit shift operations Register transfers to/from control register file (.S2 only)	Compare; Reciprocal and Reciprocal square root operations; Absolute value operations; SP → DP conversion operation (DP-Double precision, SP-Single precision)
.M Fun. unit (.M1, .M2)	16 × 16 bit multiply operation	32 × 32 bit fixed-point multiply operation floating-point multiply operation
.D Fun. unit (.D1, .D2)	Loads and stores with 5-bit and with 15-bit constant offset; 32-bit add, sub, linear and circular address calculation	Load double word with 5-bit constant offset

Data Address Paths

TMS 320C67X has two 32-bit paths LD1 for loading data from memory to register file A and 32-bit paths LD2 for loading data from memory to register file B as shown in Figure 12.2. There are two data address paths, DA1 and DA2 each connected to the .D units in both the data paths. This allows data address generated by any one path to access data to or from any register. Two cross paths (1X and 2X) allow functional unit from one data path to access a 32-bit operand from the register file on the opposite side. This allows to load 32-bit value into register file A or B and also to load two 32-bit value simultaneously into register file A and B. There are two 32-bit paths, ST1 and ST2 for storing register values from register files A and B to memory.

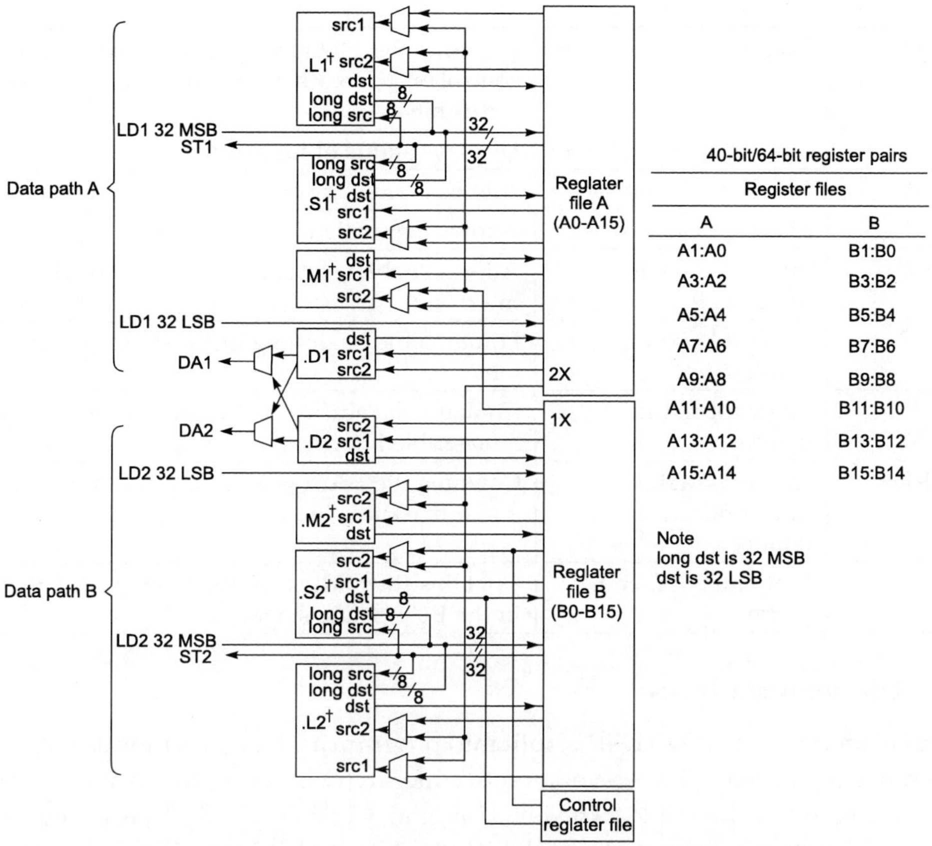

† Fixed-point instructions, these functional units execute floating-point instructions

Figure 12.2 TMS 320671X (DSP core) data paths (Courtesy: Texas Instruments)

Control Registers: Table 12.3 lists the control registers' name, abbreviation and function of each register. S2 unit can read from or write to the control register file. Each control registers is accessed by MVC instruction.

TABLE 12.3 Control Registers (Courtesy: Texas Instruments)

Abbreviation	Register Name	Description
AMR	Addressing mode register	Specifies whether to use linear or circular addressing for each of eight registers; also contains sizes of circular addressing. For each of the eight registers (A4-A7, B4-B7) that can perform linear or circular addressing, the AMR specifies the addressing mode. A 2-bit field for each register selects the address modification mode: linear (the default) or circular mode.

Contd.

CSR	Control status register	Contains the global interrupt enable bit, cache control bits, and other miscellaneous control and status bits.
IFR	Interrupt flag register	Displays status of interrupts
ISR	Interrupt set register	Allows manually setting pending interrupts
ICR	Interrupt clear register	Allows manually clearing pending interrupts
IER	Interrupt enable register	Allows enabling/disabling of individual interrupts
ISTP	Interrupt service table pointer	Points to the beginning of the interrupt service table
IRP	Interrupt return pointer	Contains the address to be used to return from a maskable interrupt
NRP	Non-maskable interrupt return pointer	Contains the address to be used to return from a non-maskable interrupt
PCE 1	Program counter, E1 phase	Contains the address of the fetch packet that is in the E1 pipeline stage

12.4.4 Internal Memory

The internal memory of C6713 DSP is split into program and data memories with two-level memory hierarchies. Two-level memory hierarchy is useful for data movement and interfacing with other DSPs. The level1 memory consists of 8 Kbyte cache and it is split into 4 Kbyte program cache and 4 Kbyte data cache. Level 2 memory consist of a 256 Kbyte memory space shared between program and data space. 192 Kbytes in L2 serves as mapped SRAM. Remaining 64 Kbytes of the 256 Kbyte in L2 memory can be configured as mapped internal memory, cache or combination of the two.

12.4.5 Peripherals

The TMS320C67X devices contains the following peripherals:

- External memory interface
- Two multichannel buffered serial ports (McBSPs)
- Two multichannel audio serial ports (McASPs)
- Two 32-bit general-purpose timers
- Enhanced direct-memory-access (EDMA) controller (16 independent channels)
- 16-bit host-port interface (HPI)
- Dedicated GPIO module with 16 pins

- Two inter-integrated circuit buses (I2C bus)
- Power down logic

The following subsections discuss the peripherals of C6713 processor.

External Memory Interface

TMS320C6713 supports interface to several types of memory which include SRAM, ROM, FIFOs, synchronous burst SRAM (SBSRAM) and synchronous DRAM (SDRAM). The external memory interface (EMIF) consists of 32-address bus to interface 4 G byte memory. Following data ordering standards are used:

- Little-endian ordering: In this standard, most significant byte has the highest address and the bytes are ordered from right to left.
- Big-endian ordering: In this standard, most significant byte has the lowest address and the bytes are ordered from left to right.

The external memory interface (EMIF) reads and writes both big and little-endian devices.

Multichannel Buffered Serial Port

TMS320C6713 supports two multichannel buffered serial ports (McBSPs). Each multichannel buffered serial port (McBSP) is based on the standard serial port interface and it provides:

- Full-duplex communication.
- Double-buffered data registers.
- Transmission and reception are supported by independent framing and clocking.
- Supports direct interface to analog interface chips (AICs), industry-standard codecs, and serially connected A/D and D/A devices.
- Supports multichannel transmission and reception of up to 128 channels.
- Supports 8, 12, 16, 20, 24, or 32 bit data formats.
- Supports both A-law and μ-law companding.
- Supports 8-bit data transfers with MSB or LSB first.
- It operates with internal programmable frequency shift clock or with external shift clock.
- It operates with programmable polarity for both data clocks and frame synchronization.
- It operates with highly programmable frame generation and internal clock.

Internal block diagram of McBSP is shown in Figure 12.3. Data communication between McBSP and the interfaced devices takes place via two different pins for transmission—data transmit (DX) pin and reception – data receive (RX) pin. 32-bit wide control registers are used to communicate McBSP with peripheral devices through internal peripheral bus. Control information in the form of clocking and frame synchronization is communicated via CLKX, CLKR, FSX, and FSR. CPU or DMA write the data to be transmitted to the data transmit register (DXR) which is shifted out to DX pin via the transmit shift register (XSR). Similarly, receive data on the DR pin is shifted into the receive shift register (RSR) and copied into the receive buffer register (RBR). RBR is then copied to data receive register (DRR), which can be read by the CPU or the DMA controller. This allows internal data movement and external data communications simultaneously.

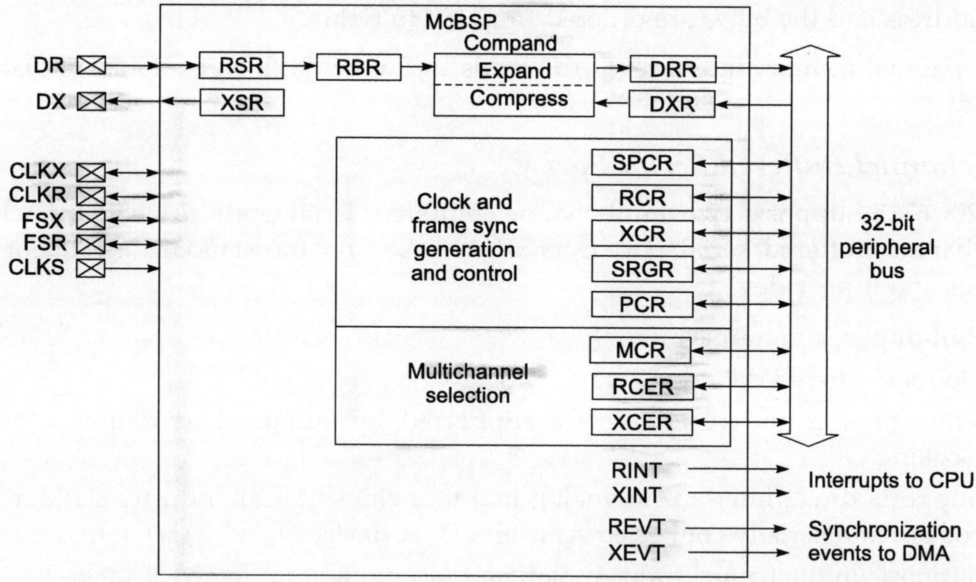

Figure 12.3 Block diagram of McBSP (Courtesy: Texas Instruments Inc.)

Other registers-receive/transmit control register (RCR/XCR), receive/transmit channel enable register (RCER/XCER), serial port control register (SPCR), pin control register (PCR), and sample rate generator register (SRGR)-support further data communication.

Multichannel Audio Serial Port

The TMS320C6713 device includes two multichannel audio serial port (McASP) interface peripherals (McASP1 and McASP0). TMS320C6713 device is capable of supporting two completely independent audio zones simultaneously. These peripherals can operate completely independently with different data formats, separate master clocks, bit clocks, and frame syncs or alternatively, transmit and receive sections may be synchronized. Each McASP consists of transmit and receive section. Each McASP module also includes a pool of 16 shift registers that may be configured to operate as either transmit data, receive data, or general-purpose I/O (GPIO).

The serial port supports the following formats:

- Burst mode
- Time-division multiplexed synchronous serial format
- Digital audio interface transmitter format.

Each McASP contains eight serial data pins which can be individually allocated to any of the two zones. The data in each zone can be transmitted and received on multiple serial data pins simultaneously and formatted in Philips Inter-IC Sound (I2S) format. The processor has sufficient bandwidth to support all 16 serial data pins transmitting a 192 kHz stereo signal. Both transmit and receive sections of the McASP support burst mode which is useful for non-audio data (for example, passing control information between two DSPs). Each McASP also includes full general-purpose I/O (GPIO) control, so any pins not needed for serial transfers can be used for general purpose. McASP peripherals have additional capability for flexible clock generation, and error detection/handling, as well as error management.

Timers

TMS 320C6713 processor has two 32-bit timers and they are used for generating pulses, for timing and counting events, interrupting CPU, and for sending synchronization events to the DMA controller.

Each timer, has a timer count register (CNT): which contains the value of the counter, has a period count register (PRD): specifies the timer's frequency, a timer control register (CTL): monitors the timer status. The input clock frequency for the timer is the CPU clock rate divided by four. The timer has an input pin (TINP) and an output pin (TOUT). The timer is clocked by an internal or by an external source. When the timer starts from reset, the value in timer count register is incremented for every clock pulse until it reaches the value of period count register, once when it reaches the value of period count register, an interrupt is generated and timer is reset. The TINP pin and TOUT pin can be used as a general-purpose input and as a general-purpose output. When it works with internal clock, the timer generates timing sequences to trigger external devices such as analog to digital converter or DMA controller, respectively. When it works with external clock, the timer can interrupt the CPU after a specified number of events and can count external events.

EDMA Controller

The enhanced direct memory access (EDMA) controller of C6713 DSP is a highly efficient data transfer engine and CPU is relieved from the burden of data transfer. The features of EDMA are as follows:

- Supports sixteen independent channels.
- Data can be transferred at the CPU clock rate.
- DMA operates independently of the CPU
- Each channel works independently with programmable priority.
- To perform both transmit and receive operations to or from memory and peripherals, single channel can be used simultaneously.
- The transfer through each channel may be either synchronized by element or by frame.
- Source and destination address registers of each channel have configurable indexes for each write and read operations. The contents of address registers can be incremented or decremented or can be changed by a programmable value or it remains constant.
- Each channel can be configured independently to transfer information in terms of bytes or 16-bit half word or 32-bit words.
- Once the block transfer is complete, each EDMA can automatically reinitialize itself for the next block transfer.

EDMA controller has 16 channels and allows the data transfer between internal memory (L2 SRAM), external memory, internal peripherals and external peripherals. EDMA controller transfers data

- Between on-chip peripherals and the memory
- Between DSP and external peripherals
- Between DSP and on-chip peripherals

EDMA controller does not use the same bus for accessing the memory, which the CPU uses for accessing operands. EDMA controller performs read/write operation from source/destination location of the memory, respectively. It also supports frame transfer and block transfer of data elements. Each EDMA channel works independently with a number of data elements per frame and number of frames per blocks. DMA registers are used to configure the DMA: address (source and destination), count reload, index, DMA global data and control registers. DMA transfers can be triggered by interrupts from internal peripherals as well as external pins. The source and destination addresses can be from internal data memory, internal program memory, internal peripheral bus, and external memory interface.

Host Port Interface

Host port interface is a 16-bit parallel port and host port device functions as a master to the interface. The host processor can directly access the CPU memory space using host port interface (HPI). CPU and the host can exchange information with internal or external memory. The host supports direct access to memory mapped peripherals. It can access the internal memory space of the device using HPI address register (HPIA) and the HPI data register (HPID).

HPI control register (HPIC) is a memory mapped register and is used to configure the interface either by the host or by the CPU.

GPIO Module

TMS 320C6713 processor provides a set of sixteen general-purpose input/output (GPIO) pins that can be configured as input or as output pin. GPIO registers are shown in Table 12.4.

TABLE 12.4 GPIO registers

Acronym	Register Name
GPEN	GPIO enable register
GPDIR	GPIO direction register
GPVAL	GPIO value register
-	Reserved
GPDH	GPIO delta high register
GPHM	GPIO high mask register
GPDL	GPIO delta low register
GPLM	GPIO low mask register
GPGC	GPIO global control register
GPPOL	GPIO interrupt polarity register
-	Reserved

To select I/O pin GPXEN bit corresponding to each pin in the GPIO enable register (GPEN) should be set. GPXDIR bit in GPIO direction register determines whether the pin is input or output. A value of '0' indicates, pin is defined as input and '1' indicates, pin is defined as output. GPXHM bit in the GPIO High Mask Register (GPHM) enables or disables a GPIO pin for generating interrupts to the CPU. When low-to-high transition occurs on the pin, GPXDH bit of GPIO delta high register (GPDH) is set else high-to-low transition occurs on the pin, GPXDH bit is reset.

Inter Integrated Circuit Bus

TMS 320C6713 processor has two inter integrated circuit bus (I2C) modules. One module may be used by the DSP, to control local peripherals ICs (DACs, ADCs, etc.), while the other may be used to communicate with other controllers in a system or to implement a user interface.

Power Down Logic

When circuit switches from one logic state to the other, most of the operating power of the CMOS logic is dissipated. Preventing this, significant power savings can be realized without losing any data or operational context. Power savings are accomplished in power-down mode 1(PD1), power-down mode 2(PD2) and power-down mode 3(PD3). In mode 1, PD1 blocks the internal clock inputs at the boundary of the CPU thus preventing most of its logic from switching and thereby effectively shutting down the CPU. In mode 2 (PD2), the entire on chip clock structure (including multiple buffers) is halted at the output of the PLL. In mode 3 (PD3), shuts down the entire internal clock tree (like PD2) and it also disconnects the external clock source (CLKIN) from reaching the PLL. Since PLL needs to be relocked, wake-up from PD3 takes longer time than wake-up from PD2.

12.5 APPLICATIONS OF TMS320C6713

TMS320C6713 includes a number of features that make it ideally suited for speech applications. Applications are listed in the Table 12.5.

TABLE 12.5 Applications

DSP Algorithms can be implemented on TMS 320c6713	Applications
Speech coding	Digital cordless telephones, secure communications, digital cellular telephones
Speech recognition	Robotics, multimedia workstations, cellular telephones
Speaker identification	Multimedia workstations, security, advanced user interface
Speech enhancement	Advanced vehicular audio, professional audio
Speech synthesis	Robotics, advanced user interface
Speech encryption	Secure communications, digital cordless telephones, digital cellular telephones
Beam forming	Navigation, radar/sonar signal intelligence

12.6 SUMMARY

Different types of speech processing chips with their features and operations have been presented in this chapter along with their applications.

EXERCISES

Review Questions

1. List salient features of SP0264 speech processor.
2. Explain Mode 0 and Mode 1 operation of SP0264.
3. List the application of RC8660 in speech processing.
4. List and explain features of RC8660.
5. List functional units of TMS 320C6713 and explain its operations.
6. What are the differences between linear and circular addressing mode in TMS 320C6713?
7. What are the peripherals supported by TMS 320C6713?
8. Explain the functions of the following registers related to Timer of 6713 processor.
   (a) Timer Control Register
   (b) Period Count Register
   (c) Timer Count Register
9. Explain the functions of mutichannel buffered serial ports (McBSPs) of 6713 processor.
10. List and explain the applications of TMS 320C6713 in speech processing.

12.4 SUMMARY

[illegible]

[illegible]

1. [illegible]
2. [illegible]
3. [illegible]
4. [illegible]

6. [illegible]
7. [illegible]
8. [illegible]
9. [illegible]
10. [illegible]

13

CHAPTER

Praat and WaveSurfer

This chapter provides insights on

- Features of Praat and WaveSurfer tools
- Recording of speech signals using Praat and WaveSurfer.
- Fundamental and formant frequency calculation

13.1 INTRODUCTION

PRAAT is a computer program for speech processing purposes. It is freely available at ***http://www.praat.org*** and versions of the program exist for all major platforms: Linux, Windows and Mac OS. The Praat has been developed by Paul Boersma and David Weenink, both at the Phonetics Institute of the University of Amsterdam. The program is used worldwide by phoneticians, phonologists, and speech researchers. Besides analysis of speech sounds, it is also used to analyze singing voices, music and even the vocalization of animals like, for example, birds, dolphins, apes, and elephants.

13.2 SOUND AND ACOUSTICS

A human who speaks, produces sounds which can be heard by anyone nearby. But there are many more sounds that we cannot hear. The sounds a human being can hear are limited to a subset of all possible sounds. For example, a dog can hear ultrasound, i.e., sounds that are too "high" for us to hear. Sounds that are too "low" for us to hear are called infrasounds. Infrasounds can travel large distances and this is why whales and elephants can communicate over very large distances using infrasounds. The descriptions "low" and "high" are connected to the physical concept of frequency. The branch of physics that studies sound is called ***acoustics***.

From acoustics the ***sound*** is defined as a mechanical disturbance from a state of equilibrium that propagates through an elastic medium. An elastic medium is capable of resuming its original shape after stretching or compressing. This disturbance may be produced in many ways like plucking a guitar string, pinching a tuning fork or by the opening and rapid closing of the vocal folds. This disturbance produces a sudden local increase in pressure, i.e., a local compression of air. Since the medium is elastic, the compression is not permanent and therefore the compressed region rebounds. In doing so, it compresses the adjacent region, which again rebounds and so on. The result of this cycle of repetition is a ***compression wave*** followed by a ***rarefaction wave*** as each region rebounds. The waves thus generated propagate through the medium with a speed that depends on the elasticity, density and the temperature of the material. The propagation velocity of a sound pressure wave in air at room temperature is approximately 340 m/s. ***But it is not the air particles in the air that carry the sound to your ear but it is the compression wave that propagates and carries the sound.*** Therefore, to study the physical aspects of sounds we don't have to study particle physics but instead we have to study waves, in particular, sound waves.

13.3 SOUND RECORDING CONDITIONS

A speaker produces a speech signal; the acoustic wave is picked up by a microphone and transduced into an electrical signal. This electrical signal is amplified and the amplified signal is recorded either in analog or in digital form. Nowadays, the recording is done in digital form only.

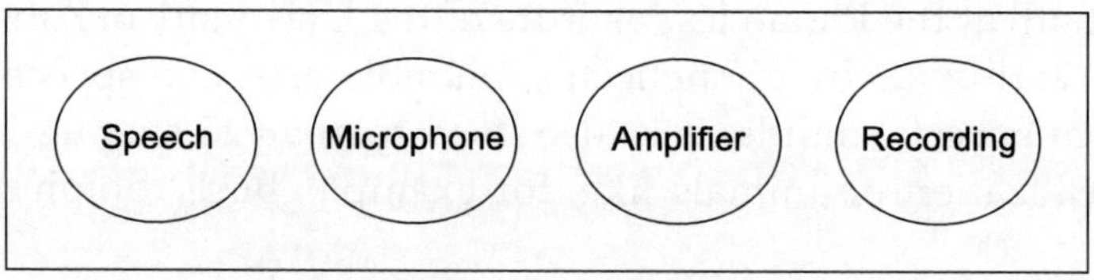

Figure 13.1 Sound recording process

The conditions at a recording session can vary as follows.

13.3.1 Recording at the Computer

The simplest environment for recording is at home with a microphone directly attached to a computer. It needs a good microphone and a sound card. But there is no control over the environment. The possible sources of noise are, for example, somebody enters the room, a car honks outside or one of the neighbours starts drilling a hole. Another source of noise is the computer itself: a fan might start to blow or a disk might spin up or the electrical circuit of the computer interferes with the circuitry of the input of the sound card.

For recording, connect the microphone with the sound card in the computer. Make sure that the microphone is selected as the only sound input device and any other input devices like the CD or line-in are deselected. Then select the microphone and deselect all other devices for the Windows XP mixer. While recording it is best to deselect all outputs to avoid possible reverberations between the loudspeaker in the computer and the microphone.

Now with the sound streams in place, use Praat for actual recording. Choose from the "New" menu the "Record mono Sound..." option: the sound recorder pops up, it. The mono channel is automatically set and check "Microphone" and the "44100 Hz" boxes. The actual recording starts as soon as the "Record" button is clicked and stops when click "Stop". If click "Save to list" a new sound object appears in the list of objects. It is possible to inspect the new sound in the sound editor and save it to a file on disk. If you close the sound recorder without having saved the sound the recording will be lost.

13.3.2 Recording in a Studio

This is clearly the preferred way because we have almost complete control over the recording environment. Environmental noise, the main disturbing factor in making a recording, can be minimized here.

13.3.3 Recording in the Field

This might be at extreme places like somewhere in the Amazon basin where may have to record the sounds of a very uncommon Indian language, or in the tundra of Siberia where the last sounds of some extinguishing language have to be recorded or in a Chinese city making recordings for a language change project. These kinds of field recordings clearly limit the amount of control on the recording environment and they also limit the size and weight of our recording equipment.

13.3.4 Guidelines for Recording Speech

- Try to avoid making recordings in reverberant rooms (a church is very reverberant).
- Try to avoid making recordings at places where uncontrollable noises like traffic noise, airplanes flying over, trains passing by, other people interrupting, etc., might turn up.
- Write the recording protocol beforehand and inform the speaker about how and what you want from him.
- Arrange a comfortable reading/speaking position for the speaker. If he/she needs to read from paper, take care that no paper crackling should take place while reading and the speaker only moves the paper during breaks. These

sounds get unnoticed during the recording but afterwards, they will sound awfully loud.

- To avoid large intensity variations in the recording, the distance from the speaker's mouth to the microphone should remain as constant as possible.
- Beware of lip noises: many speakers get dry lips during talking, their lips start sticking together and then they open with a smacking sound that you probably won't notice during the recording, only afterwards when you listen with headphones. Try to avoid this from happening by having a glass of water.
- Take care to comfort the speaker, he/she should speak as naturally as possible. If you record read-aloud speech then embed the words or sentences you want recorded in other words and sentences.
- Try to avoid repeating rhythmic patterns in consecutive sentences: speakers will adapt their speaking rate to this rhythm.
- Test extensively the sound level before making the recording. Try to avoid that your recordings are either too loud or too soft.
- Always start with the speaker saying her name, date, and time of the recording.
- Legal concerns. Depending on the use of the recorded material the speaker has to give his/her written consent. This means that you have to state explicitly how the speech is going to be used and/or stored or distributed.

13.4 SOUND EDITOR IN PRAAT

Sound editor, is the window that appears when clicked the Edit command of a selected sound.

The sound editor shows vertically the sound amplitude as a function of the time which is displayed horizontally. The most important parts of the editor are:

1. **Title bar:** It shows, from left to right, the identification id of the sound, the type of object being edited, i.e., "Sound" and the name of the sound being edited. This is typical for the title bar of all editors in Praat: they always show the id, the type of object and the name of the object, respectively.
2. **Menu bar:** Besides options for displaying the sound amplitudes we can display various other aspects of the sound like the spectrogram, pitch, intensity, formants and pulses.
3. **The sound display area:** It shows the sample "amplitudes" as a function of time. The two numbers at the left near the top and the bottom line of the sound display area show the maximum and the minimum sound amplitudes, respectively.

4. **Rows of rectangles:** It represents parts of the sound with their durations. Five rectangles are displayed and each one when clicked plays a different part of the sound.

   In case of a selection which is indicated in the display area, a rectangle also appears above the sound display area. In this rectangle the duration and the inverse duration are displayed.

5. **Global selections:** After clicking the "all" button complete signal is displayed in the display area. The "in", "out" or "bak" button functions as zoom in or zoom out or back. By successively clicking "in" you will see ever more detail of the sound. This goes on and on until the duration of the display area becomes smaller than the sample time. In this case there will be no signal drawn anymore. By successively clicking "out" you will get less details. This goes on until the complete sound can be drawn within the display area. Further zooming out is not possible.

6. **Cursor:** The cursor position is indicated by a vertical red dashed line. Its time position is written above the display area.

7. **Horizontal scrollbar:** We can either drag the scrollbar or click in the part not covered by it to change the part of the sound that is displayed.

8. **Grouping feature:** It only applies if you have more than one time-based editor open on the same sound. For example, if you make a selection in one of the grouped editors, the selection is replicated in all the editors. If you scroll in one of them, all other editors scroll along.

9. A number will display the average value of the sound part in this window.

13.5 SOUND ANALYSIS IN PRAAT

The fundamental limitation of the analysis tool is that the speech is an unpredictable signal. The practical solution is to model the speech signal as a slowly varying function of time. By slowly varying we mean that during intervals of say 5 to 25 ms the speech characteristics hopefully don't change too much and we consider them to be almost constant. The parts of the speech signal where this model holds best is for monophthong vowels. This is because the articulators do not move as fast here as in other parts of speech (for example, during the release of a plosive the articulators move faster).

Therefore, to analyze a speech sound we cut up the sound into small segments and analyze each interval separately and pretend it has constant characteristics. The analysis of a whole sound is split up and becomes a series of analyses on successive intervals. Such intervals are termed analysis intervals and all sound analyses give the opportunity to set the duration of the ***analysis intervals***.

In the upper left part, the first part of a sound signal and in the upper right part the first part of the analysis results. The next lines visualize the cutting up of the sound into successive segments. Each segment is analyzed in the rectangular block labelled "Analysis". The results of the analysis are stored in an analysis frame and the analysis frame is stored in the output object. What happens in the "Analysis" block depends on the particular type of analysis and the contents of the analysis frames also depends on the analysis. For example, a pitch analysis will store pitch candidates and a formant analysis will store formants. Before the analysis the following three parameters have to be specified:

1. **Window length** is the duration of such a segment. This duration will hold for all the segments in the sound. For pitch measurements window length is derived from the desired lowest pitch.

   There is no one optimal window length that fits all circumstances as it depends on the type of analysis and the type of signal being analyzed. For example, to make spectrograms one often chooses either 5 ms for a wideband spectrogram or 40 ms for small band spectrogram. For pitch analysis one often chooses the lowest frequency of 75 Hz and this implies a window length of 40 ms. If you want to measure lower pitches the window length has to increase.

2. **Time step** determines the amount of overlap between successive segments. If the time step is much smaller than the window length we have more overlap. If time step is larger than the window length, we have no overlap at all. In general, we like to have at least 50% overlap between two succeeding frames and we will choose a time step smaller than half the window length.

3. **Window shape:** This parameter determines how a segment will be cut from the sound. With the help of windowing function can make the sound segment's amplitudes to start and end smoothly.

Now the analysis algorithm scheme can be summarized as follows:

1. Determine the values of the three parameters windowLength, timeStep and windowShape from values specified by the user. The duration of the sound and the values of these three parameters together determine how many times we have to perform the analysis on separate segments. This number therefore equals the number of times we have to store an analysis frame, we call it *number of frames*.
2. Calculate the midpoint (t_0) of the first window.
3. Copy the windowed sound segment, centred at t_0 with duration window length from the sound and analyse it.

4. Store the analysis frame in the output object.
5. If this is the last frame then stop. Else calculate a new $t_0 = t_0$ +time step and start again at 3. Once the analysis has finished, a new object will appear in the list of objects, for example, pitch or formant or spectrogram.

13.6 SPEECH ANALYSIS WITH PRAAT SOFTWARE (GENERAL)

The following steps will give a clear picture of speech analysis.

Step 1: Run the Praat software then two windows, Praat objects and Praat picture are displayed.

Step 2: In Praat object window, →new→record mono sound then recording window. Then record the sound and save it. Close the window.

Step 3: In Praat object window go to object list → select the recorded sound → view and edit option. Then recorded sound wave and spectrum window are displayed. Then select the recorded sound portion only.

Step 4: From the previous window go to the select option and click on it. Then only the selected recorded sound will be displayed.

Step 5: Selection of window in the analysis of spectrogram: Then go to spectrogram → spectrogram settings then will get option to select the window length.

Step 6: Select window length of short duration, then go to Pitch→Get pitch→mean Pitch value is displayed. Select widow length of long duration, then go to Formant→First formant→First formant frequency is displayed. Similarly, get the second, and third formant frequencies.

Step 7: In the above window, go to Spectrum →extract visible spectrogram. Then the spectrogram of the sound gets listed in the object window.

Step 8: Select the portion of the sound, go to sel and get clear visible waveform.

Step 9: Go to file →draw visible sound then the selected portion of sound wave is seen in the picture window.

Step 10: In picture window go to edit→ copy to clip board then the waveform is copied which can be pasted into word file.

Step 11: In Praat Objects window, in object list select the spectrogram and go to draw option and click on it. Then the spectrogram will display.

Step 12: In spectrogram picture window, go to margins→marks→to the left. Then marks for frequency are indicated in the picture. Then copy and paste it into the word file.

Note: After recording the sound in Sound recorder window save it as .WAV file. Use *'wavread'* command to read it in MATLAB.

13.7 EXAMPLES

VOWELS=FRONT=>I

202.58258524230666 Hz (mean pitch in SELECTION)

187.46326942876564 Hz (minimum pitch in SELECTION)

243.78332460734785 Hz (maximum pitch in SELECTION)

969.7642104444544 Hz (mean F1 in SELECTION)

2119.097146885455 Hz (mean F2 in SELECTION)

3207.8794540298572 Hz (mean F3 in SELECTION)

4427.427862184191 Hz (mean F4 in SELECTION)

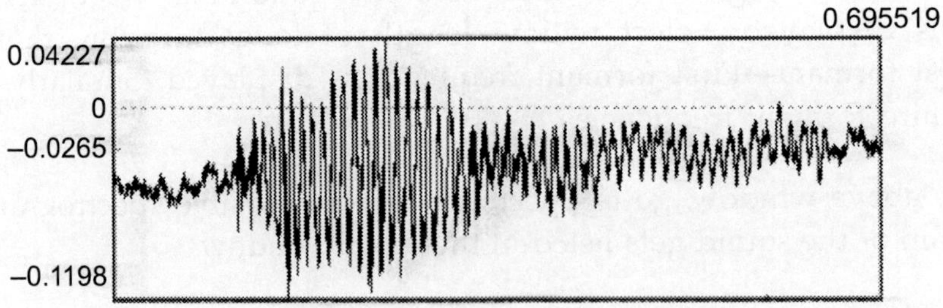

VOWELS=MID=>A

218.60557703239812 Hz (mean pitch in SELECTION)

209.5071766182528 Hz (minimum pitch in SELECTION)

247.40902931509854 Hz (maximum pitch in SELECTION)

628.6121759813302 Hz (mean F1 in SELECTION)

2019.5697218212822 Hz (mean F2 in SELECTION)

3174.8880407913416 Hz (mean F3 in SELECTION)

4108.318815754196 Hz (mean F4 in SELECTION)

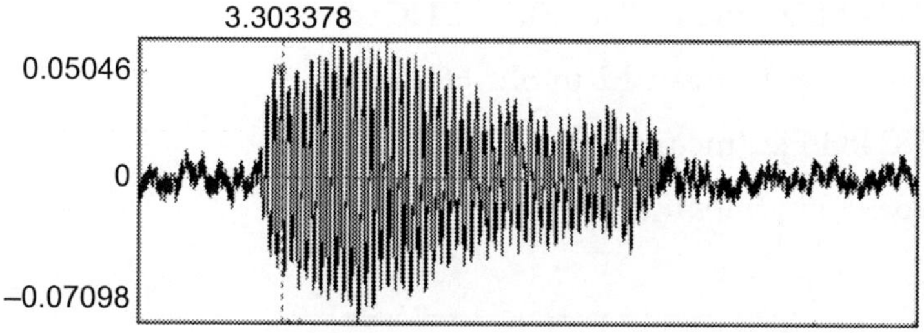

VOWELS=BACK=>O

232.36041528334982 Hz (mean pitch in SELECTION)

212.70141518594806 Hz (minimum pitch in SELECTION)

243.81553820954676 Hz (maximum pitch in SELECTION)

733.1585923326607 Hz (mean F1 in SELECTION)

1508.7311815381847 Hz (mean F2 in SELECTION)

2940.233439400575 Hz (mean F3 in SELECTION)

4063.6811121345704 Hz (mean F4 in SELECTION)

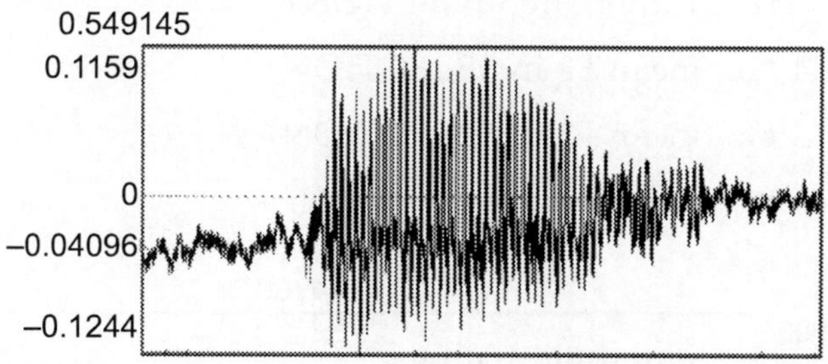

DIPTHONGS=>EI

200.17673414221252 Hz (mean pitch in SELECTION)

107.10664257322617 Hz (minimum pitch in SELECTION)

249.74801653767958 Hz (maximum pitch in SELECTION)

645.0774498791449 Hz (mean F1 in SELECTION)

2225.1231708716277 Hz (mean F2 in SELECTION)

3127.4972644187483 Hz (mean F3 in SELECTION)

4096.221782526393 Hz (mean F4 in SELECTION)

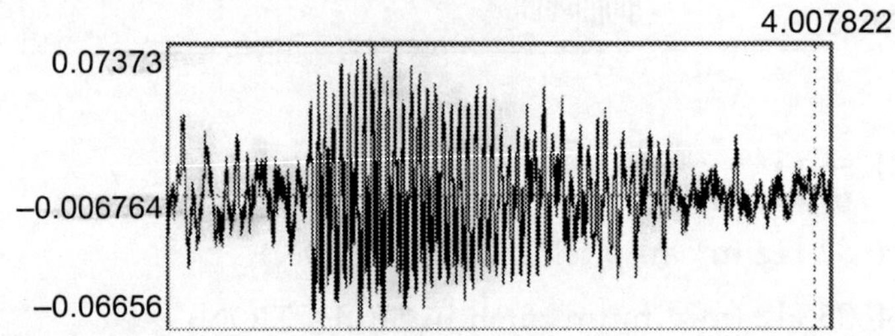

SEMIVOWELS=>LIQUIDS=>L

228.1860849074359 Hz (mean pitch in SELECTION)

207.8823076477542 Hz (minimum pitch in SELECTION)

236.96631972377565 Hz (maximum pitch in SELECTION)

579.8342708962033 Hz (mean F1 in SELECTION)

1750.2232514583334 Hz (mean F2 in SELECTION)

2868.2416430420612 Hz (mean F3 in SELECTION)

4021.381486623768 Hz (mean F4 in SELECTION)

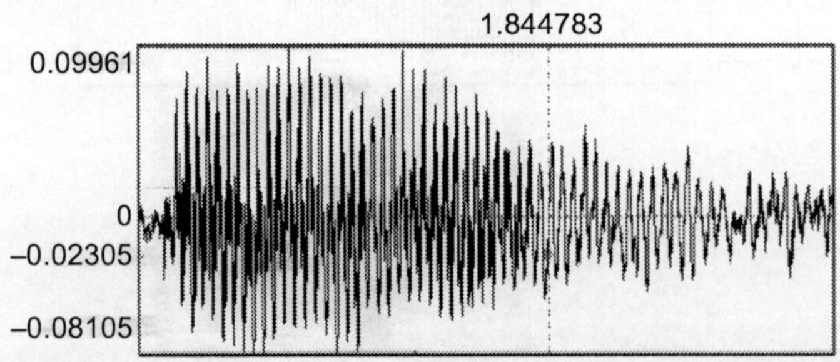

SEMIVOWELS=>GLIDES =>R

210.75901763700767 Hz (mean pitch in SELECTION)

202.67649806898135 Hz (minimum pitch in SELECTION)

223.25595063585737 Hz (maximum pitch in SELECTION)

922.6810133043324 Hz (mean F1 in SELECTION)

1579.1176227793087 Hz (mean F2 in SELECTION)

2631.7187706882683 Hz (mean F3 in SELECTION)

4003.2608128384877 Hz (mean F4 in SELECTION)

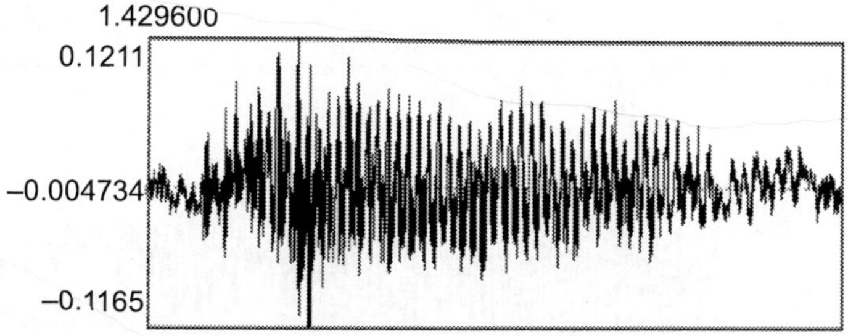

CONSONANTS=>NASAL=>M

228.91548707264755 Hz (mean pitch in SELECTION)

96.15491140984047 Hz (minimum pitch in SELECTION)

255.37489793262768 Hz (maximum pitch in SELECTION)

686.3567155907301 Hz (mean F1 in SELECTION)

1878.6348149329958 Hz (mean F2 in SELECTION)

2949.9401094388368 Hz (mean F3 in SELECTION)

4014.432378375391 Hz (mean F4 in SELECTION)

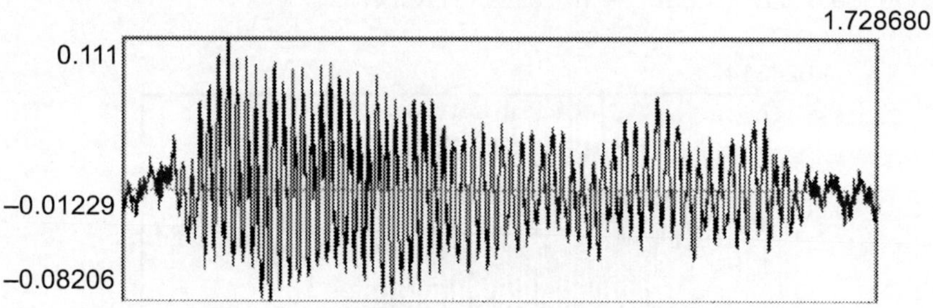

CONSONANTS=>STOPS=>VOICED=>B

229.53422059780866 Hz (mean pitch in SELECTION)

214.05375448357645 Hz (minimum pitch in SELECTION)

255.37211011873566 Hz (maximum pitch in SELECTION)

669.9869170193003 Hz (mean F1 in SELECTION)

1832.6238088972052 Hz (mean F2 in SELECTION)

3002.747949375926 Hz (mean F3 in SELECTION)

3970.3414950786346 Hz (mean F4 in SELECTION)

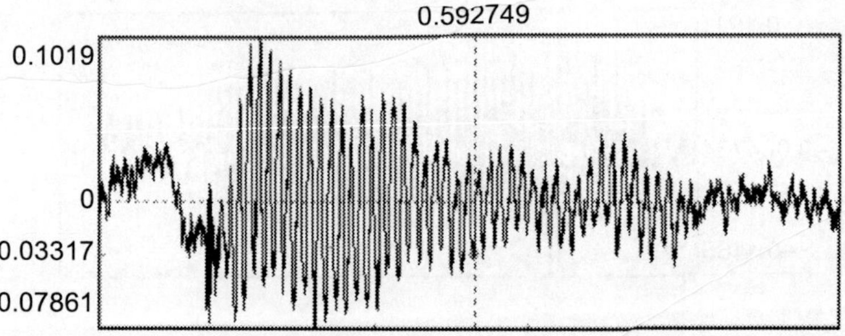

CONSONANTS=>STOPS=>UNVOICED=>P

240.48084467203572 Hz (mean pitch in SELECTION)

212.88349898654218 Hz (minimum pitch in SELECTION)

258.05565246594443 Hz (maximum pitch in SELECTION)

526.3286126055336 Hz (mean F1 in SELECTION)

1800.1409643736079 Hz (mean F2 in SELECTION)

3020.6574968578493 Hz (mean F3 in SELECTION)

3905.700235683097 Hz (mean F4 in SELECTION)

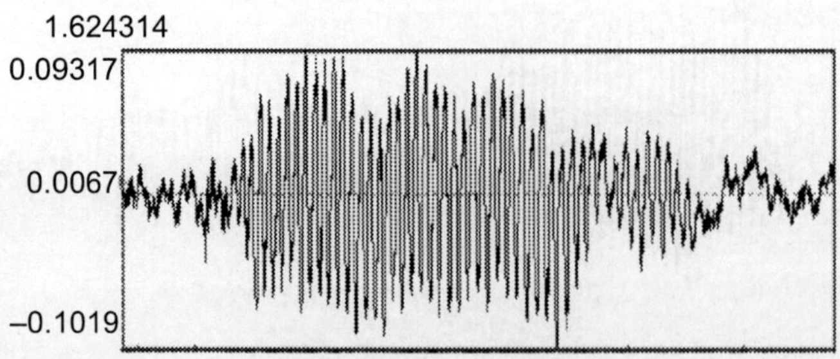

CONSONANTS=>FRICATIVES=>VOICED=>Z

200.14250076855333 Hz (mean pitch in SELECTION)

82.39220272138842 Hz (minimum pitch in SELECTION)

243.28662234367465 Hz (maximum pitch in SELECTION)

803.7392144551764 Hz (mean F1 in SELECTION)

1779.2408584904938 Hz (mean F2 in SELECTION)

2817.373303759259 Hz (mean F3 in SELECTION)

4047.9341975121934 Hz (mean F4 in SELECTION)

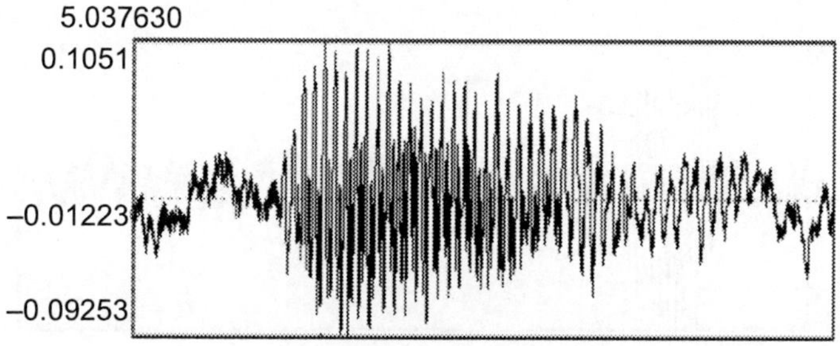

CONSONANTS=>FRICATIVES=>UNVOICED=>F

223.25942656295837 Hz (mean pitch in SELECTION)

78.58917258496469 Hz (minimum pitch in SELECTION)

249.81366201311477 Hz (maximum pitch in SELECTION)

784.609809991432 Hz (mean F1 in SELECTION)

1784.8418265068112 Hz (mean F2 in SELECTION)

2874.281875636057 Hz (mean F3 in SELECTION)

3995.081623758115 Hz (mean F4 in SELECTION)

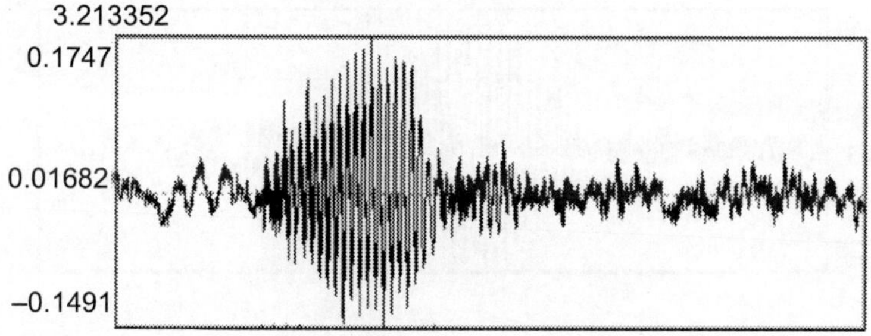

CONSONANTS=>WHISPER=>H

194.7598115920048 Hz (mean pitch in SELECTION)

81.99042683713762 Hz (minimum pitch in SELECTION)

254.63747824587458 Hz (maximum pitch in SELECTION)

980.3059197426745 Hz (mean F1 in SELECTION)

2203.5487961469194 Hz (mean F2 in SELECTION)

3304.1119462664888 Hz (mean F3 in SELECTION)

4281.708167672829 Hz (mean F4 in SELECTION)

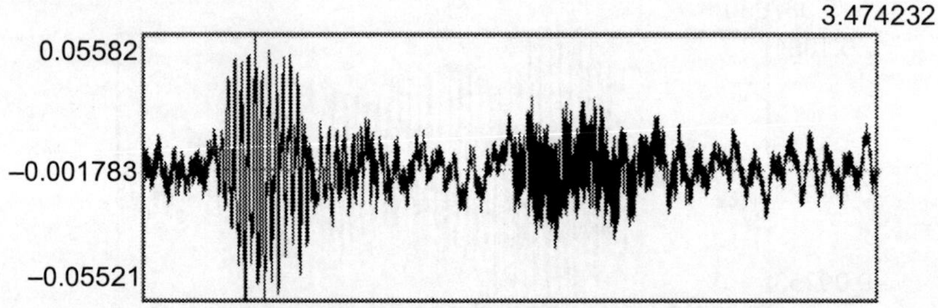

CONSONANTS=>AFFRICATES=>J

237.62663612135745 Hz (mean pitch in SELECTION)

221.03344284002642 Hz (minimum pitch in SELECTION)

273.16928889431296 Hz (maximum pitch in SELECTION)

772.3223180539234 Hz (mean F1 in SELECTION)

1977.2843648588594 Hz (mean F2 in SELECTION)

3020.3908132874244 Hz (mean F3 in SELECTION)

3974.00558003694 Hz (mean F4 in SELECTION)

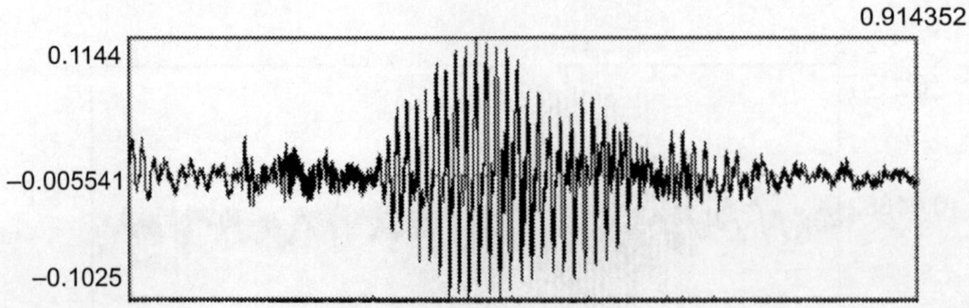

13.8 SOUND EDITOR IN WAVESURFER

WaveSurfer is an audio editor widely used for studies of acoustic phonetics. It is a simple but fairly powerful program for interactive display of sound pressure waveforms, spectral sections, spectrograms, pitch tracks and transcriptions. It can read and write a number of transcription file formats used in industrial speech research. WaveSurfer is an open source tool for sound visualization and manipulation. Typical applications are speech/sound analysis and sound annotation/transcription. WaveSurfer may be extended by plug-ins as well as embedded in other applications.

WaveSurfer has a simple but powerful interface. The basic document you work with is a *sound*. When WaveSurfer is first started, it contains an empty sound. You can load a sound file from disk or start recording, using the tape recorder like controls in the upper right corner. You can add any number of sounds (using *File* | *New* or File | *Open*). Each sound has a small arrow-shaped button in the upper left corner, that can be used to collapse the view of the sound into only a blue titlebar.

WaveSurfer is a **free software**, distributed under a permissive free software licence. www.speech.kth.se/wavesurfer/

Features

- Customizable – users can create their own configurations. Localization support.
- Basic audio editing operations, such as excision, copying, pasting, zero-crossing adjustment.
- Effects such as fading, normalization, echo, inversion, reversal, replacement with silence.
- Extensible – new functionality can be added through a plug-in architecture.
- Embeddable – WaveSurfer can be used as a widget in custom applications.
- Transcription file formats – reads, and writes HTK (and MLF), TIMIT, ESPS/Waves+, and Phondat. Support for encodings and Unicode.
- Multi-platform – Microsoft Windows, Mac OS X, Linux, Solaris, HP-UX, FreeBSD, and IRIX.

Popup Menus

Right-clicking on the different parts of the interface brings up a popup menu with choices that are applicable to that part. (Use Ctrl-Mousebutton on the Macintosh.)

WaveBar

The WaveBar is used to navigate in a sound file. The left mouse button is used for scrolling and the middle mouse button (or shift-left button) is used for zooming. If your mouse is equipped with a *scrollwheel*, you can use this for scrolling and zooming in the sound under the mouse pointer: Spinning the wheel up & down will scroll, pressing Ctrl and spinning the wheel will zoom.

Play/Record

Use the tape recorder controls to play and record audio. If a current selection exists only that segment is played. You can use the space bar to toggle between play and pause. Right-clicking on the play button brings up a popup menu with different play options.

Adding Panes

For more sophisticated tasks, a single waveform view of the whole sound might not be sufficient. You might want to zoom in to look at small portions, you might want other representations such as spectrograms or pitch-curves, or maybe you have a multichannel sound and want to look at each channel individually. To allow for this and many other tasks, WaveSurfer gives you the possibility of adding *panes*. A pane is a window stacked on top of the WaveBar that contains waveform, spectrogram, pitch-curve, time axis, or transcription or something else. (In fact, through the use of WaveSurfer's plug-in architecture, a pane can contain pretty much anything you can think of!)

To add a new pane, you right-click on the WaveBar or any existing pane where you want to insert the new pane, select ***Create Pane***, and choose what type of pane you want.

Unlike the WaveBar, a pane will not necessarily display the whole sound. Rather it will display a portion of the sound that is specified in the WaveBar. Think of the WaveBar as an overview and the pane as a variable magnifying glass.

A sound can contain any number of panes (as long as they fit on the screen) They will be stacked on top of each other and always be aligned in time.

Once the pane is created it can be resized vertically by dragging the handle below the pane. (Some panes such as ***transcription*** and ***time axis*** cannot be resized.) You can also control many properties of the pane by selecting *Properties* | *Pane* from the popup menu (right-click).

Pane Properties Dialog

It is possible to modify various properties of a pane by selecting *Popup* | *Properties*. Depending on the specific type of pane, there will be different tabs in the dialog. For example, it will allow you to modify the color scale and analysis parameters of a spectrogram.

Configurations

For a certain task, you might want a specific set-up of panes, with specific properties. WaveSurfer comes with a set of ready-to-use pane configurations, for transcription, speech analysis, etc. You can easily define your own configurations. Once you have created a set of panes and set their properties as needed, you simply choose *Save Configuration* from the popup menu. Once the configuration is saved, it will appear as a choice when you select *File* | *New*. It is possible to change the pane configuration used for a sound by selecting *Popup* | *Apply Configuration*. A configuration does not only specify what panes should be visible, but also their sizes and everything else that can be specified in the *Popup* | *Properties* dialog. The configuration files reside in ~/.wavesurfer/1.8/configurations/ and under some circumstances it might be useful to edit them by hand. For example, the configuration *n-waveforms* have been created in this way. Examine that configuration in a text editor to get an idea of how this can be done.

Selection

All panes share the current selection. Selections are made in the standard way, press-drag-release mouse left mouse button. The selection can be adjusted by dragging the bars marking the end points or by pressing shift and clicking mouse left mouse button. Clicking the play button causes the marked region to be played. The end points of the current selection can be adjusted to the nearest zero crossing using *Edit* | *Zero Cross Adjust*.

Basic Editing and Processing

The *Edit* menu contains basic editing functions such as cut, copy, and paste. The *Transform* menu contains transformation functions for conversions, amplification, normalization, etc. On the *View* menu zoom functions can be found. The *Master Sound* option makes it possible for one window to control the zoom-level and scroll position of other windows.

Sound Conversion

You can convert properties such as frequency, number of channels, and sample encoding format using the dialog in *Popup* | *Properties* | *Sound*.

Printing

WaveSurfer creates postscript files for printing.

Preferences Dialog

The preferences dialog (*File* | *Preferences*...) is used to specify global properties of WaveSurfer. These include:

- commands used to print and preview
- key-bindings
- interpretation of raw sound data from filename extension
- method of sound storage management
- maximum recording time
- automatic scroll during playback
- temporary work directory
- maximum zoom-in level

Mixer dialog (Unix versions)

The mixer dialog (*File* | *Mixer*...) is used to control volume and select output/input jacks.

Sound Files

WaveSurfer can read a number of sound file formats including WAV, AU, AIFF, MP3, CSL, and SD. It can also save files in several formats, including WAV, AU, and AIFF. There are separate plug-ins to handle Ogg/Vorbis and NIST/Sphere files. For raw sound files (headerless PCM, etc.) WaveSurfer tries to guess some properties and displays a dialog giving the user a chance to modify these. It is possible to associate the filename extension with these settings for future use. WaveSurfer reads files by analyzing their headers regardless of filename extension. When saving files, the file format to use is decided from the filename extension specified, e.g., to save a file in AU format specify a filename like myfile.au.

Sound Analysis

WaveSurfer can be used to visualize and analyze sound in several ways. The standard analysis plug-in can display Waveform, Spectrogram, Pitch, Power or Formant panes.

Many properties of these panes can be adjusted using *Popup* | *Properties*. In addition, a spectrum window can be opened using *Popup* | *Spectrum Section*. Special control windows are available for waveforms and spectrograms, which allows the user to make quick modifications. A special enlarged waveform window, which is centred on the cursor, can be opened using *Popup* | *Waveform Blow-up*. This is useful for detailed inspection and adjustment.

Plotting Time Aligned Data

The visualization of data related to a sound file is handled by the dataplot plug-in. Pitch, power and formants are examples of data that this plug-in can be used to plot. It can also be used to visualize other time aligned data, for example, output from other programs. The plug-in plots tabulated numerical ASCII data. The data values should be contained in a text file, formatted in rows with one or more columns to be plotted. One row for each point in time, starting with the first row typically at 0.0 seconds and equally spaced according to the frame interval option. The name of this text file should be the same as the sound file except for the extension. The default extension is .txt, which can be changed in the property pane. Optionally, a spectrogram or a waveform can be drawn as a backdrop. The data can be plotted either as continuous curves or using dots. For large number of data points the dots style might be slow. The data values can be edited by simply dragging them with the mouse.

Transcription

WaveSurfer has many facilities for transcribing sound files. Transcription is handled by a dedicated plug-in and it's associated pane type.

Use the properties, dialog, to specify which label file that should be displayed in a transcription pane. It is usually practical to create a special configuration for a certain combination of sound and transcription files, specifying file properties such as filename extension, label format, and character encoding. There are many options to control appearance and editing functionality.

Depending on the transcription file format additional options might be available. Unicode characters are supported if using the source version of WaveSurfer, in order to keep the binary versions small. See the Tcl language documentation for a list of available character encodings. The transcription plug-in is used in combination with format handler plug-in which handles the conversion between file formats and the internal format used by the transcription plug-in.

The standard popup menu has additional entries for transcription panes. *Popup | Load Transcription* and *Popup | Save Transcription* are used to load and save transcription files. Label editing is straightforward, simply click where you want to insert a label and type it in. The label fields are user-configurable and used to insert a label directly at the cursor position. You can either drag time boundaries using the mouse or use Ctrl-r/Ctrl-l to right/left justify boundaries with the cursor. It is possible to specify where WaveSurfer should look for transcription files using *Properties | Trans1 | Label file path* entry, or *Properties | HTK | Master Label File*. The latter is used in the case of HTK transcriptions contained in an HTK MLF-file with immediate definitions. HTK label files need to have both start and end times specified. Choosing *File* | *Save* will save a modified transcription in addition to any sound changes.

Command Line Switches

WaveSurfer recognizes the following command line switches

-play

Start playing as soon as the sound is loaded.

-config *configuration*

Use the *configuration* and do not query the user for one. WaveSurfer uses the first matching configuration it can find. If no one matches, a wavebar will be displayed.

-filelist *file*

Load the file list contained in *file* (text file with one file name per line). This list will be displayed in the Chooser dialog for quick access in large file sets.

-master *filename*

When loading several sound files at once, specify which sound's wavebar will control the rest. For example, **wavesurfer.tcl -config Waveform -master ex1.wav ex1. wav ex2.wav**, will make the first sound's wavebar control the second one as well. Same function can be done by choosing *View* | *Master Sound* | *ex1.wav.*

-tfa *filename*

Search the specified WaveSurfer Transcription File Archive. Several options can be specified and they will apply to the corresponding transcription pane counting from the top.

-mlf *filename*

Search the specified HTK Master Label File (only for HTK format transcription panes). Several options can be specified and they will apply to the corresponding transcription pane counting from the top.

-filelist *file*

Load the file list contained in *file* (text file with one file name per line). This list will be displayed in the Chooser dialog for quick access in large file sets.

-debug *level*

Generate a debug trace which is useful when trying to track down bugs. Use with a value of 6 for the *level* parameter in order to generate the trace file *wsurf.log*.

Site Wide Installation

WaveSurfer can be installed on a central server. Several users can share configurations and plug-ins by copying them from each other and installing them in ~/.wavesurfer/1.8/configurations/ or ~/.wavesurfer/1.8/plugins/. From a system administration point of view a better solution is to keep such files in common directories and to use the environment variables WSCONFIGDIR and WSPLUGINDIR to point to these. Windows administrators can use the registry entry HKEY_LOCAL_MACHINE\SOFTWARE\CTT\WAVESURFER\1.8 and set the keys WSCONFIGDIR and WSPLUGINDIR.

Large Sound Files

Very large sound files can be handled by choosing "keep on disk" in *Preferences* | *Miscellaneous* | *Sound storage*. This choice will make WaveSurfer access sound data on disk as needed instead of loading them all into memory. The first time a file is opened it is scanned and its waveform shape is computed and stored on disk. Next time the file is opened, WaveSurfer will only load this pre-computed file, which gives almost instantaneous access. Note that currently WaveSurfer cannot edit sound files in this mode.

Key Bindings

It is possible to customize most key/event bindings in the preferences and pane properties dialogs. Here are some examples of how such event bindings might look

- ButtonPress-3
- Shift-Control-ButtonPress-1
- Control-space
- F1
- Alt-a

13.9 SUMMARY

This chapter discusses software tools used to record and analyze various speech signal features quantitatively and qualitatively.

Programs

PROGRAM 1

The input sequence x(n)=(1,2,3,4) is applied to DSP system, with unit sample response h(n)=(2,3). What is the resulting output sequence?

```
clc;
clear all;
close all;
xn=input('Enter the input sequence:');
hn=input('Enter the system response:');
yn1=conv(xn,hn) %Convolution result
yn2=conv(hn,xn) %Convolution result
figure;
stem(yn1);
title('Convolution result of yn1');
xlabel('n samples');
ylabel('amplitude');
figure;
stem(yn2);
title('Convolution result of yn2');
xlabel('n samples');
ylabel('amplitude');
```

OUTPUT

```
Enter the input sequence:[1 2 3 4]
Enter the system response:[2 3]
```

```
yn1 =

     2     7    12    17    12

yn2 =

     2     7    12    17    12
```

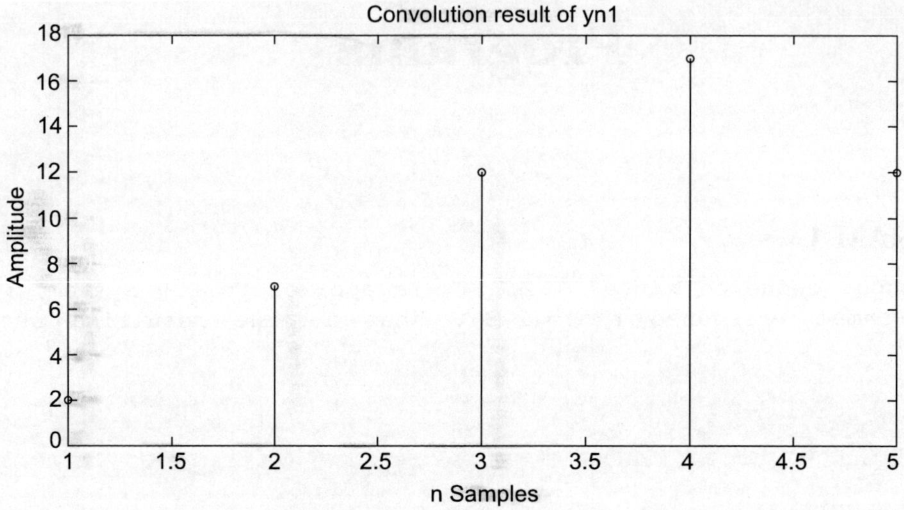

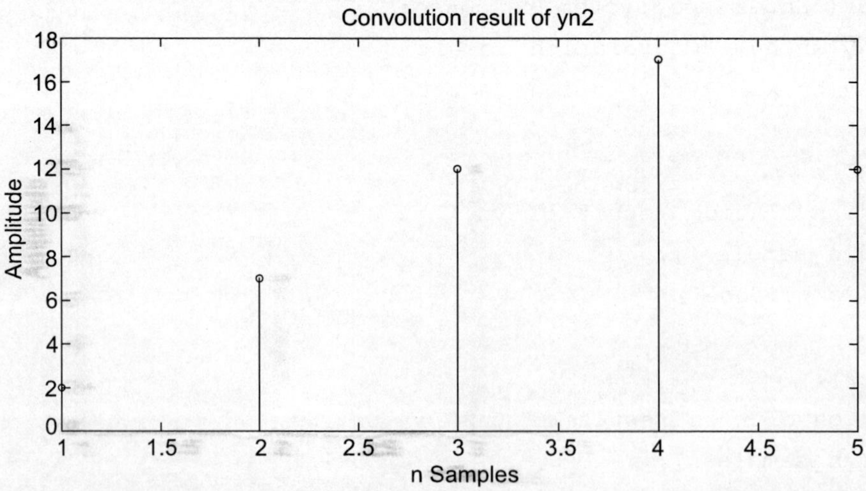

PROGRAM 2

The input sequence x(n)=(1,1,-2,1,1) is applied to a DSP system, with unit sample response h(n)=(1,4). What is the resulting output sequence of the processor?

```
clc;
clear all;
close all;
xn=input('Enter the input sequence:');
hn=input('Enter the system response:');
yn=conv(xn,hn) %Convolution result
figure;
stem(yn);
title('Convolution result');
xlabel('n samples');
ylabel('amplitude');
```

OUTPUT

```
Enter the input sequence:[1 1 -2 1 1]
Enter the system response:[1 4]
yn =
      1       5       2      -7       5       4
```

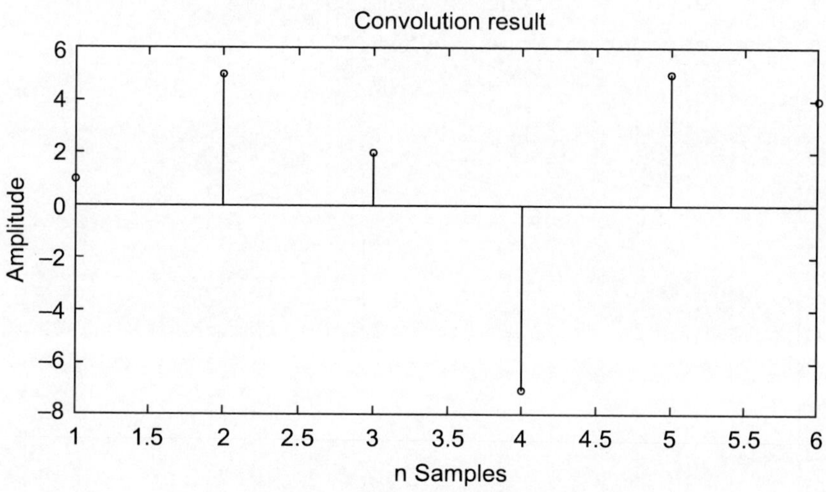

PROGRAM 3

The input sequence x(n)=(3,1,2) is applied to a DSP system, having two subsystems with unit sample response hn1 = (5,8,1) and hn2 = (1,1) connected in cascade. What is the overall impulse response and resulting output sequence of the processor?

```
clc;
clear all;
close all;
xn=input('Enter the input sequence:');
hn1=input('Enter the 1st response of the system:');
hn2=input('Enter the 2nd response of the system:');
hn=conv(hn1,hn2)
yn=conv(xn,hn)
figure;
%stem(yn);
%title('Convolution result of yn');
%xlabel('n samples');
%ylabel('amplitude');
```

OUTPUT

```
Enter the input sequence:[3 1 2]
Enter the 1st response of the system:[5 8 1]
Enter the 2nd response of the system:[1 1]
hn =
     5    13     9     1
yn =
    15    44    50    38    19     2
```

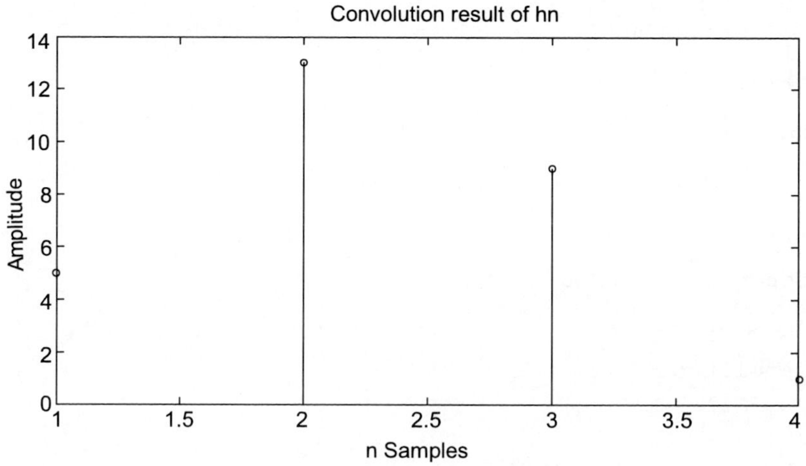

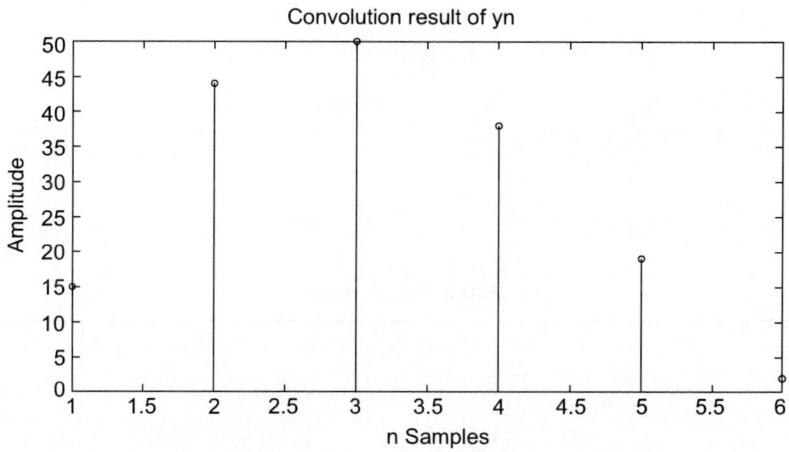

PROGRAM 4

The input sequence x(n)=(1,2,4,8) is applied to a DSP system, having two subsystems with unit sample response hn1 = (1,1,1)and hn2 = (2,2,2) connected in parallel. What is the overall impulse response and resulting output sequence of the processor?

```
clc;

clear all;

close all;

xn=input('Enter the input sequence:');

hn1=input('Enter the 1st response of the system:');
```

```
hn2=input('Enter the 2nd response of the system:');
hn=hn1+hn2
yn=conv(xn,hn)
%figure;
%stem(yn);
%title('Convolution result');
%xlabel('n samples');
%ylabel('amplitude');
```

OUTPUT

```
Enter the input sequence:[1 2 4 8]
Enter the 1st response of the system:[1 1 1]
Enter the 2nd response of the system:[2 2 2]
hn =
     3     3     3
yn =
     3     9    21    42    36    24
```

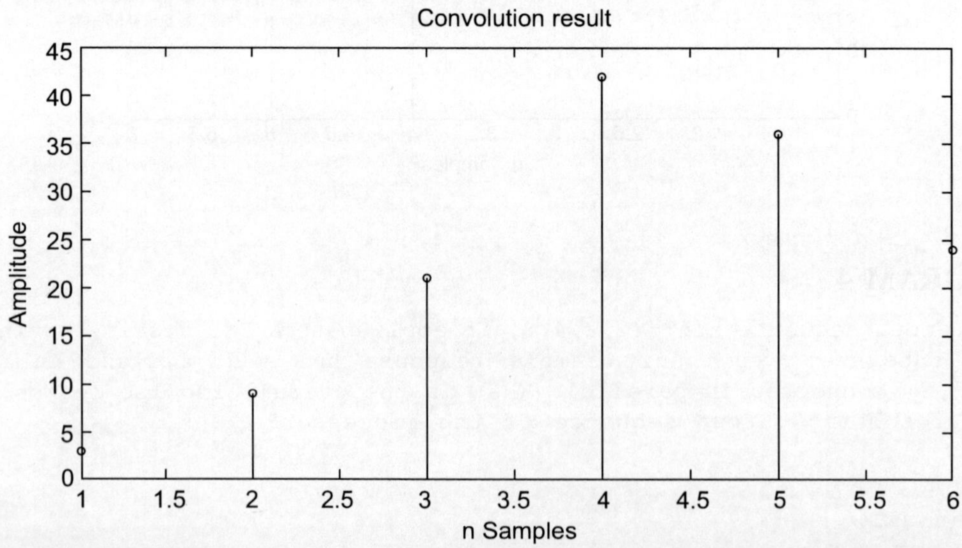

PROGRAM 5

For input sequence x(n)=(1,2,3,4) is applied to a discrete time processor with output y(n)=(2,7,12,17,12). What is the system response h(n)?

```
clc;
clear all;
close all;
xn=input('Enter the input sequence:');
yn=input('Enter the output sequence:');
hn=deconv(yn,xn) %Deconvolution result
%figure;
%stem(hn);
%title('Deconvolution result');
%xlabel('n samples');
%ylabel('amplitude');
```

OUTPUT

```
Enter the input sequence:[1 2 3 4]
Enter the output sequence:[2 7 12 17 12]
hn =
     2     3
```

PROGRAM 6

For input sequence x(n)=(1,1,-2,1,1) is applied to a discrete time processor with output y(n)=(1,5,2,-7,5,4). What is the system response h(n)?

```
clc;
clear all;
close all;
xn=input('Enter the input sequence:');
yn=input('Enter the output sequence:');
hn=deconv(yn,xn) %Deconvolution result
%figure;
%stem(hn);
%title('Deconvolution result');
%xlabel('n samples');
%ylabel('amplitude');
```

OUTPUT

```
Enter the input sequence:[1 1 -2 1 1]
Enter the output sequence:[1 5 2 -7 5 4]
hn =
     1     4
```

PROGRAM 7

Determine the correlation of sequence x(n)=(2,1,2,3) and y(n)=(2,1,2,3)

```
clc;
clear all;
close all;
xn=input('Enter input sequence:');
yn=input('Enter output sequence:');
rxy=xcorr(xn,yn)
stem(rxy);
title('Crosscorrelation');
xlabel('samples');
ylabel('amplitude');
```

OUTPUT

```
Enter input sequence:[2 1 2 3]
Enter output sequence:[2 1 2 3]
rxy =
     6     7    10    18    10     7     6
```

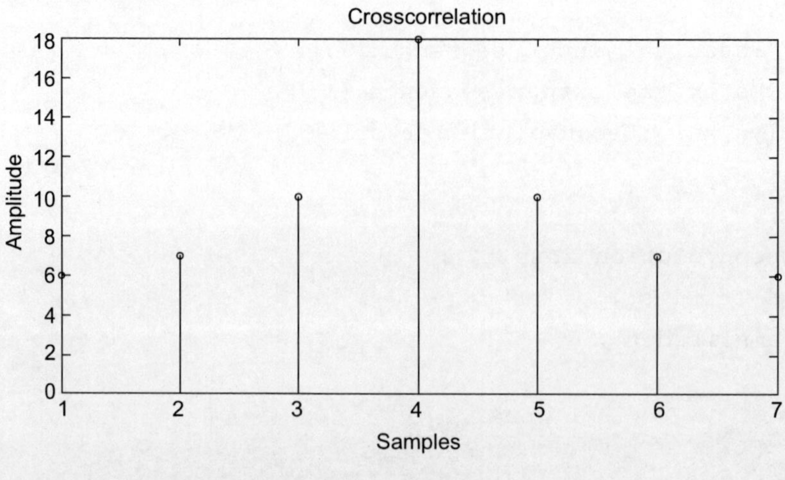

PROGRAM 8

```
Determine the correlation of sequence x(n)=(1,2,3,1) and
y(n)=(1,1,1,1)
clc;
clear all;
close all;
xn=input('Enter input sequence:');
yn=input('Enter output sequence:');
rxy=xcorr(xn,yn)
stem(rxy);
title('Crosscorrelation');
xlabel('samples');
ylabel('amplitude');
```

OUTPUT

```
Enter input sequence:[1 2 3 1]
Enter output sequence:[1 1 1 1]
rxy =
 1.0000    3.0000    6.0000    7.0000    6.0000    4.0000    1.0000
```

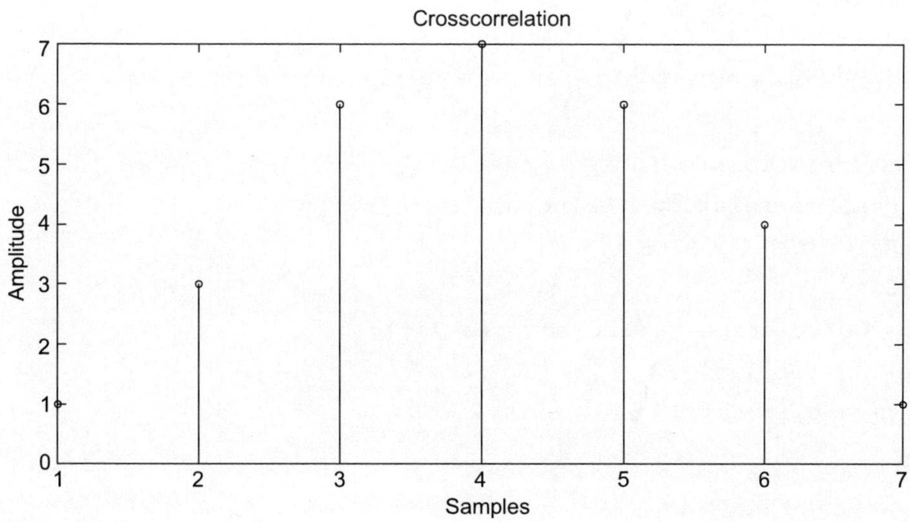

PROGRAM 9

```
(A) Compute correlation coefficient of the sequence x(n)= (4,2,3)
and y(n)=(4,2,3)
```

```
clc;
clear all;
close all;
A=input('enter 1st sequence:');
B=input('enter 2nd sequence:');
Xm=mean(A);
Ym=mean(B);
[R,C]=size(A);
N=C;
v1=0;
v2=0;
%finding the denominator
for k=1:1:N
v1=v1+(A(k)-Xm)^2;
v2=v2+(B(k)-Ym)^2;
end
DENOMINATOR=sqrt(v1*v2);
%finding the numerator
NUMERATOR=0;
for k=1:1:N
     NUMERATOR=NUMERATOR+(A(k)-Xm)*(B(k)-Ym);
end
%final correlation coefficient result
disp('final correlation coefficient result=');
Z=NUMERATOR/DENOMINATOR
%stem(Z);
%title('Correlation coefficient result');
%xlabel('n samples');
%ylabel('amplitude');
```

OUTPUT

```
enter 1st sequence:[4 2 3]
enter 2nd sequence:[4 2 3]
final correlation coefficient result=
Z =
     1
```

(B) Compute correlation coefficient of the sequence x(n)= (4,2,3) and y(n)=(4,2,3) using MATLAB Command.

```
clc;
clear all;
close all;
xn=input('Enter input sequence:');
yn=input('Enter output sequence:');
r=corrcoef(xn,yn)
stem(r);
title('Correlation coefficient');
xlabel('samples');
ylabel('amplitude');
```

OUTPUT

```
Enter input sequence:[4 2 3]
Enter output sequence:[4 2 3]
r =
      1     1
      1     1
```

PROGRAM 10

If x(n)={δ(n), δ((n-2)}.Find DFT of the periodic signal x(n) with N=4.

```
clc;
clear all;
close all;
N=input('Enter the length of the sequence:');
xn=input('Enter the sequence:');
Xk=fft(xn,N);
disp('FFT of the given sequence');
disp(Xk);
%subplot(2,1,1);
%stem(xn);
%xlabel('n samples');
%ylabel('amplitude');
%title('X(n)');
%subplot(2,1,2);
```

```
%stem(Xk);
%xlabel('n samples');
%ylabel('amplitude');
%title('FFT of the sequence');
```

OUTPUT

```
Enter the length of the sequence:4
Enter the sequence:[1 0 1 0]
FFT of the given sequence
     2     0     2     0
```

PROGRAM 11

If X(k)=(2,0,2,0), find the periodic signal x(n) with N=4.

```
clc;
clear all;
close all;
N=input('Enter the length of the sequence:');
xn=input('Enter the sequence:');
Xk=ifft(xn,N);
disp('IFFT of the given sequence');
disp(Xk);
subplot(2,1,1);
stem(xn);
xlabel('n samples');
ylabel('amplitude');
title('X(n)');
subplot(2,1,2);
stem(Xk);
xlabel('n samples');
ylabel('amplitude');
title('IFFT of the sequence');
```

OUTPUT

```
Enter the length of the sequence:4
Enter the sequence:[2 0 2 0]
IFFT of the given sequence
     1     0     1     0
```

PROGRAM 12

If x(n) = (2,1,0,1,3,1,5,4),find X(k) by using decimation in time FFT algorithm.

```
clc;
clear all;
close all;
N=input('Enter the length of the sequence:');
xn=input('Enter the sequence:');
Xk=fft(xn,N);
disp('FFT of the given sequence');
disp(Xk);
%subplot(2,1,1);
%stem(xn);
%xlabel('n samples');
%ylabel('amplitude');
%title('X(n)');
%subplot(2,1,2);
%stem(Xk);
%xlabel('n samples');
%ylabel('amplitude');
%title('FFT of the sequence');
```

OUTPUT

```
Enter the length of the sequence:8
Enter the sequence:[2 1 0 1 3 1 5 4]
FFT of the given sequence
  Columns 1 through 6
17.0000 + 0.0000i    1.1213 + 7.1213i    0.0000 + 3.0000i   -3.1213
- 2.8787i   3.0000 + 0.0000i  -3.1213 + 2.8787i
  Columns 7 through 8
   0.0000 - 3.0000i    1.1213 - 7.1213i
```

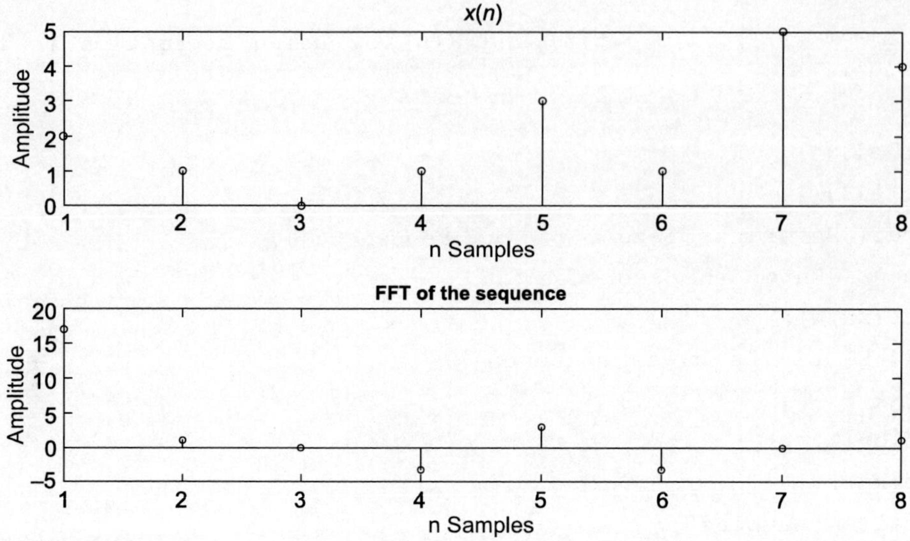

PROGRAM 13

```
If  X(k)=(17,1.12+j7.12,j3,-3.121-j2.87,3,- 3.12+j2.87,
-j3,1.12-j7.12). % Find x(n) by using decimation in time IFFT
algorithm.
clc;
clear all;
close all;
N=input('Enter the length of the sequence:');
xn=input('Enter the sequence:');
Xk=ifft(xn,N);
disp('IFFT of the given sequence');
disp(Xk);
%subplot(2,1,1);
%stem(xn);
%xlabel('n samples');
%ylabel('amplitude');
%title('X(n)');
%subplot(2,1,2);
%stem(Xk);
%xlabel('n samples');
%ylabel('amplitude');
%title('IFFT of the sequence');
```

OUTPUT

```
Enter the length of the sequence:8
Enter the sequence:[17, 1.12+7.12i, 3i, -3.121-2.87i, 3, -3.12+2.87i, -3i, 1.12-7.12i]
IFFT of the given sequence
  Columns 1 through 6
1.9999 + 0.0000i   0.9983 - 0.0001i   0.0025 + 0.0001i   0.9991 - 0.0001i   3.0001 + 0.0000i   1.0017 + 0.0001i
 Columns 7 through 8
   4.9975 - 0.0001i   4.0009 + 0.0001i
```

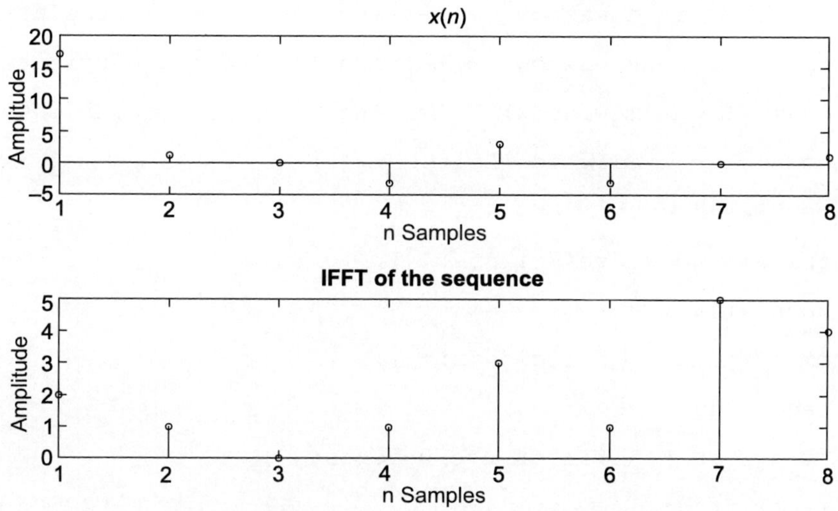

PROGRAM 14

Design Low pass FIR filter for N=5 and Wc=0.6 radian.

```
clc;
clear all;
close all;
N = input('Enter the length of the filter: '); % order of iir filter
Wc = input('Enter the cut off frequency in radians: ');% Cut off frequency
n = 0 : N-1;
```

```
tau = (N-1)/2;
eps = 0.001; % To avoid divide by zero, we use correction factor
Hd = (sin(Wc * (n - tau + eps))) ./ (pi * (n - tau + eps));
disp('For LOWPASS FIR filter-->')
disp('H(d) = ');
disp(Hd);
Wd_rect_window = (window('rectwin', N)'); % take the transpose to
bring column vector to a row vector
Wd_hann_window = (window('hann', N)'); % hanning window
Wd_hamm_window = (window('hamming', N)'); % hamming window
Hn_rect = Hd.*Wd_rect_window; % rectangular fir filter coefficients
Hn_hann = Hd.*Wd_hann_window; % Hanning fir filter coefficients
Hn_hamm = Hd.*Wd_hamm_window; % Hamming fir filter coefficients
disp('Rectangular window coefficients: ');
disp(Wd_rect_window)
disp('Hanning window coefficients: ');
disp(Wd_hann_window)
disp('Hamming window coefficients: ');
disp(Wd_hamm_window)
disp('Rectangular FIR Filter coefficients: ')
disp(Hn_rect);freqz(Hn_rect);
title('Frequency Response of  Rectangular FIR filter');
xlabel('Normalized frequency');
ylabel ('Magnitude');
disp('Hanning FIR Filter coefficients: ')
disp(Hn_hann);
figure
freqz(Hn_hann);
title('Frequency Response of  Hanning FIR filter');
xlabel('Normalized frequency');
ylabel ('Magnitude');
```

```
disp('Hamming FIR Filter coefficients: ')
disp(Hn_hamm);
figure
freqz(Hn_hamm);
title('Frequency Response of  Hamming FIR filter');
xlabel('Normalized frequency');
ylabel ('Magnitude');
disp('The end')
```

OUTPUT

```
Enter the length of the filter: 5
Enter the cut off frequency in radians: 0.6
For LOWPASS FIR filter-->

H(d) =

    0.1484    0.1798    0.1910    0.1797    0.1483

Rectangular window coefficients:

     1     1     1     1     1

Hanning window coefficients:

         0    0.5000    1.0000    0.5000         0

Hamming window coefficients:

    0.0800    0.5400    1.0000    0.5400    0.0800

Rectangular FIR Filter coefficients:

    0.1484    0.1798    0.1910    0.1797    0.1483

Hanning FIR Filter coefficients:

         0    0.0899    0.1910    0.0899         0

Hamming FIR Filter coefficients:

    0.0119    0.0971    0.1910    0.0970    0.0119

The end
```

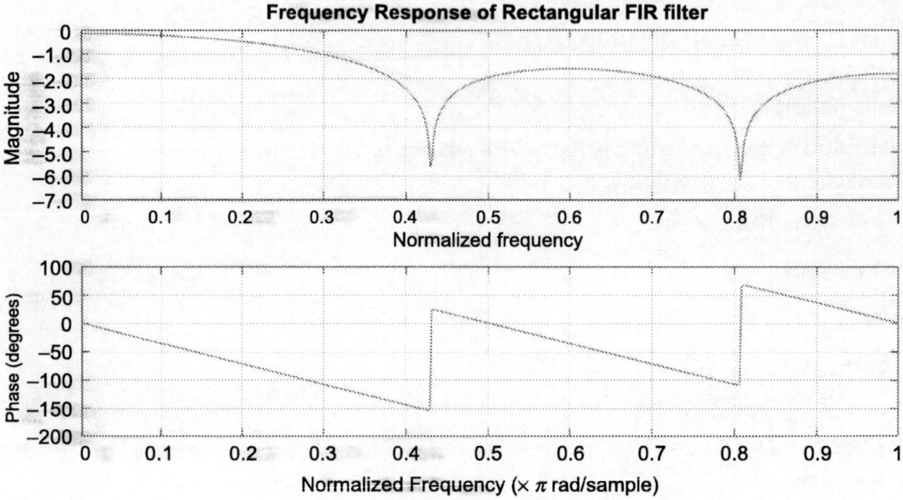
Frequency Response of Rectangular FIR filter
Magnitude
Normalized frequency
Phase (degrees)
Normalized Frequency (× π rad/sample)

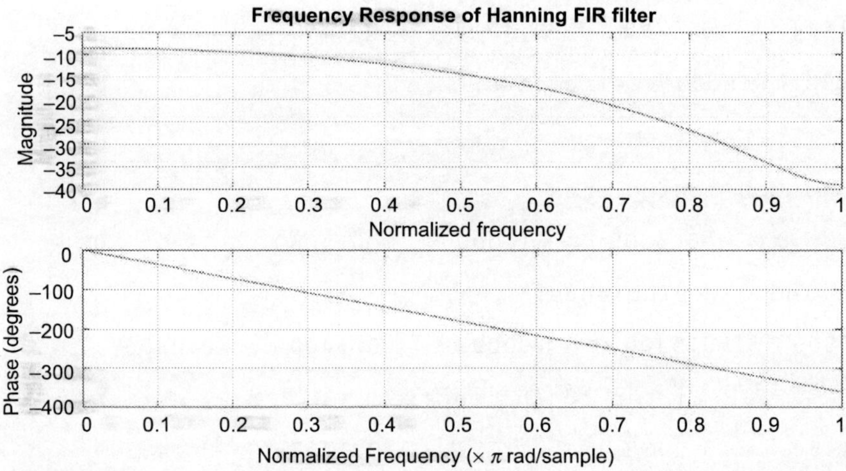
Frequency Response of Hanning FIR filter
Magnitude
Normalized frequency
Phase (degrees)
Normalized Frequency (× π rad/sample)

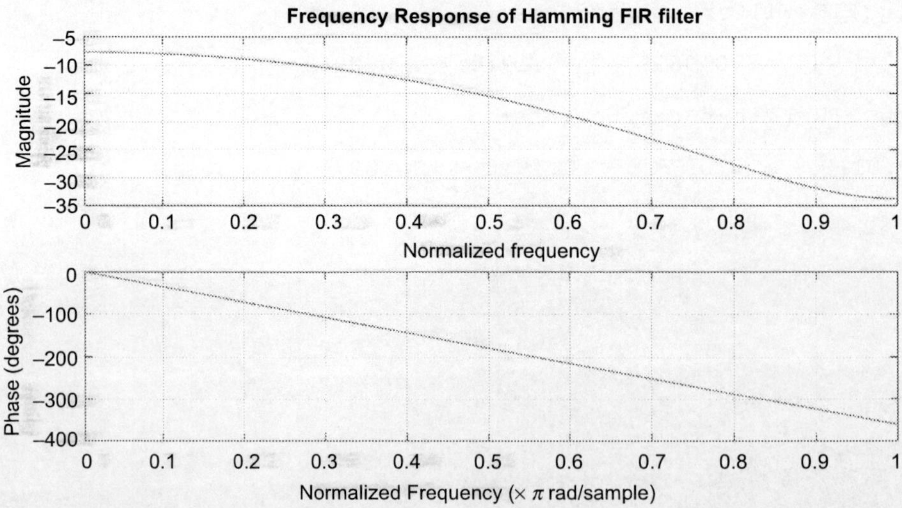
Frequency Response of Hamming FIR filter
Magnitude
Normalized frequency
Phase (degrees)
Normalized Frequency (× π rad/sample)

PROGRAM 15

Design High pass FIR filter for N=5 and Wc=0.6 radian

```
clc;
clear all;
close all;
N = input('Enter the length of the filter: '); % order of iir filter
Wc = input('Enter the cut off frequency in radians: ');% Cut off
frequency
n = 0 : N-1;
tau = (N-1)/2;
eps = 0.001; % To avoid divide by zero, we use correction factor
Hd = (sin(Wc * (n - tau + eps))) ./ (pi * (n - tau + eps));
disp('For LOWPASS FIR filter-->')
disp('H(d) = ');
disp(Hd);
Wd_rect_window = (window('rectwin', N)'); % take the transpose to
bring column vector to a row vector
Wd_hann_window = (window('hann', N)'); % hanning window
Wd_hamm_window = (window('hamming', N)'); % hamming window
Hn_rect = Hd.*Wd_rect_window; % rectangular fir filter coefficients
Hn_hann = Hd.*Wd_hann_window; % Hanning fir filter coefficients
Hn_hamm = Hd.*Wd_hamm_window; % Hamming fir filter coefficients
disp('Rectangular window coefficients: ');
disp(Wd_rect_window)
disp('Hanning window coefficients: ');
disp(Wd_hann_window)
disp('Hamming window coefficients: ');
disp(Wd_hamm_window)
disp('Rectangular FIR Filter coefficients: ')
disp(Hn_rect)
disp('Hanning FIR Filter coefficients: ')
disp(Hn_hann)
```

```
disp('Hamming FIR Filter coefficients: ')
disp(Hn_hamm)
disp('The end')
```

OUTPUT

```
Enter the length of the filter: 5
Enter the cut off frequency in radians: 0.6
For HIGHPASS FIR filter-->
H(d) =
   -0.1489   -0.1788    0.8090   -0.1807   -0.1478
Rectangular window coefficients:
     1     1     1     1     1
Hanning window coefficients:
         0    0.5000    1.0000    0.5000         0
Hamming window coefficients:
    0.0800    0.5400    1.0000    0.5400    0.0800
Rectangular FIR Filter coefficients:
   -0.1489   -0.1788    0.8090   -0.1807   -0.1478
Hanning FIR Filter coefficients:
         0   -0.0894    0.8090   -0.0904         0
Hamming FIR Filter coefficients:
   -0.0119   -0.0965    0.8090   -0.0976   -0.0118
The end
```

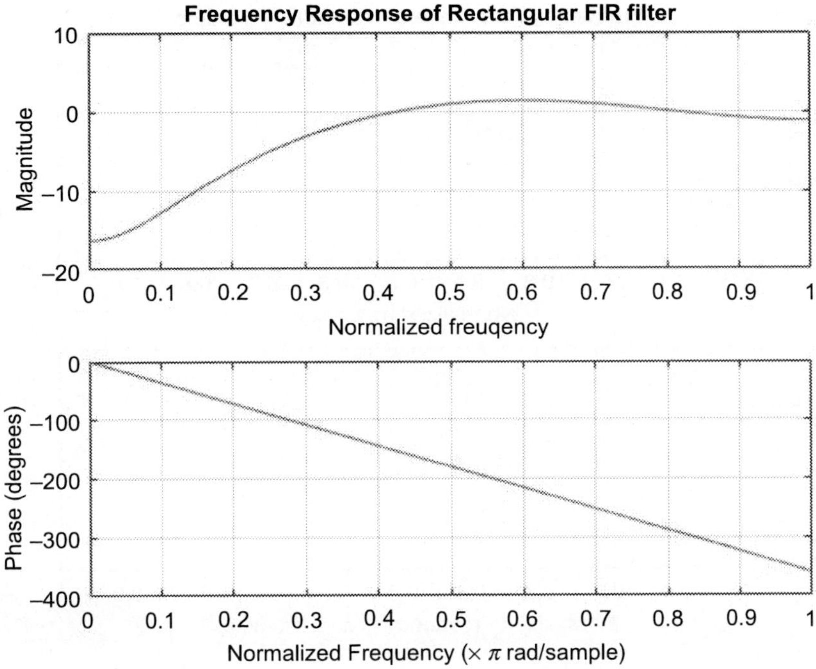
Frequency Response of Rectangular FIR filter
Magnitude
Normalized freuqency
Phase (degrees)
Normalized Frequency (× π rad/sample)

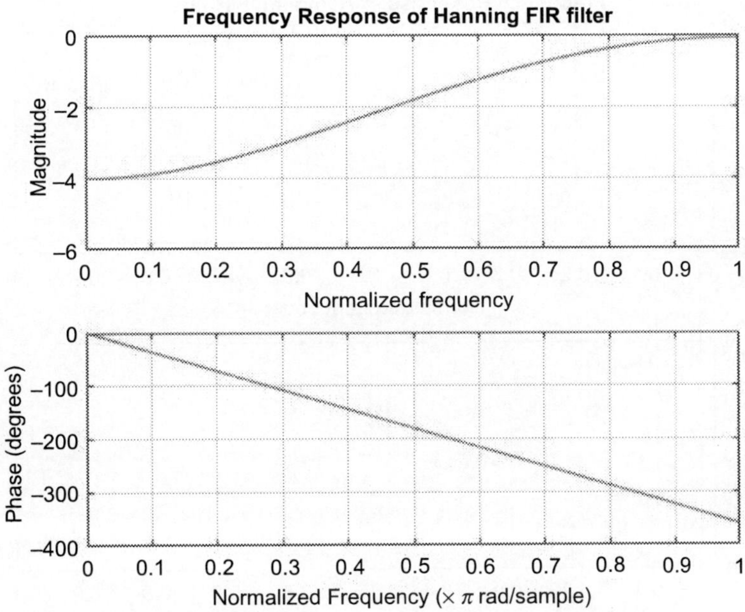
Frequency Response of Hanning FIR filter
Magnitude
Normalized frequency
Phase (degrees)
Normalized Frequency (× π rad/sample)

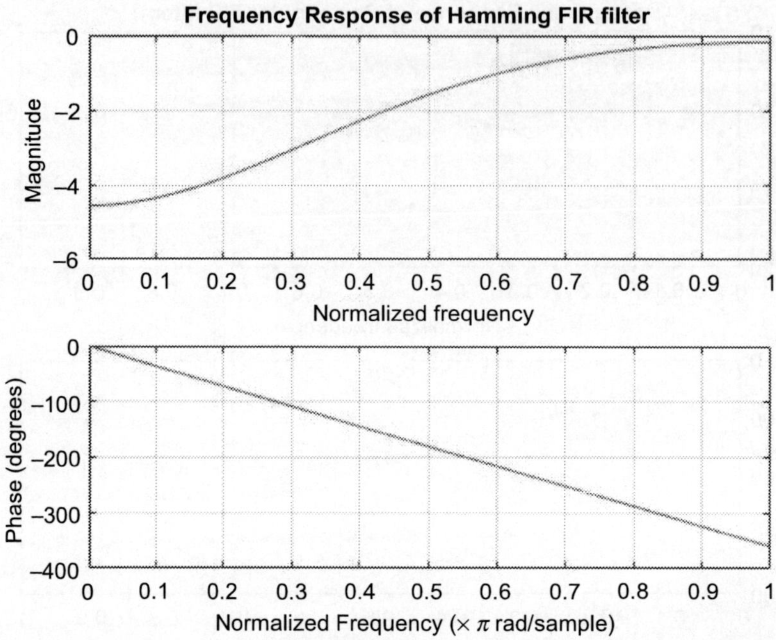

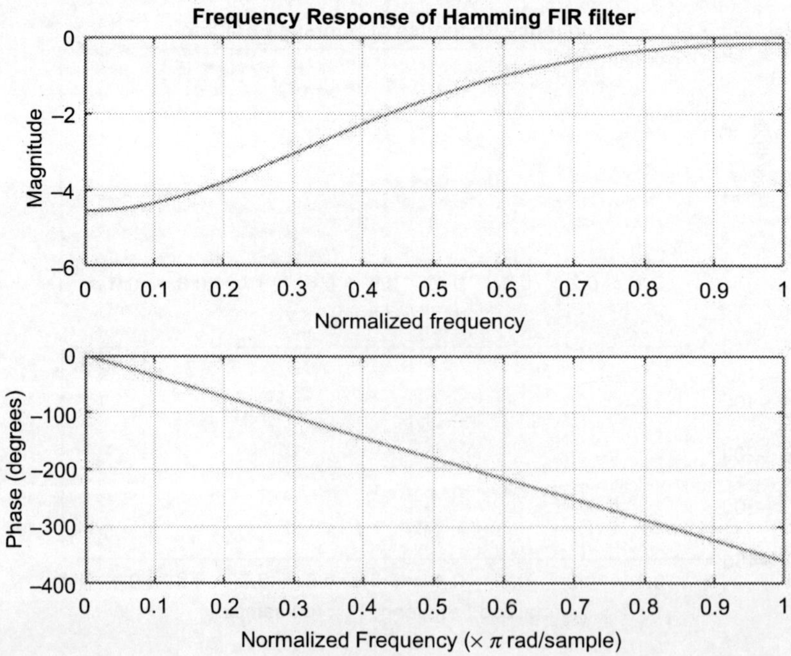

PROGRAM 16

```
Design Band pass FIR filter for N=7, Wcl=0.4 radian and Wcu=0.5
radian.
```

```
clc;
clear all;
close all;
N = input('Enter the length of the filter: '); % order of iir filter
Wcl = input('Enter the lower cut off frequency in radians: ');
%Lower Cut off frequency
Wcu = input('Enter the upper cut off frequency in radians: ');
%Upper cut off frequency
n = 0 : N-1;
tau = (N-1)/2;
eps = 0.001; % To avoid divide by zero, we use correction factor
Hd = (sin(Wcu * (n - tau + eps)) - sin(Wcl * (n - tau + eps)))
./ (pi * (n - tau + eps));
disp('For BANDPASS FIR filter-->')
disp('H(d) = ');
disp(Hd);
Wd_rect_window = (window('rectwin', N)'); % take the transpose to
bring column vector to a row vector
Wd_hann_window = (window('hann', N)'); % hanning window
Wd_hamm_window = (window('hamming', N)'); % hamming window
Hn_rect = Hd.*Wd_rect_window; % rectangular fir filter coefficients
Hn_hann = Hd.*Wd_hann_window; % Hanning fir filter coefficients
Hn_hamm = Hd.*Wd_hamm_window; % Hamming fir filter coefficients
disp('Rectangular window coefficients: ');
disp(Wd_rect_window)
disp('Hanning window coefficients: ');
disp(Wd_hann_window)
disp('Hamming window coefficients: ');
disp(Wd_hamm_window)
disp('Rectangular FIR Filter coefficients: ')
disp(Hn_rect)
disp('Hanning FIR Filter coefficients: ')
disp(Hn_hann)
disp('Hamming FIR Filter coefficients: ')
```

```
disp(Hn_hamm)
disp('The end')
```

OUTPUT

```
Enter the length of the filter: 7
Enter the lower cut off frequency in radians: 0.4
Enter the upper cut off frequency in radians: 0.5
For BANDPASS FIR filter-->
H(d) =
 0.0070    0.0198    0.0287    0.0318    0.0286    0.0197    0.0069
Rectangular window coefficients:
     1     1     1     1     1     1     1
Hanning window coefficients:
     0    0.2500    0.7500    1.0000    0.7500    0.2500         0
Hamming window coefficients:
 0.0800    0.3100    0.7700    1.0000    0.7700    0.3100    0.0800
Rectangular FIR Filter coefficients:
 0.0070    0.0198    0.0287    0.0318    0.0286    0.0197    0.0069
Hanning FIR Filter coefficients:
     0    0.0049    0.0215    0.0318    0.0215    0.0049         0
Hamming FIR Filter coefficients:
 0.0006    0.0061    0.0221    0.0318    0.0221    0.0061    0.0006
The end
```

PROGRAM 17

Design Band reject FIR filter for N=5, Wcl=0.35 radian and Wcu=0.65 radian.

```
clc;
clear all;
close all;
N = input('Enter the length of the filter: '); % order of iir filter
Wcl = input('Enter the lower cut off frequency in radians: ');
```

```
%Lower Cut off frequency
Wcu = input('Enter the upper cut off frequency in radians: ');
%Upper cut off frequency
n = 0 : N-1;
tau = (N-1)/2;
eps = 0.001; % To avoid divide by zero, we use correction factor
Hd = (sin(pi * (n - tau + eps)) - (sin(Wcu * (n - tau + eps)) -
sin(Wcl * (n - tau + eps)))) ./ (pi * (n - tau + eps));
disp('For BANDREJECT FIR filter-->')
disp('H(d) = ');
disp(Hd);
Wd_rect_window = (window('rectwin', N)'); % take the transpose to
bring column vector to a row vector
Wd_hann_window = (window('hann', N)'); % hanning window
Wd_hamm_window = (window('hamming', N)'); % hamming window
Hn_rect = Hd.*Wd_rect_window; % rectangular fir filter coefficients
Hn_hann = Hd.*Wd_hann_window; % Hanning fir filter coefficients
Hn_hamm = Hd.*Wd_hamm_window; % Hamming fir filter coefficients
disp('Rectangular window coefficients: ');
disp(Wd_rect_window)
disp('Hanning window coefficients: ');
disp(Wd_hann_window)
disp('Hamming window coefficients: ');
disp(Wd_hamm_window)
disp('Rectangular FIR Filter coefficients: ')
disp(Hn_rect)
disp('Hanning FIR Filter coefficients: ')
disp(Hn_hann)
disp('Hamming FIR Filter coefficients: ')
disp(Hn_hamm)
disp('The end')
```

OUTPUT

```
Enter the length of the filter: 5
Enter the lower cut off frequency in radians: 0.35
Enter the upper cut off frequency in radians: 0.65
For BANDREJECT FIR filter-->
H(d) =
   -0.0514    -0.0825     0.9045    -0.0845    -0.0503
Rectangular window coefficients:
     1     1     1     1     1
Hanning window coefficients:
         0    0.5000    1.0000    0.5000         0
Hamming window coefficients:
    0.0800    0.5400    1.0000    0.5400    0.0800
Rectangular FIR Filter coefficients:
   -0.0514    -0.0825     0.9045    -0.0845    -0.0503
Hanning FIR Filter coefficients:
         0   -0.0413    0.9045   -0.0422         0
Hamming FIR Filter coefficients:
   -0.0041    -0.0446     0.9045    -0.0456    -0.0040
The end
```

PROGRAM 18

Design a digital low pass butterworth filter for cut-off frequency Fc = 1000 Hz,N=2 and sampling frequency Fs = 10,000 Hz.

```
clc;
clear all;
close all;
N = input('Enter the length of the filter: '); % order of iir filter
Fc = input('Enter the cut off frequency: '); % cut off frequency
Fs = input('Enter the sampling frequency: '); % sampling frequency
Wc=(2*pi*Fc)/Fs % cut off frequency in radians
[b a]=butter(N,Wc/pi,'low') % b and a are coefficients of H(z)
freqz(b,a);
```

```
title('frequency response of the filter');
xlabel('frequency in radians');
ylabel('gain in dB');
```

OUTPUT

```
Enter the length of the filter: 2
Enter the cut off frequency: 1000
Enter the sampling frequency: 10000
Wc =
     0.6283
b =
     0.0675     0.1349     0.0675
a =
     1.0000    -1.1430     0.4128
```

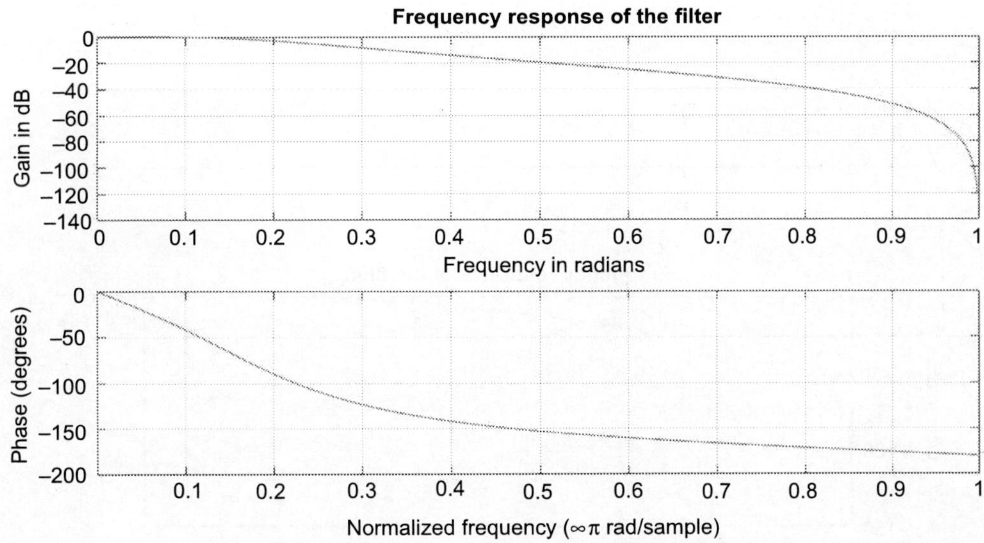

PROGRAM 19

For cut-off frequency Fc=50 Hz and sampling frequency Fs=150 Hz, design a second order digital high pass butterworth filter.

```
clc;
clear all;
close all;
```

```
N = input('Enter the length of the filter: '); % order of iir filter
Fc = input('Enter the cut off frequency: '); % cut off frequency
Fs = input('Enter the sampling frequency: '); % sampling frequency
Wc=(2*pi*Fc)/Fs % cut off frequency in radians
[b a]=butter(N,Wc/pi,'high') % b and a are coefficients of H(z)
freqz(b,a);
title('frequency response of the filter');
xlabel('frequency in radians');
ylabel('gain in dB');
```

OUTPUT

```
Enter the length of the filter: 2
Enter the cut off frequency: 50
Enter the sampling frequency: 150
Wc =
    2.0944
b =
    0.1551   -0.3101    0.1551
a =
    1.0000    0.6202    0.2404
```

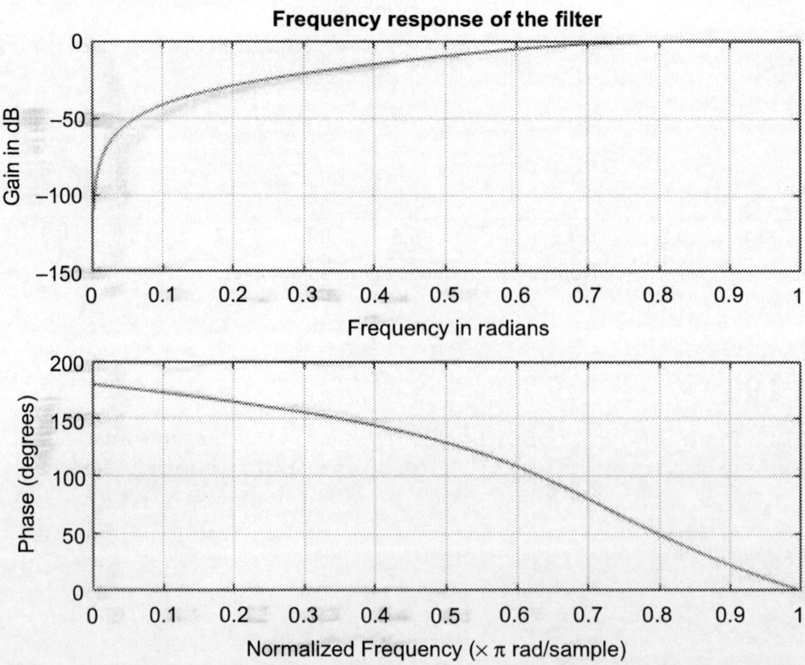

PROGRAM 20

Design digital band pass butterworth filter for cut-off frequency Fcl=70 Hz,Fcu=200 Hz,N=1 and sampling frequency Fs=500 Hz.

```
clc;
clear all;
close all;
N = input('Enter the length of the filter: '); % order of  filter
Fs = input('Enter the sampling frequency: '); % sampling frequency
Fcl = input('Enter the lower cut off frequency: '); % lower cut
off frequency
Fcu = input('Enter the upper cut off frequency: '); % upper cut
off frequency
Wcl = (2*pi*Fcl)/Fs% cut off frequency in radians
Wcu = (2*pi*Fcu)/Fs% cut off frequency in radians
[b,a] = butter(N,[Wcl/pi Wcu/pi]) % b and a are coefficients of H(z)
freqz(b,a);
title('frequency response of the filter');
xlabel('frequency in radians');
ylabel('gain in dB');
```

OUTPUT

```
Enter the length of the filter: 1
Enter the sampling frequency: 500
Enter the lower cut off frequency: 70
Enter the upper cut off frequency: 200
Wcl =

    0.8796

Wcu =

    2.5133

b =

    0.5157         0   -0.5157

a =

    1.0000    0.1773   -0.0314
```

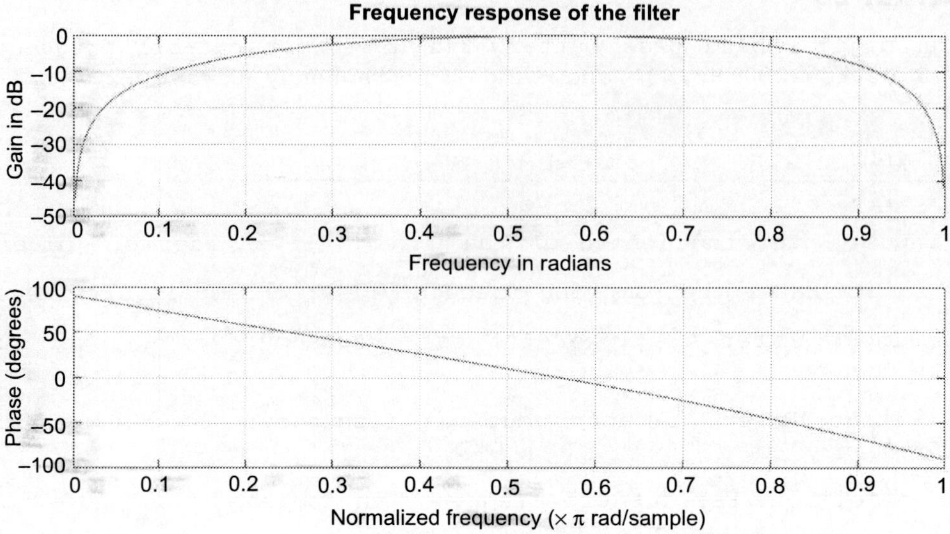

PROGRAM 21

```
%Program to calculate zero crossings of a voiced segment
clc;
clear all;
close all;
[m,Fs]=wavread('sp01.wav');
m=m(12000:13000);
plot(m);
xlabel('Sample number');
ylabel('Amplitude');
title('Plot of speech segment for 1000 samples');
% Read and plot first 1000 samples.
y=0;
for k=10:999,
if (m(k)>0) && (m(k+1)<0)
    y=y+1;
else y=y;
end
end
disp('number of zero crossings for voiced segment=');
disp(y);
```

COMMAND WINDOW RESULTS

```
number of zero crossings for voiced segment=
     69
```

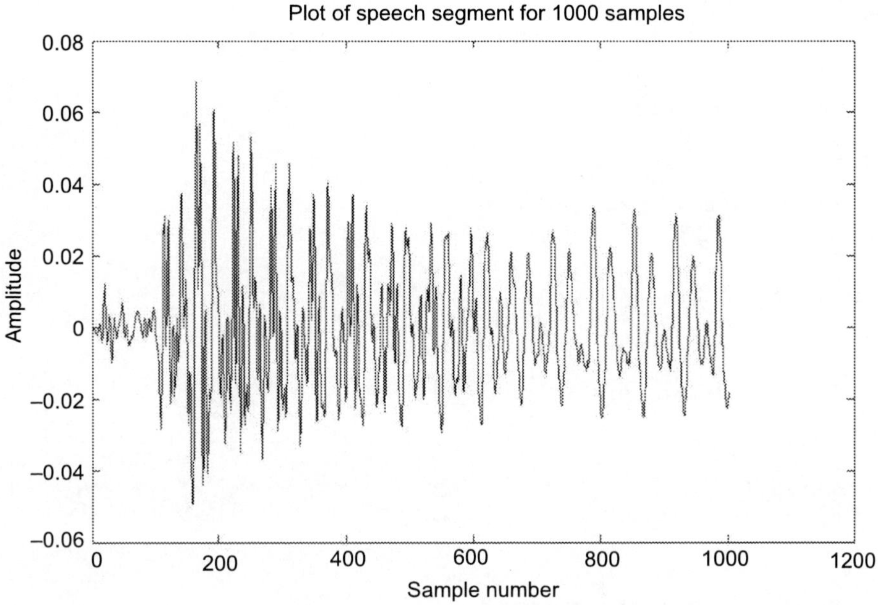

PROGRAM 22

```
% Program to plot short time energy of a signal
clc;
clear all;
close all;
[signal,fs]=audioread('sp01.wav');
L=length(signal);
t=[1:L]/fs;
% To plot signal in time domain
figure(1);
plot(t,2*signal)
title('Time Domain');
xlabel('Time in sec');
ylabel('Amplitude in volts');
n=300;
```

```
w=hamming(n);
overlap=n-1;
frames=buffer(signal,n,overlap,'nodelay'); %To form frame of speech
of length 'n'
w1=diag(w); %To form diagonal matrix with window sequence as
diagonal
sigwindowed=w1*frames;
STenergy=sum(sigwindowed.^2);
%To find short time energy
figure(2);
delay=(n-1)/2;
plot(t(delay+1:end - delay),STenergy, 'r');
title('Short Time Energy');
xlabel('Time in sec');
ylabel('Energy');
hold off
```

COMMAND WINDOW RESULTS

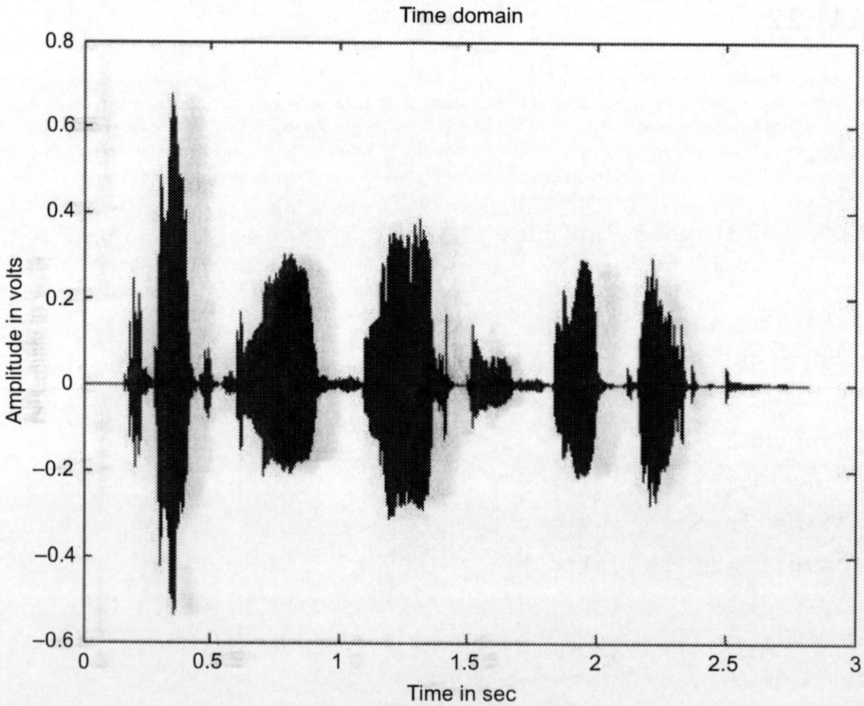

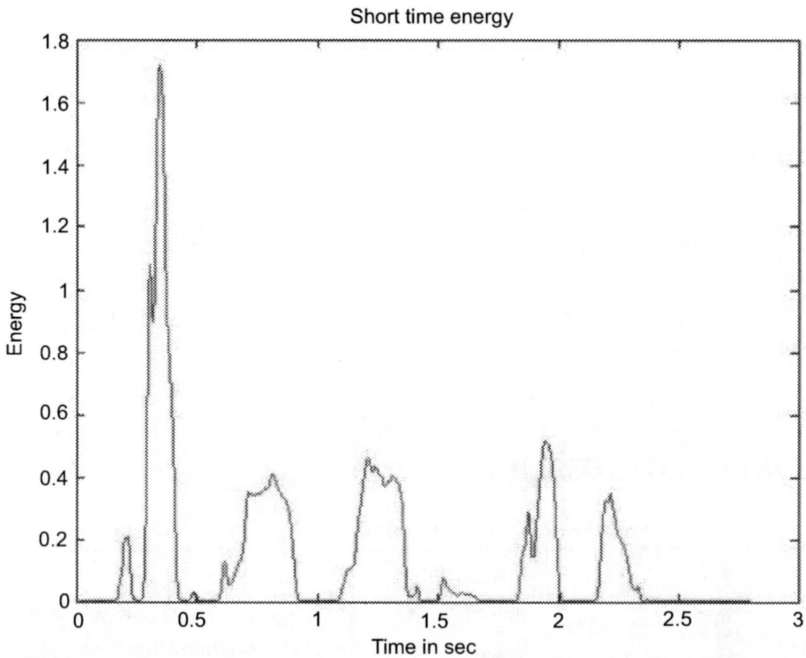

PROGRAM 23

```
% Program to plot Speech signal, Short-time energy, and Average
amplitude
clc;
close all;
clear all;
[SpeechSignal,Fs]=wavread('sp01.wav');
window_length=401;
window_Overlap=400;
window_Hamm=hamming(window_length);
FrameingSignal=buffer(SpeechSignal,window_length,window_
Overlap,'nodelay');
WindowedSignal=diag(sparse(window_Hamm))*FrameingSignal;
EnergyST=sum(WindowedSignal,1);
Magnitude_Avg=sum(FrameingSignal,1);
t=0:length(SpeechSignal)-1/Fs;
subplot(1,1,1);
plot(t,SpeechSignal/max(abs(SpeechSignal)));
```

```
title('speech:Expect nothing appreciate everything');
xlims=get(gca,'xlim');
hold on;
delay=(window_length-1)/2;
plot(t(delay+1:end-delay),EnergyST/max(EnergyST),'g');
plot(t(delay+1:end-delay),Magnitude_Avg/max(Magnitude_Avg),'r');
xlim(xlims);
legend('Speech','Short-time Energy','Average Magnitude');
xlabel('Time(sec)');
hold off;
```

COMMAND WINDOW RESULTS

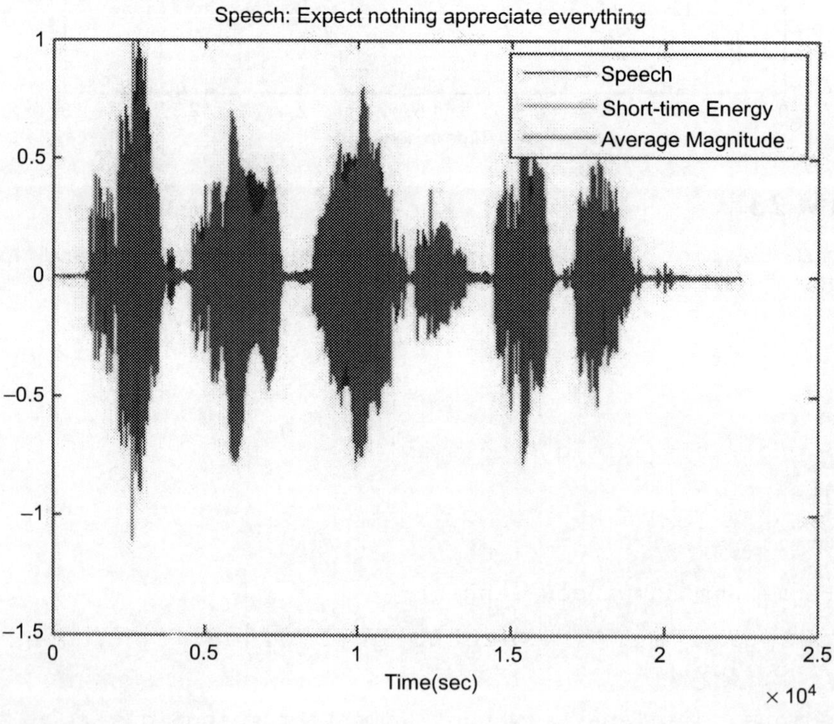

PROGRAM 24

```
% Program to plot Short-time autocorrelation function
clc;
clear all;
close all;
[SpeechSignal,fs]= wavread('sp01.wav'); % Get a section of vowel
disp('L=');
```

```
disp(length(SpeechSignal));
ss1=SpeechSignal(12095:12700); % Speech segment
% Calculating short-time autocorrelation
[ac1,lags1]=xcorr(ss1);
figure;
plot(lags1,ac1);
legend('Window Length:700 samples');
title('Short-Time Autocorrelation Function');
grid on;
```

COMMAND WINDOW RESULTS

```
L=
    22529
```

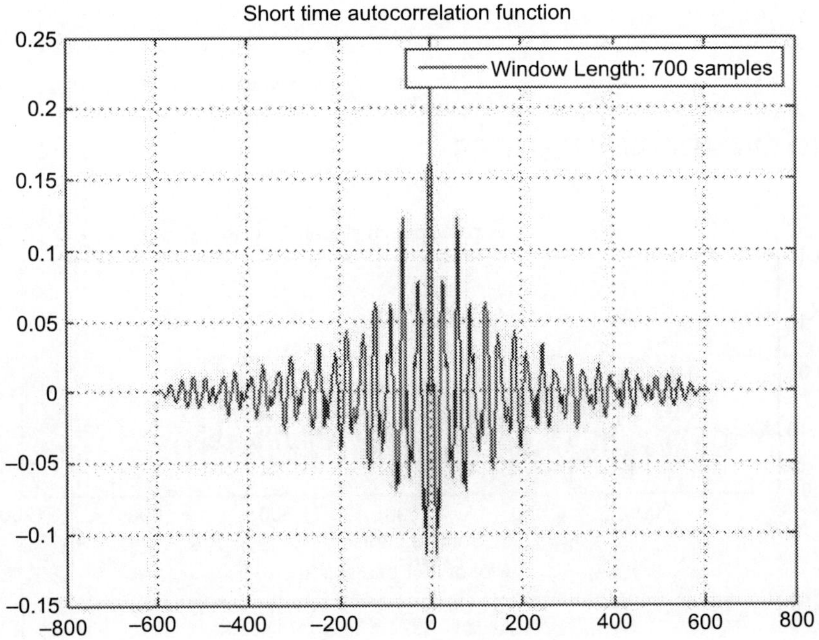

PROGRAM 25

```
% Program to find FFT of the signal.
clc;
clear all;
```

```
close all;
[m,Fs]=wavread('sp01.wav');
m=m(12000:13000);
subplot(2,1,1);
plot(m);
title('Part of Speech signal');
xlabel('Sample number');
ylabel('Amplitude');
b=abs(fft(m));
subplot(2,1,2);
plot(b,'r');
title('Plot of FFT of a signal');
xlabel('Frequency');
ylabel('Amplitude');
disp(b);
```

COMMAND WINDOW RESULTS

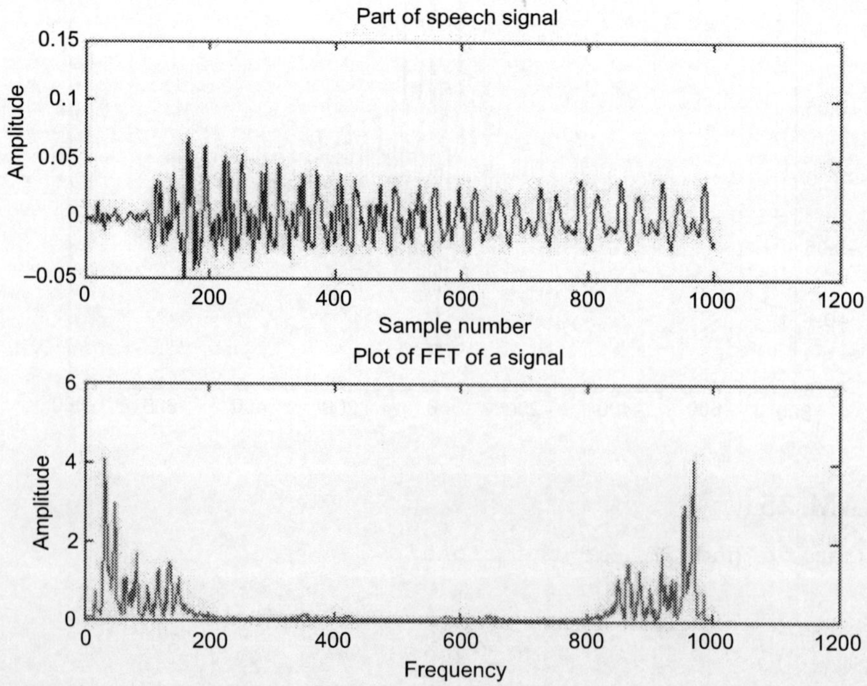

PROGRAM 26

```
%Program to find correlation of the signal
clc;
clear all;
close all;
[m,Fs]=wavread('sp01.wav');
m=m(12000:13000);
subplot(2,1,1);
plot(m);
title('Plot of voiced part of a signal');
xlabel('Sample number.');
ylabel('Amplitude');
for k=1:400,
    sum(k)=0;
end
    for k=1:400,
for i=1:45,
    sum(k)=sum(k)+(m(i)*m(i+k));
    sum(k)=sum(k)/45;
end
end
subplot(2,1,2);
plot(sum,'m');
title('Plot of correlation of a signal');
xlabel('Sample number');
ylabel('Correlation');
```

COMMAND WINDOW RESULTS

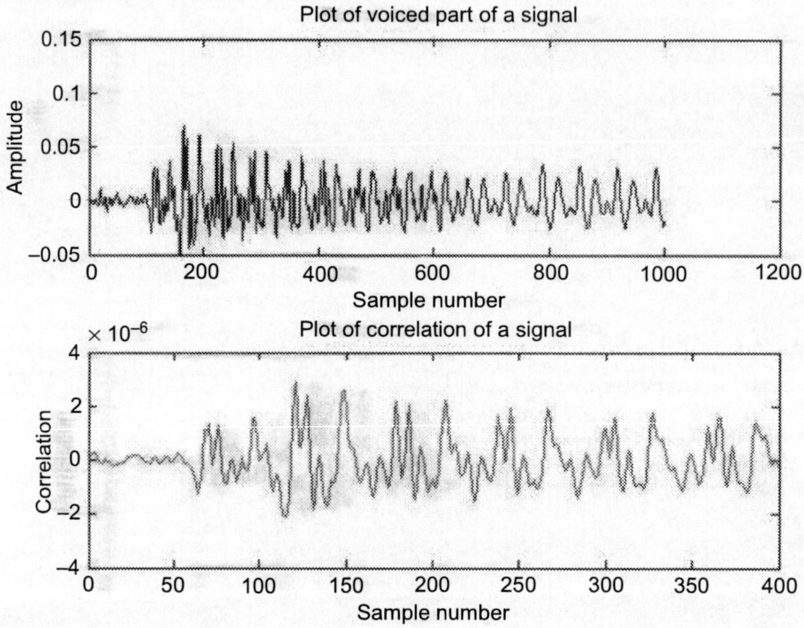

PROGRAM 27

```
% Program to display spectrograph
clc;
close all;
clear all;
%Long window in time-narrowband
[y,fs]=wavread('sp01.wav');
%window-time
time_window_narrowband=0.04;
time_overlap_narrowband=0.001;
%window-samps
window_narrowband=time_window_narrowband*fs;
no_overlap_narrowband=time_overlap_narrowband*fs;
nfft_narrowband=1024;
window=window_narrowband;
no_overlap=no_overlap_narrowband;
nfft=nfft_narrowband;
subplot(2,1,1)
```

```
specgram(y,nfft,fs,window,no_overlap);
xlabel('time(sec)');
ylabel('frequency(Hz)');
%Spectrogram,wideband,short window in time
%window-time
t_window_wideband=0.004;
t_overlap_wideband=0.001;
%window-samps
Window_wideband=t_window_wideband*fs;
no_overlap_wideband=t_overlap_wideband*fs;
nfft_wideband=8192;
window=Window_wideband;
no_overlap=no_overlap_wideband;
nfft=nfft_wideband;
subplot(2,1,2)
specgram(y,nfft,fs,window,no_overlap);
xlabel('time(sec)');
ylabel('frequency(Hz)');
```

COMMAND WINDOW RESULTS

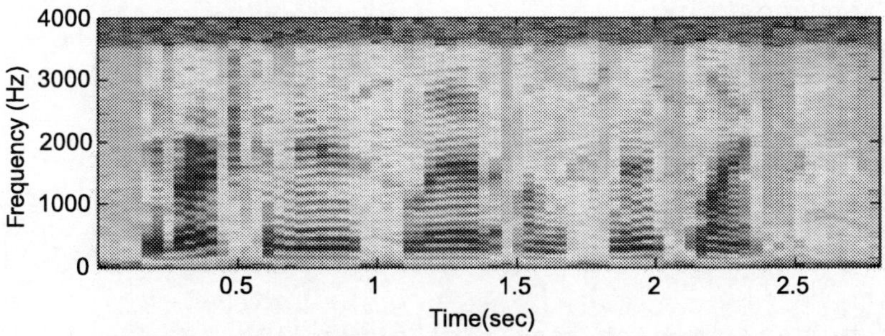

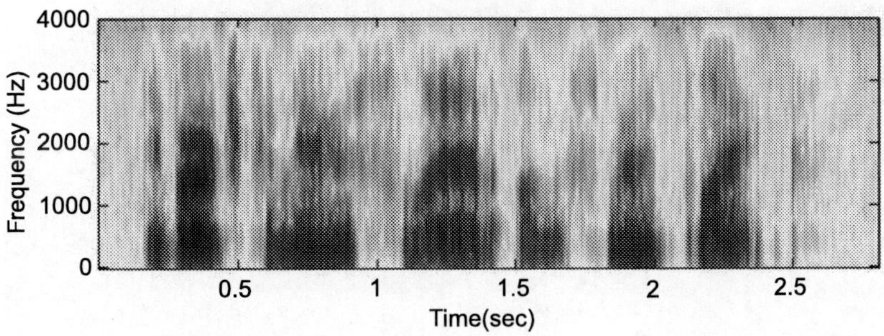

PROGRAM 28

```
% Program to plot formant frequency
clc;
clear all;
close all;
[m,Fs]=wavread('sp01.wav');
m=resample(m,10000,Fs); % Resample to 10,000 Hz
Fs=10000;
t=(0:length(m)-1)/Fs; % Plot waveform
subplot(2,1,1);
plot(t,m);
title('Original speech waveform');
xlabel('Time');
ylabel('Amplitude');
ncoeff=2+Fs/1000; %To get linear predictive filter
z=lpc(m,ncoeff);
[h,f]=freqz(1,z,512,Fs); % To Plot frequency response
subplot(2,1,2);
plot(f,20*log10(abs(h)+eps),'r');
title('Formant frequency');
xlabel('Frequency');
ylabel('Gain');
% To find frequencies by root solving
% To find roots of polynomial
r=roots(z);
% Only look for roots>0 Hz up to fs/2
r=r(imag(r)>0.01);
f_freq=sort(atan2(imag(r),real(r))*Fs/(2*pi));
% Converting to Hz and sort
for i=1:length(f_freq)
    fprintf('Formant %d frequency %0f\n',i,f_freq(i));
end
```

COMMAND WINDOW RESULTS

Formant 1 frequency 379.708928
Formant 2 frequency 812.860283
Formant 3 frequency 1443.710479
Formant 4 frequency 1909.926959
Formant 5 frequency 2732.763785
Formant 6 frequency 3302.644275

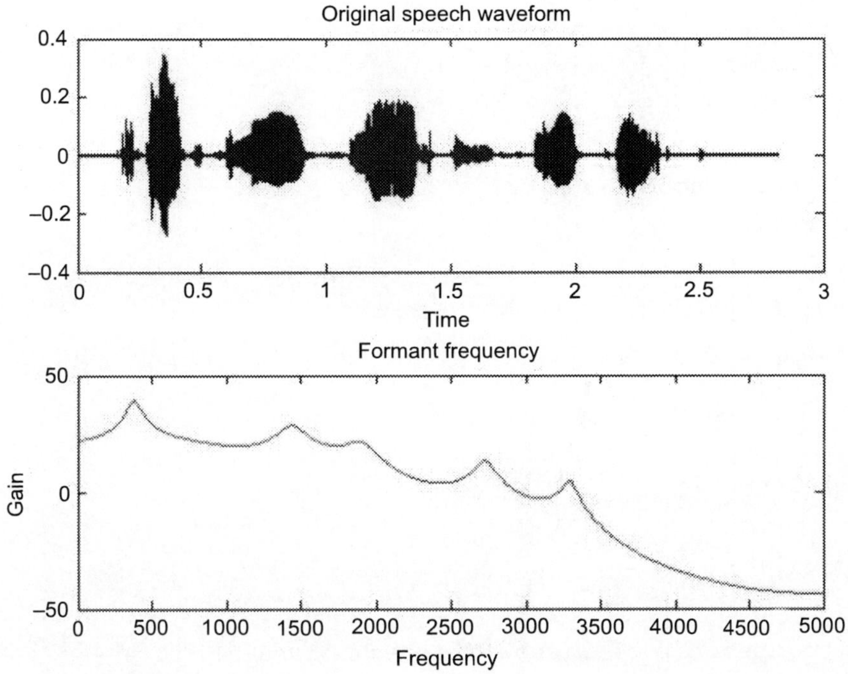

PROGRAM 29

```
% Program to plot log spectrum and Cepstrum of speech signal
clc;
clear all;
close all;
[m,fs]=wavread('sp01.wav');
m=m(12000:13000);
```

```
subplot(2,1,1);
plot(m);
title('Plot of part of Speech signal');
xlabel('Sample Number');
ylabel('Amplitude');
window=window(@hamming,900);
for i=1:900
    y(i)=m(i)*window(i);
end
subplot(2,1,2);
plot(y,'r');
title('Plot of windowed part of speech signal');
xlabel('Sample Number');
ylabel('Amplitude');
%the signal is passed via Hamming window and displayed.
a_fft=abs(fft(y));
C=log(a_fft);
figure;
for i=1:499
    d(i)=C(i);
end
subplot(2,1,1);
plot(d);
title('Log Spectrum of windowed speech signal');
xlabel('Frequency in Hz');
ylabel('Amplitude in dB');
e=abs(ifft(C));
subplot(2,1,2);
plot(e,'r');
title('Cepstrum of speech signal');
xlabel('Quefrency');
ylabel('Amplitude in dB');
```

COMMAND WINDOW RESULTS

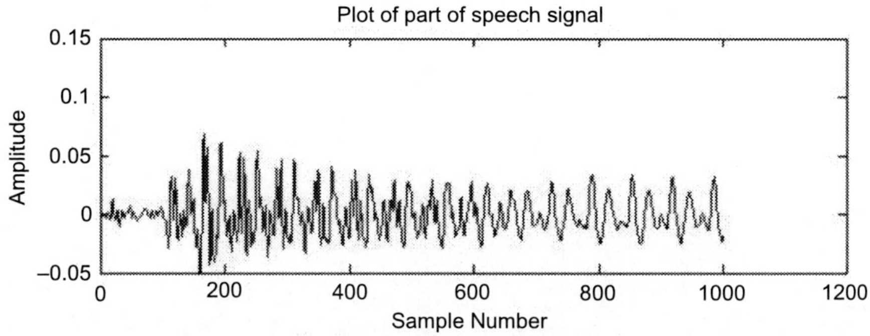

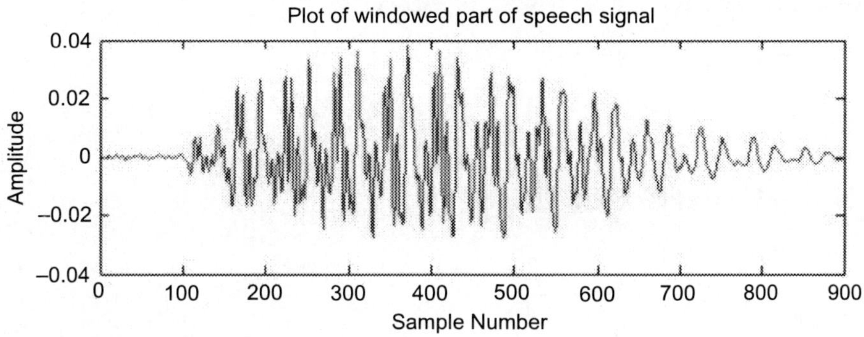

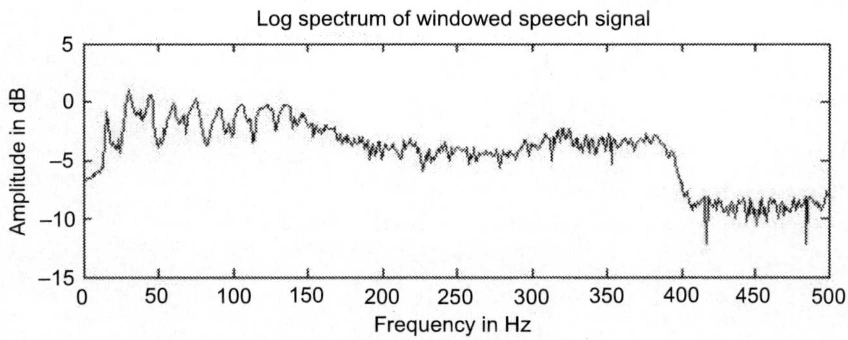

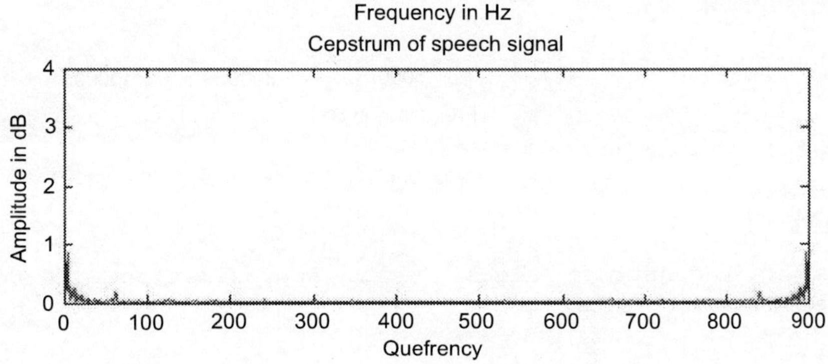

PROGRAM 30

```
% Program to convert frequency to Mel Scale
clc;
clear all;
close all;
for k=1:11050,
    z(k)=2595*log(1+k/700);
end
plot(z);
title('Frequency in Hz to Mel scale');
xlabel('Frequency in Hz');
ylabel('Mel frequency');
```

COMMAND WINDOW RESULTS

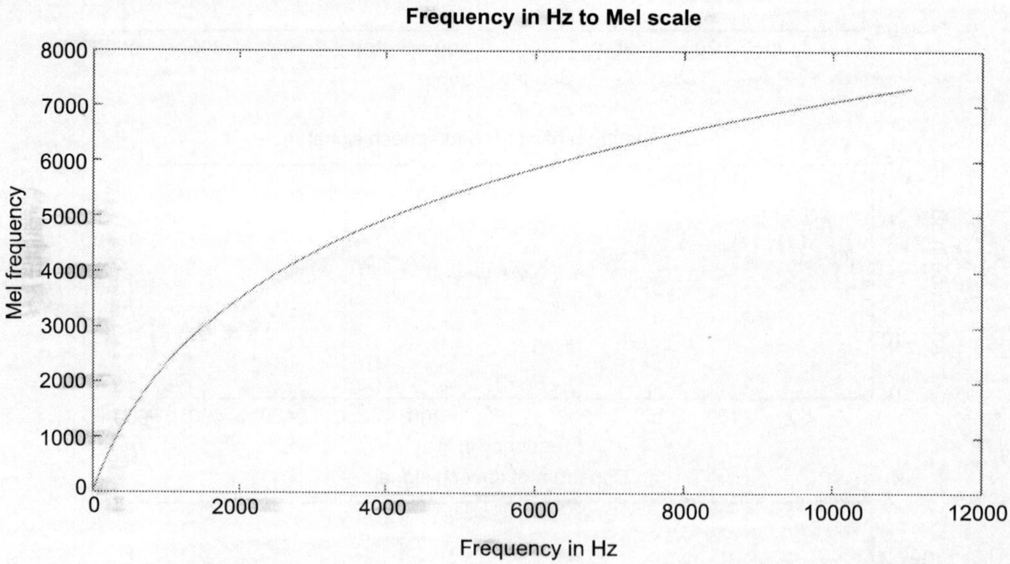

PROGRAM 31

```
%Program to reconstruct signal using Approximation and Detailed
Coefficients of wavelets

path='C:\1-1.wav';

sig=audioread(path);
```

```
% Obtain the DWT down to level 5 with the |'db5'| wavelet.
[a,d] = dwt(sig,'db5');
x=idwt(a,[],'db5');%Reconstrution using Approximation Coefficients
x1=idwt(a,d,'db5');%Reconstrution using Appr and Detailed Coefficients
subplot(4,1,1)
plot(sig); title('Original Signal');
subplot(4,1,2);
plot(a); title('Approximation Coefficients');
subplot(4,1,3);
plot(x);title('Reconstrution using Approximation Coefficients');
subplot(4,1,4);
plot(x1);title('Reconstrution using Appr and Detailed Coefficients');
```

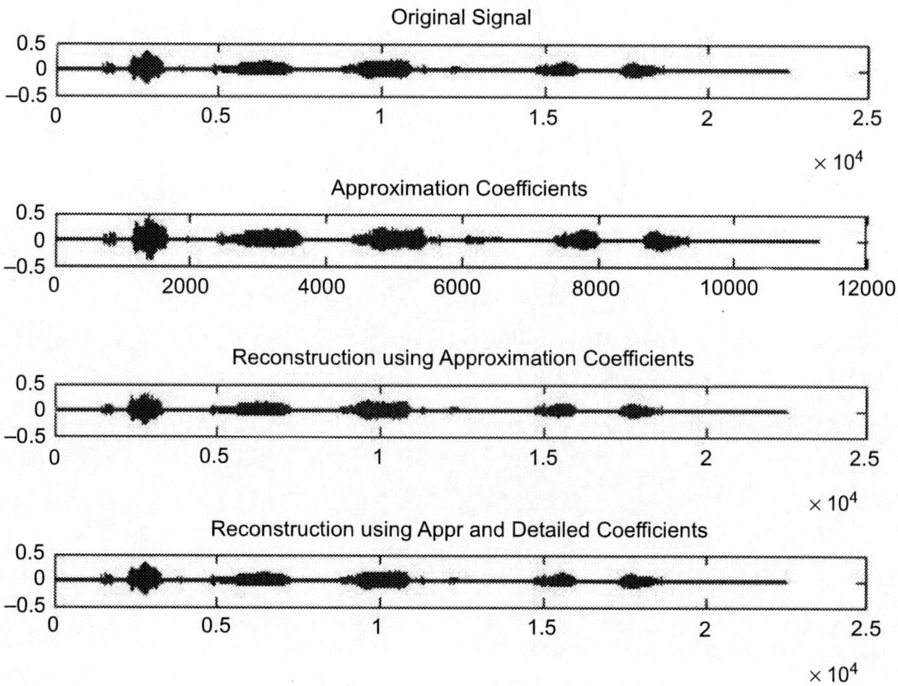

PROGRAM 32

```
%Program to cancel 60 Hz noise in corrupted speech signal with 60
Hz noise using LMS algorithm
```

```
clc; clear all; close all;
fs= 200; % Sampling frequency
s=wavread('sp23.wav',[2000  4000]);
L = length(s);
i = [1 : L]';
x1 = sin(2*pi*60* i/fs); % 60 Hz Noise as a Reference input
x=0.2*x1;
p=x+s;% 60 Hz Noise added with signal
N=input('enter filter order N= ');
mu=input('enter mu= '); % to define convergence parameter
% adaptive algorithm for training the filter co-efficients
h = zeros ( N  , 1 ) ;
  for n = N : L
      X = x(n:-1:n-N+1) ;
      xh(n)= h' * X;
      e(n) = p(n) - xh(n) ;
      h = h + 2*(mu * X * e(n)) ;
   end
figure(1);
subplot(311) ;plot(p);ylim([-0.5  +0.5]); title('Corrupted Speech
signal with 60 Hz noise');
subplot(312) ;plot(x);ylim([-0.5 +0.5]); title(' 60 Hz reference
noise ');
subplot(313) ;plot(e); ylim([-0.5  +0.5]);title(' speech signal
after filtering');
z=fft(r');
z1=abs(z);
f=(0:L-1)*199/L;%frequency scale
figure(2),subplot(211),plot(f,z1),ylim([0  15]);title('Magnitude
spectrum of the signal before filtering');
z2=fft(e);
z3=abs(z2);
f=(0:L-1)*199/L;
```

```
subplot(212),plot(f,z3),ylim([0  15]);title('Magnitude  spectrum  of
the  signal  after  filtering');
```

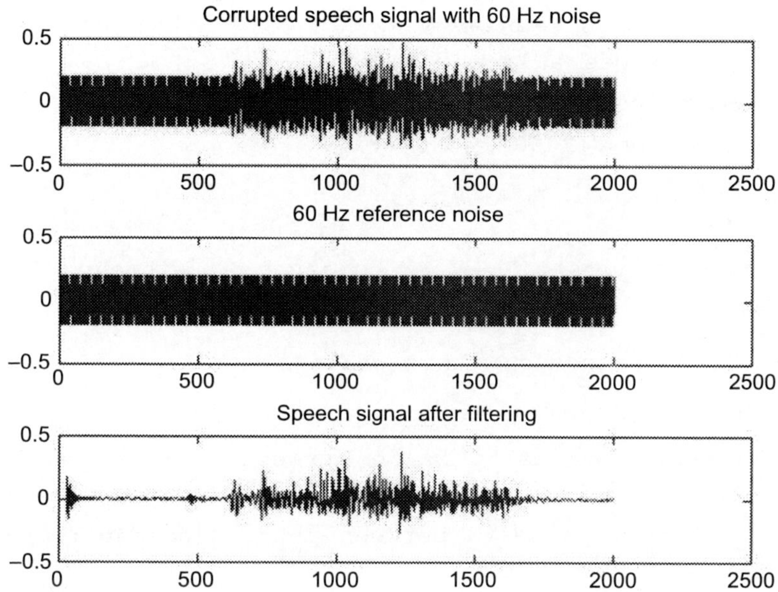

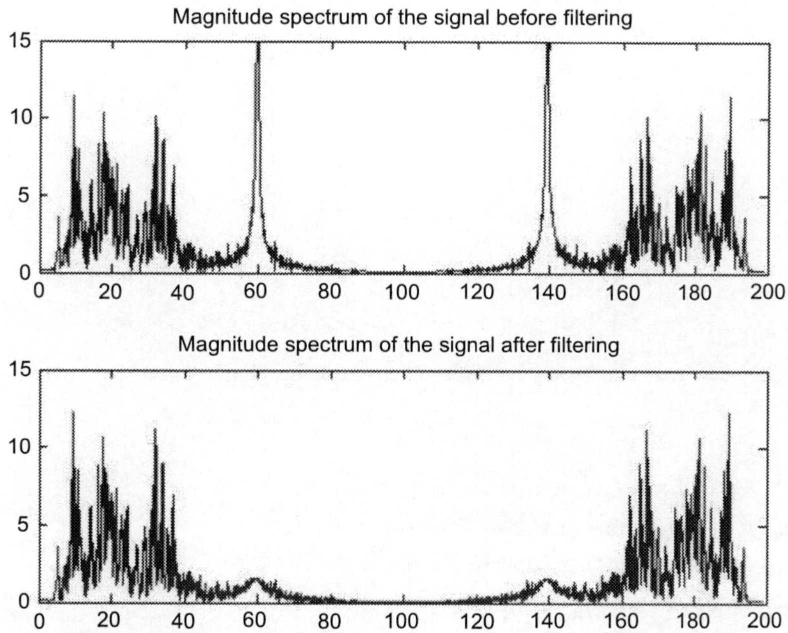

PROGRAM 33

```
% Program to cancel 60 Hz noise in corrupted speech signal with
60 Hz noise using RLS algorithm
clc; clear all; close all;
fs= 200; % Sampling frequency
s=wavread('sp23.wav',[2000  4000]);
L = length(s);
i = [1 : L]';
x1 = sin(2*pi*60* i/fs); % 60 Hz Noise as a Reference input
x=0.2*x1;
d=x+s;% 60 Hz Noise added with signal
N=input('enter filter order N= ');
delta=input('Enter the delta value between 0 to 1 delta=');% to
compute p(0)
lambda=input('Enter the value between 0 to 1 lambda=');%forgetting
factor
P = diag((ones(N,1)/delta));
Lx = length(x);
[m,n] = size(x);
if (n>m)
    x = x';
end
x = [zeros(N-1,1); x];
h = zeros(N,1);
itr=1;% defines number of iterations(more the number of iterations
less is the error)
for j=1:itr
     for i = 2:Lx
          X = x(i+N-1:-1:i);
          Pi = P*X;
          g = Pi/(lambda + X'*Pi);
          xh(i,1) = h'*X;
          a(i,1) = d(i,1) - xh(i,1);
          h = h + g*a(i,1);
          P = (P - (g*(X')*P))/lambda;
     end
```

```
end
z=fft(d);
z1=abs(z);
f=(0:L-1)*199/L;%frequency scale
figure,subplot(211),plot(f,z1),ylim([0    15]);title('Magnitude
spectrum of the signal before filtering');
z2=fft(a);
z3=abs(z2);
f=(0:L-1)*199/L;
subplot(212),plot(f,z3),ylim([0  15]);title('Magnitude spectrum of
the signal after filtering');
figure(2);
subplot(311) ;plot(d);ylim([-0.5  +0.5]); title('Corrupted Speech
signal with 60 Hz noise');
subplot(312) ;plot(x);ylim([-0.5  +0.5]); title(' 60 Hz reference
noise');
subplot(313) ;plot(a);ylim([-0.5  +0.5]); title(' speech signal
after filtering');
```

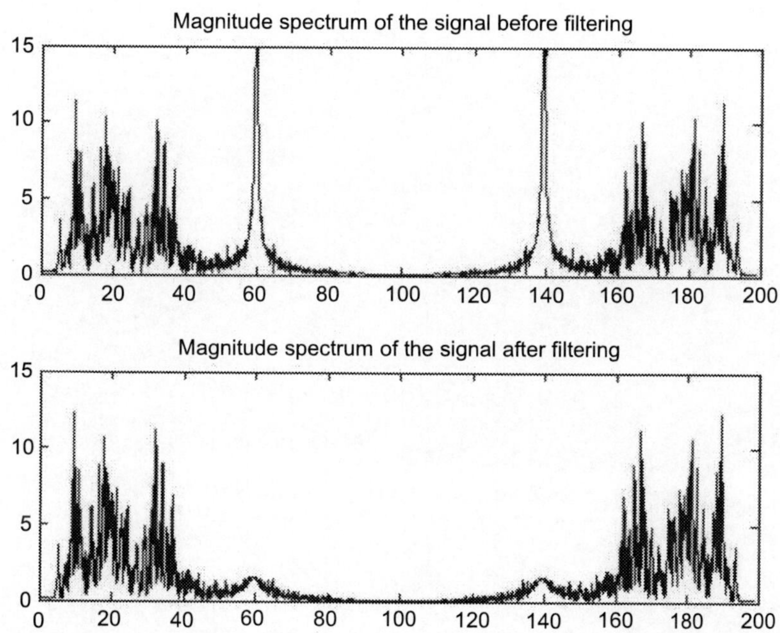

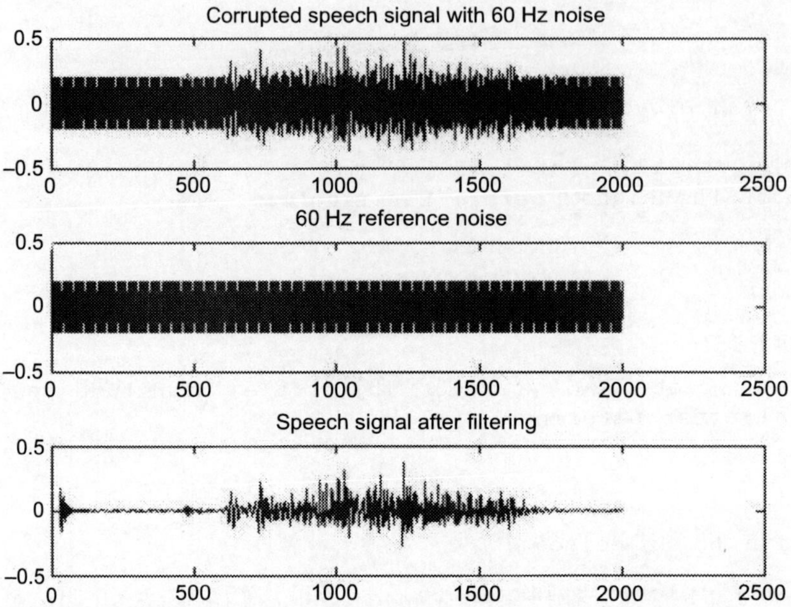
Corrupted speech signal with 60 Hz noise
0.5
0
−0.5
0
500
1000
1500
2000
2500
60 Hz reference noise
0.5
0
−0.5
0
500
1000
1500
2000
2500
Speech signal after filtering
0.5
0
−0.5
0
500
1000
1500
2000
2500

Index